THE END OF THE CERTAIN WORLD

THE END OF THE CERTAIN WORLD

The Life and Science of MAX BORN

THE NOBEL PHYSICIST WHO IGNITED THE QUANTUM REVOLUTION

NANCY THORNDIKE GREENSPAN

A Member of the Perseus Books Group
New York

Published by Basic Books,
A Member of the Perseus Books Group

Books published by Basic Books are available at special discounts for bulk purchases in the United States by corporations, institutions, and other organizations. For more information, please contact the Special Markets Department at the Perseus Books Group, 11 Cambridge Center, Cambridge MA 02142, or call (617) 252-5298 or (800) 255-1514, or e-mail special.markets@perseusbooks.com.

Set in 10.5-point Minion by the Perseus Books Group

Library of Congress Cataloging-in-Publication Data
Greenspan, Nancy Thorndike.
The end of the certain world : the life and science of Max Born : the Nobel physicist who ignited the quantum revolution / Nancy Thorndike Greenspan.
p. cm.
Includes bibliographical references and index.
ISBN 0-7382-0693-8 (alk. paper)
1. Born, Max, 1882–1970. 2. Physicists—Germany—Biography. I. Title.
QC16.B643G74 2005
530'.092—dc22 2004021809

05 06 07 / 10 9 8 7 6 5 4 3 2 1

For Irene,
whose devotion to her father's memory
inspired this book.

CONTENTS

INTRODUCTION

EIGHT YEARS AGO, I MET IRENE BORN NEWTON-JOHN WHEN WE WERE BOTH STAYING WITH her daughter Olivia in California. During a long weekend, she told of her father's discovery of quantum mechanics, the family's exile from Nazi Germany, and the personalities of her father's many famous students. Seventy years later, she could still see the Mephisto-like eyes of Edward Teller and hear the Bach concertos for two pianos performed by her father, Max Born, and his young assistant, Werner Heisenberg. She recalled Albert Einstein's visits to their home when she was young and how special they were. The story was fascinating, as was the woman; but Irene had one regret: No one had written a biography of her father, who had been so instrumental in the quantum revolution, the most important scientific advance of the twentieth century.

After I arrived home, Irene's story bounced around in my head. The first half of this century had always held a particular absorption for me: the political reverberations of World War I, the creative explosion in science, literature, and the arts during the 1920s, the global Depression, the profound changes in Germany, World War II, the bomb. I realized that Max Born's life was an intimate portrait of all these events. I called Irene to ask if I could write his biography. She agreed. A few months later in London, I met her brother, Professor Gustav Born, the executor of his father's papers—who straightaway became Gustav. His gracious consent for me to use the papers and write the biography gave me access to the rich trove of family documents stored at the University of Edinburgh.

The thousands of letters, diaries, and photos there, a largely untapped historical source, chronicle not only the life of Max Born and his family, but also the scientific and political developments of Germany and Europe that profoundly affected him. They are by turns fascinating and troubling. Perhaps the most disconcerting find was a letter to Born signed by Adolf Hitler, thanking him for his service to the University of Göttingen—from which the Nazis suspended him in 1933 because he was Jewish. Later I learned that the signature was by autopen, but the impact was only slightly lessened.

The Edinburgh papers reveal Born's personal side, in both relationships and thoughts, and complement the 10,000 or so letters in his professional correspondence that exist in public archives. (Max Born and his wife, Hedi, were prodigious letter-writers.) In all of these, the importance of religion, war, and scientific pursuit stand out as factors crucial to his development, but paradoxically: an assimilated German-Jew, Born had no interest in formal religion, yet it defined the direction of much of his life; a pacifist, he shaped the scientific background of many of the inventors of the atomic bomb. As for his own theoretical search for fundamental truths, it gave him optimism for the future and escape from the present. He was dependent on his work for his strength and sense of purpose and proud and excited when he found answers. Modest about his accomplishments, Born saw former assistant Werner Heisenberg receive credit for some of his own ideas, and he saw other ones simply subsumed into the quantum mechanical theory known as the Copenhagen Interpretation.

The intimate side of his life revealed a paradox as well: a wife he desperately needed, but whom he described in his later years as the cross he had to bear. Their relationship was sometimes tortuous for him, but he chose to pursue it.

As I had anticipated, Born's life contained the elements that drew me and made his story so compelling. Especially satisfying was that, in spite of the hardships through which Born journeyed, he persisted and ultimately prevailed. In old age, when he had graciously accepted his fate, he unexpectedly received the recognition that he had striven for and that he deserved. Life does not usually resolve itself in such a way, but it is pleasing when it does.

Nancy Greenspan
Bethesda, MD

PROLOGUE

DECEMBER 1953, EDINBURGH, SCOTLAND

One Sunday evening, Max Born was relaxing upstairs in his sitting room listening to the radio. It was 9:15 and the soft voice just beginning his talk belonged to J. Robert Oppenheimer, Born's doctorate student in Germany twenty-seven years earlier. Oppenheimer was in London to give six lectures on quantum physics for the BBC's prestigious Reith Lecture Series.

Born had enjoyed the first three and was anticipating the fourth entitled "Atom and Void in the Third Millennium." This evening's topic covered Born's particular field of research. As the speaker began, he described the excitement of the quantum discoveries in the 1920s along with the illustrious physicists associated with them except for Born. Oppenheimer gave a lucid explanation of the discovery but passed over his old teacher without a single mention. Oppenheimer's omission did not come from ignorance; indeed, it was this specific research that originally had drawn him to work with Born in Germany.

Oppenheimer told his radio audience that when researchers investigated the position of a particle, they never found it where they thought they would, nor did they see it spread out. He explained, "The spreading of the waves in space did not mean that the electron itself spread; it meant that the probability or likelihood of our finding the electron, when we look for it, spread as the wave does." This fundamental concept, so simply put by Oppenheimer, changed the nature of science, creating the transition from the deterministic world of Newtonian mechanics to the modern statistical world of quantum theory. Born discovered the principle and originated the shift.

Born's forty-five year career of shaping the theoretical physicists of the future had reached its end. A couple of weeks earlier, his department at the University of Edinburgh had presented him with a collection of articles from colleagues and students—Albert Einstein, Erwin Schrödinger, Theodore von Kármán, Alfred Landé, Louis de Broglie, David Bohm—to honor a scientific legacy that included major contributions in quantum theory, solid-state physics, optics, special relativity, and field theory. It was time for Born to take account of his life. Oppenheimer's omission felt all the more pointed in juxtaposition.

Born thought about the lecture for a few days then wrote to Oppenheimer. It was December 11, 1953, Born's seventy-first birthday. Twenty years and a day earlier, his friends Paul Dirac and Erwin Schrödinger, and his former young assistant, Werner Heisenberg, had received the Nobel Prize for their contributions to quantum theory. Born, who had intensified the search for a solution to the quantum mystery and played a major role in the revolutionary new theory, did not.

Born started off in the traditional European manner with a number of gracious remarks then came to the point.

> I was particularly happy about your emphasis of the importance of the statistical interpretation of quantum mechanics that I inaugurated 27 years ago, but I cannot conceal my disappointment that you did not mention my part in the matter, although you have quoted others, like Bohr, Heisenberg, etc. I am now very old and beyond ambition or greed for honours. I have been silent for these 27 years, but I think that I now can at least ask the question: Why is it that my participation, or may I say leadership, in this development from the mechanistic to the modern way of thinking is almost everywhere neglected? It began when in 1934 [sic] Heisenberg alone got the Nobel Prize, for work done in collaboration with me and, to some degree, with Jordan. He did not know at the time (1925) what matrices were; but a short time later the expression "Heisenberg Matrices" was introduced. This I can understand, for who could disentangle the distribution of 3 collaborators, except these themselves? But with the statistical interpretation of the wave function the matter is quite different. I found violent opposition from Heisenberg, who in a letter called my ideas a "treason against the spirit of matrix mechanics." This letter is lost in the Nazi upheaval during my emigration, but Heisenberg has frankly acknowledged the correctness of my memory.

A week later, Oppenheimer replied that he had kept the names to a minimum to reduce confusion. Trying to soften the slight, he added, "I am one of the last to forget your part in these affairs, for I was one of those who learned of the discoveries more or less as you made them."

The day Born wrote the letter, he and his wife moved out of their home of seventeen years and went to stay in a hotel before leaving Edinburgh for good. Up until that day, it had been a long journey through petty squabbles and world wars, quantum mechanics and atomic explosions. Now they were going back to Germany—where it had all started.

ONE

A KIND OF SHELL

MAX WANTED A PAIR OF LOVEBIRDS. HE KNEW HIS CHANCES WERE SLIM—HIS FATHER already had said no. Their city apartment in Breslau could not accommodate the smell and noise of pet birds. In early December, though, shortly before his fifth birthday, Max overheard from the nursery an encouraging conversation in the drawing room. "I got the couple of love-birds yesterday for Max," said Fräulein Wedekind, his grandmother's secretary. "I hope he will like them."

Max glowed.

On the morning of December 11, Max ran eagerly into the drawing room. He saw a table covered with presents—toys, books, and cakes—but where was the birdcage? Overcome with disappointment, his eyes welled up and he cried quietly. Then he realized that Grandmother had not yet arrived. Surely she would bring the lovebirds! When she appeared empty-handed, Max burst into sobs. His family watched in bewilderment at the distraught little boy. Finally, Fräulein Wedekind took him into a corner and prodded the full story from him. He had misunderstood, she explained—the lovebirds were a Christmas present for his uncle, Max Kauffmann.

Max hid his embarrassment amid the family laughter. Years later, recalling this public revelation of his greed and eavesdropping, he noted, "It was the first experience which taught me to be reserved. Many others followed."

Young Max's shyness and sensitivity were reinforced by a deeper loss than a pair of lovebirds. "There was no mother in whom I could confide, and who could restore my confidence."

Gretchen Kauffmann Born had died a year and a half earlier, in summer 1886. Pregnant with her third child, she had fled the heat of Breslau for the cooler air and forest beauty of Tannhausen, her parents' country estate. In this sanctuary southwest of Breslau, she hoped to find relief from the searing pain of gallstones. Max and his two-year-old sister, Käthe, came with her.

At Tannhausen, curly-headed Maxel and giggly Kätchen (as the family called them) scampered about with a gang of seven or eight cousins, running on short plump legs through the park and gardens surrounding the house. Watching them was a pleasure long awaited by Gretchen. She had spent the previous year at the mineral springs in Karlsbad, hoping that soothing waters would ease her suffering. For months she had only been able to enjoy her children and their antics through her husband Gustav's letters. He had lovingly told her how the serious Maxel was always careful not to muss his clothes when splashing about with a watering can in the flower bed or when playing coachman to Gustav's unruly horse. If "horse" were not lined up with the reins properly, Maxel would yell, "Horse not in order." Gustav described the happy and lively Kätchen snatching Maxel's toys and making him search for them; and he lamented, "Ach, that the Mama cannot see the joyful hunt."

As a physician, Gustav knew the risks of gallstones. Besides acute pain, they caused inflammation and intense pressure that medical treatment in 1886 could do little to abate. His hopes for Gretchen's recovery and emptiness without her filled his letters to her.

> I have such a longing for my wife, to see her again with me healthy and blooming, that I am not quite capable of saying it. For nearly a year, I am so out of my mind, and so often I have feared who knows what. Look after yourself properly and conscientiously obey the rules. And before saying your prayers before meals, say to yourself: Lead me not into temptation with fatty and acidic foods; let me have plenty of meat, white bread, light vegetables and much wine.

Thus he cautioned before continuing,

> I am alone again and so return my soul to you. I have a dream, a truly bold dream, a truly gigantic dream that comes to fruition when I have my wife and children, healthy, cheerful, and contented around me, and I myself have a permanent position which allows me to be independent. When will these bold dreams be fulfilled? Now enough of dreams. (That dream is my main vice anyway.) We defend ourselves bravely in the present.

Gustav Born and Gretchen (for Margarethe) Kauffmann met in Breslau in 1880. He had come to Breslau six years earlier to be an assistant researcher in the anatomy department at the university. His love was embryology; he was beginning a serious investigation of hybridization, artificially inseminating frog eggs. First he had followed his father's wish and completed a medical degree, making two generations of Born physicians, unusual for Jewish families in the late nineteenth century. Gustav's father, Marcus Born, had been a district physician in the small German town of Görlitz. When he died suddenly, Gustav came to the university.

Gustav was a poet and amateur thespian with a sensitive face framed by curly dark hair and a full beard and mustache that accentuated the sparkle of his eyes when he was excited—eyes that to his young nephew Hans seemed "on fire." Much sought after by the young ladies of Breslau, he had captivated Gretchen Kauffmann at once. Gustav in turn was charmed by the warmth and intelligence that lit up her strong, pretty features. Having grown up in an atmosphere of music and culture, she was a talented pianist and complemented his poetic temperament.

Gretchen's father, Salomon, was director of the Meyer-Kauffmann Textilwerke A. G., a family-owned textile manufacturing company with numerous spinning, weaving, and dying plants located in Silesia, then part of Germany. In the Kauffmanns' eyes, at least, their wealth and social position outshone the educational and professional achievements of the solidly middle-class Born family. Gretchen's parents were not pleased by the prospect of marrying their oldest daughter to a junior researcher at the university. Shortly after Gustav and Gretchen's engagement was announced, Gustav's sister Selma called on the Kauffmanns. She was shown into a large first-floor apartment in a Florentine palazzo-style building located on one of Breslau's most prestigious squares. Frau Marie Kauffmann, with hair piled high in a bun and mouth sternly set, greeted her, and then, according to Selma, "regally rustled out of the room."

That meeting was a defining moment for the Born and Kauffmann families. To Selma Born, Marie Kauffmann forever remained a self-glorified person with whom she wanted little interaction. Only Gretchen, whom Selma saw as an "amiable, lovely personality," saved that first day.

On a splendid August afternoon in 1881, Gretchen and Gustav married at Tannhausen. To make sure their daughter would maintain her social position, the Kauffmanns provided the new couple with a large annual allowance along with access to all the family trappings: Tannhausen, their summer estate; Kleinburg, a grand weekend retreat near Breslau; the family box at the symphony; dinner parties with renowned musicians; and a handsome apartment. Gustav chafed under this heavy-handed generosity but felt helpless to object, especially since Gretchen's fragile health benefited from her family's largesse. Bound by love, Gustav and Gretchen created what Selma called "a house in the sun" that shone for five years—especially with the birth of Max in 1882 and then Käthe in 1884.

Then on August 29, 1886, as the sun began to set over the Weistritz River at the valley's base, Gretchen Born died at Tannhausen. Gustav showed his grief and love in the poem he wrote for her gravestone.

There remains a comfort which lifts us up,
What you have planted in me with your
wonderful spirit of love and goodness,
That endures on and on and that works and lives.

After Gretchen's death, Gustav withdrew into his research.

For a while, a young Polish Catholic girl who had been Kätchen's wet nurse looked after the children. She was soon replaced by a governess—a stern spinster incapable of filling the emotional void left by their mother's death. Later, Max would recall, "There was never anybody with whom we children could take refuge with our little sorrows and joys, as other children did with their mothers." As with the lovebirds, he buried his frustration, loneliness, and suffering deep inside himself. No one made things better, and no one explained about the world, leaving him frequently confused as a boy.

Behind the governess stood the imposing figure of Grandmother Kauffmann: brisk, controlling, and overprotective. Almost every day the children would hear her carriage arrive, see her sweep into the house, then listen quietly as she gave the governess her orders for the day. Friends, health care, and little gifts—nothing escaped her attention. Gustav buried himself in his work, too tired to argue with either woman about the children's daily care.

In the early years after their mother's death, the Born children had few friends apart from one another. Grandmother, fearing draughts because Max had had an asthma attack, restricted their activities except for playing with their cousins. Max and Käthe became each other's best companions, even creating their own language to confound the adults.

When Max reached school age, his grandmother demanded that he be tutored at home, due to his fragile health. She hired Herr Böhr—a friendly man with a "full resounding voice"—as tutor. Böhr's first challenge was to get the timid boy to talk. With coaxing he eventually succeeded, and by the end of the year Max was shouting out answers with delight. His grandmother decided Max was healthy enough to go to Wankel's Knabenschule, an elementary school for boys, the next year.

Afternoons and weekends were playtimes with his cousins. Once a week he played with the three Jacobi boys, sons of Aunt Selma and Dr. Joseph Jacobi, at Grandmother Born's house. Sometimes they also visited each other's houses; on other days, there were cousins from the Kauffmann side of the family. Gretchen Kauffmann Born had had a brother and two sisters, and their father, Salomon, came from a large family that provided plenty of playmates for Max and Käthe. Not liking their values, however, Selma kept her boys at home when Max was with his other cousins.

The Kauffmann cousins often gathered at their retreat in Kleinburg, a budding suburb a few kilometers south of Breslau. On sunny days, Grandfather's carriage would collect the children and their nannies and drive them out to the private parklike garden, with a large orchard, greenhouse, vegetable garden, tennis courts, stables, and playground. The children romped, filched fruit from Grandfather's favorite trees, and played soldiers: Max as General Helmut von Moltke, hero of the Franco-Prussian War, and cousin Hans Schäfer as the great chancellor Otto von Bismarck, unifier of Germany, led the others in mock battle.

These lively afternoons in Kleinburg were a prelude to summer at Tannhausen, where life was idyllic for Max. An assortment of cousins roamed through the forest where the "Teutons" fought the "Romans" using twigs for swords and pinecones for missiles.

Years later, when Max and Käthe were young adults, they climbed to a large outcropping in the woods called the Haunstein. From there they could see all that Max remembered as paradise—the river Weistritz meandering past quaint villages, the fields and meadows, the hills and forest. Age had not diminished its glory in Max's eye. He told his sister, "If one traveled the world, one would never see a more beautiful place than this."

Summers also brought Max cherished time with his father. After a long winter of work, Gustav would go trekking with friends in the Tirolean or Swiss Alps before joining the children at Tannhausen. There Max and the cousins excitedly anticipated nature walks through the "dark, mysterious spruce forests" with Gustav, whom Max described as "an intimate friend of all living creatures."

During one summer, Gustav returned from the Tirol with a rare alpine salamander for Max. He explained to his son the animal's need for special care and feeding, and for a couple of weeks, Max carefully followed his instructions. One night, however, he forgot. The next morning he found the salamander dead. His father did not scold him but rather used the event to illustrate the meaning of life and death. As Max later wrote, "From that moment I suddenly knew what it meant to make another creature suffer and die."

From his father, Max learned his outlook on life, the basic principles for assessing life's most complex issues:

> I am certain that his deepest conviction was that the most fundamental rules of ethics can just as well, or even better, be derived from a study of nature, including man, than from religious tradition. For he believed that these rules, during the progress of evolution which culminates in man's civilization, become intrinsic parts of our soul. . . . He did not say "never kill"—he killed many animals in his biological research— . . . but nevertheless he imbued in

me a deep respect for the right of every living creature to enjoy its span in the light of the sun.

What Max took from his father was that life without struggle and killing is an illusion. Nature did not follow the Christian rule. To survive, animals must kill each other, and humans must kill animals. The Christian rule came in to stop humans from killing each other—the difference being that humans "remember, understand, and imagine; that means, *our* fear is not the terror of the moment, as that of a deer chased by a dog, but it is the clear vision of the cruel *future*."

Years later, Max accepted a friend's invitation to a rabbit hunt. He expected to observe, but his friend persuaded him to shoulder a gun. As the two walked along, a hare jumped out in front of them, and the friend yelled "Fire!" Max did, purposely aiming too high, but then he "heard a scream from some living creature. I do not know whether it was that hare or another animal, but I had certainly hit some living being, and there was no way to find it in the woods and end its pain." He never raised a gun again, and he never forgot the scream.

Gustav recognized Max's sensitive nature as akin to his own. Aside from his children, his most passionate attention went into his work, a quest for understanding the nature of life. He developed a method for constructing wax-plate models of biological organs that allowed him to demonstrate the structure of the mammalian heart and to discover a septum in the upper part of the atrium. His investigations of factors affecting sex ratio in the birthrate of frogs showed that poor nutrition favored the development of females. He experimented with the regenerative capacity of tadpoles by cutting up various species while they were still supplied with yolk, creating artificial unions and observing differences in growth and development. A "newspaper" published yearly by the university's young lecturers spoofed Gustav's efforts:

Two frogs made love in a pond,
Born waited in vain for their spawn,
Inseparable they did seem,
But both were masculine.

Gustav indulged his love for biology at serious emotional cost. From his start at the Institute of Anatomy in 1874, he had been under the direction of Professor Karl Hasse. The German laws of 1871 that legislated complete equality of Jews apparently had not diminished Hasse's anti-Semitism. He treated Gustav with disdain, ridiculing him in public and hampering his advancement. While Gustav looked on in chagrin, Hasse would reexamine students in the dissection lab in order to double-check Gustav's assessments. When Gretchen was still alive, Gustav had written to her, "The professor torments me . . . [I will] pursue the matter with Hasse himself to show him clearly and sharply how very much he damages me." Unfortunately, Hasse, like many other Germans, was not to be moved.

The "truly bold dream" Gustav had written Gretchen about included finding a position at another university, but opportunities for Jewish scientists were scarce. At the University of Halle, officials told him that employment would be conditional on his being baptized. They felt that having one Jewish scientist on the staff—a physiologist named Bernstein—was enough. Gustav refused the offer. He thought that religion should remain private. He would not renounce his religion for convenience, even though he did not adhere to its beliefs. After all, his father and grandfather had not converted in the face of anti-Semitic pressures.

Indeed, Marcus Born, Gustav's father, had faced a similar decision. After establishing his first medical practice in the small Prussian-Polish town of Kempen, he had riled his Polish patients by supporting Germany in a political dispute and they abandoned him. Throughout a three-year hunt for another medical position—during which he worried constantly about supporting his wife and infant son—conversion would have simplified matters considerably. Instead, he only searched harder.

Marcus made this decision even though he was not religious, describing himself as "a bit of a heathen" to his fiancée, Fanny Ebstein, before their marriage in 1849. The designation was by way of explaining that he could not tolerate Jewish dietary laws, such as the separation of meat and dairy products, or accept the concept of a wrathful God.

> I do not love the very old gray selfish ruler . . . who gets yellow with anger when he gets the smell of buttered chicken up his nose. Either he does not know what is good or he begrudges it. Father Zeus did not bother about the fleshpots or milkpots of mankind. He looked at their strengths as in Hermes or Theseus . . . and lifted them up to himself and the other gods in heavenly Olympus. . . . What do you think of this lovely region in which God is not a strange terrible unfathomable might on the other side of the stars, but lives in you and in everything around you?

As Marcus's son, Gustav grew up on this philosophy. Their family by this time had resettled in the German town of Görlitz, which had very few Jews and no noticeable anti-Semitism. His parents belonged to a private club whose members were for the most part Gentiles. The household was nonkosher; Gustav went to the local Gymnasium rather than to a religious school. "The loveliest feast" of the year was Christmas, complete with presents and a tree that the four children decorated with their own handiwork of gilded apples, long paper chains, and strings of raisins and almonds. For the children, at least, assimilation was a fact. Unusual for most Jewish families, they unselfconsciously thought of themselves as Germans whose religious background was incidentally Jewish.

Not surprisingly, Gustav brought these same values into his own family, raising Max and Käthe to be freethinkers with respect to religion. Still, he wanted them to know the Bible because so much of western culture was infused with its images. With no Jewish instruction at Wankel's Knabenschule, Max took the school Bible class taught by a Lutheran parson. On the first day, the parson asked Max to recite the Lord's Prayer. Max proudly began to recite the version learned from his sister's Polish Catholic wetnurse years earlier. The parson interrupted him several times to correct his wording to correspond to the Protestant form. From then on, the other boys in the class subjected Max to merciless teasing. The effect on him was profound, "the beginning of a deep resentment against religion."

In the years after Gretchen's death, Gustav met and courted a few women. His quest did not find him a second great love, but he still wanted to remarry so that his children would have a mother. Friends thought they had the perfect match for him: a good woman in her late twenties—approaching spinsterhood—who could both care for his children and provide a dowry that would end his financial dependence on Gretchen's parents. Gustav got to know Bertha Lipstein through letters, an acceptable practice at a time when some marriages were still arranged. In July 1892, six years after Gretchen's death, they wed. Max was nine and a half; Käthe was eight.

The new Mrs. Born came from Königsberg in East Prussia. Bertha Lipstein was one of a dozen children of a wealthy timber merchant who had started out in the Russian part of Poland and moved west to escape violent pogroms. Max found his stepmother a kind and honest person. His cousin Helene Schäfer thought "she was not charming or beautiful like the first, but had a splendid character." Later on her granddaughter described her as very kind, shy, and lacking in social skills.

To Marie and Salomon Kauffmann, however, Bertha was a poor substitute for Gretchen. Max later summarized their reaction. "How could Gustav, the lover of the fair and charming Margarethe Kauffmann, marry this unattractive, shy, and reserved spinster!" Their dismay, however, was not based solely on the new bride's lack of charm and beauty; there was a more complex side: Bertha's strong East Prussian (i.e., "Jewish") accent unmistakably linked her to the ghetto and to ethnic Russian and Polish Jews flocking to western cities such as Breslau. Bertha's conspicuous ethnicity seems to have threatened the Kauffmanns' sense of belonging to mainstream German society. Their attitude may have rubbed off on Max; some of the Lipsteins later had the impression that he too would have liked Bertha to be "more German."

The crucial milestone of legal equality for Jews in 1871 did not create a truly open society any more than it ended the anti-Semitism of Gustav's boss. Barriers remained even for those Jews who wanted to assimilate fully into local and national customs, who simply wanted to be regarded as Germans, just like their Protestant and Catholic neighbors. Striving for parity, they willingly abandoned any vestige of their religious background, embraced German culture and higher education, accumulated wealth, and, perhaps most importantly, tried to erase the signs of their ethnic roots. Boarding school in Switzerland, however, had not succeeded in erasing Bertha's accent.

Both the Kauffmanns and Borns were as assimilated as German society would allow. "They believed they differed from their neighbours only in respect of religion," wrote Max, "for which they had hardly any use, and that they were Germans by culture and custom." This identity was so strong that the late nineteenth-century novel *Soll und Haben* (*Debit and Credit*)—a story with sharply drawn, negatively stereotyped Jews who contrasted with the story's noble German merchants—was a favorite in their households as it was in other assimilated ones. Max read it as a boy and enjoyed it. Rereading the novel as an adult, he was struck by the contradictions. "[The families] obviously considered themselves as belonging to the class of noble German merchants, and having nothing to do with the 'Galician' small traders with their 'Yiddish' dialect and low standard of education."

Russian and Polish Yiddish accents and ethnic clothing were, for some assimilated Jews, an embarrassment, a reminder of the recent past and a liability in a society rife with anti-Semitism. Yet the Borns heartily welcomed Bertha into their family. They were generally more welcoming than the Kauffmanns, but they also had something in common: not only a reasonably similar background, but a shared dislike of Gustav's intrusive, self-important in-laws.

Gustav's father, Marcus, had grown up in the small Polish town of Lissa. His childhood was typical of Jewish experience of the 1820s. He lived with his parents and six brothers and sisters in the town's Jewish quarter, which was considerably apart from non-Jews. Marcus went to the *cheder*—a school often crammed into a room in the teacher's house—where he mainly studied Torah and the Talmud. His father was a trader whose life was largely at the mercy of the governing authority, which taxed the Jews excessively and did not allow them to move to another town, own real estate, or educate their children at the local Gymnasium.

But as the Jews of Lissa knew, this disconcerting situation was far better than the constant fear felt by those who lived further east, where random violence by the peasantry or the more systematic looting, burning, and raping by a Cossack-instigated pogrom were a constant threat. The prosperity of Lissa, a market town in the province of Posen dependent on its Jewish traders, tended to shield Marcus's family and others from such life-threatening rampages.

Except for a brief interlude under Napoleon, the Prussians had controlled this region of Poland since the country's partition between Prussia, Austria, and Russia in 1795. After the partition, the Prussian government found itself with a large Jewish population that was culturally, physically, and economically separate from the Poles and, because of governmental economic restrictions, considerably impoverished. It was a population the government did not want, so it set about concocting schemes to force the poorest Jews to migrate from the province. When population flows proved too fluid and borders too porous, the Prussian government gave up this approach and decided that what it could not get rid of, it would change. The official policy became one of "Germanization," or cultural assimilation, a way to make the Jews seem less foreign and threatening.

Jewish assimilation was a gradual process. By 1830, government policy allowed Marcus and his brothers to attend the German Gymnasium in town. Moreover, in 1835, his father, who was modestly successful at his trade, became a naturalized citizen of the state of Prussia. A German census taker, walking through the dusty streets of the Jewish quarter with printed form in hand, had knocked on the family's door to ask six important questions: Do you speak German in professional activities? Have you adopted a family name? Is your lifestyle irreproachable? Have you been a permanent resident of the Duchy of Posen since 1815? What is your occupation? Finally, and most important, are you economically secure? In that year only 5 percent of Posen's Jews fulfilled the six criteria necessary to earn citizenship. For those special few, the new status opened the door to Germany. Marcus was able to study medicine at the University of Berlin and left the Jewish quarter of Lissa far behind.

As the list of census questions shows, last names were an important factor to the government. Before the 1795 Prussian takeover, most Jews did not have family names, referring to themselves by the patronymic "son of" (or "ben" in Hebrew, as in Isaac ben Solomon). To the Prussians, last names were essential, not only for cultural conformity but as a system for tracking Jews. At first, Jews simply chose a name based on a descriptor such as a trade (Schneider for tailor) or a physical trait (Schwartz for black). In some towns, though not in Lissa, local clerks contrived to sell more desirable names such as Rosenthal (valley of roses) to the highest bidder and sadistically assigned humiliating names such as Affenkraut (monkey weed) to less affluent Jews.

For reasons unknown, Marcus's family had taken the name Buttermilch. Marcus later wrote to a friend that his father had lived honorably with the name for years, but he felt differently. "How is it possible to be called Buttermilch?" he wrote. "I am surprised how I managed to put up with such a sweet-sour name." And so, without their father's knowledge, the five sons petitioned the royal magistrate in Lissa for a change of name. On February 14, 1842, the Buttermilch sons became known as Born, German for "rejuvenating spring." They were now, as Marcus said, "one syllable men" in the High German tradition.

Gretchen Kauffmann Born's ancestors came from the other side of the Oder River from Prussian Poland—in Prussia itself. In the eighteenth century, throughout the German states, local authorities forbade Jews to live within the walls of towns and cities. Only with the local ruler's approval were a few exceptions granted. Having successfully petitioned the crown for special privileges in Frankfurt am Oder, Gretchen's ancestors established a business there in white goods—linens and bedding—in the early 1700s. In addition to moving inside the city wall, after much argument they were permitted to buy a home, largely because no one else would buy the deteriorating building. The next few generations managed to amass wealth in spite of the restrictions on professions, property ownership, and residence. As the nineteenth century began, the descendants of Chajim ben Jekuthiel Kauffmann Präger became known as Kauffmann (German for businessman).

Gretchen's grandfather—Salomon's father—left the family business in Frankfurt am Oder, crossed the river into Silesia, and set up a small white goods shop in Schweidnitz. By 1843, he had expanded to Breslau, where he sent his oldest son, nineteen-year-old Salomon, as the firm's representative. Well-connected by rail, Breslau had an active market that supported a thriving textile smuggling business over the Polish border, and Salomon's office prospered. His business skills had sufficiently matured by 1851 that the firm sent him to London's Great Exhibition to display woven goods as one of more than 13,000 exhibits. There he marveled at spectacular fountains, fireworks, and most extraordinary of all, the Crystal Palace—a dazzling feat of Victorian architecture, more than 1 million square feet with 300,000 sheets of glass. But for Salomon, even the Crystal Palace was eclipsed by the superiority of English woven goods produced by mechanized looms.

On his recommendation, the firm replaced old-fashioned handlooms with water-powered, mechanized ones. Within a year, they were well on their way to real prosperity, shipping white goods throughout Prussia and the provinces that were equal in quality to those produced in Britain. By the time Gustav Born met Margarethe Kauffmann in 1880, her father and his brothers owned four textile factories in various parts of Silesia and were one of Breslau's wealthiest families.

German law and social structure had allowed Jews only a few narrow routes out of the ghetto. The Kauffmanns had successfully taken that of trade and business. Marcus Born had taken another—medicine. Law would have worked equally well. In entering academia, however, Gustav Born broke with custom and practice. Universities were state institutions, and professors were state employees. The state did not want to hire Jews; and if it did, it did not promote them.

For Gustav, marriage to Bertha severed the two most irritating of the Kauffmann strings: Grandmother's reign over the children—for which Max was pleased—and Grandfather's domineering financial control. First on Gustav's agenda was to move out of the apartment belonging to the Kauffmanns and into one on the other side of town. About a year later, Gustav and Bertha further consolidated their independence by having a son of their own, whom they named Wolfgang.

Bertha, who had no more interest in the Kauffmanns than they in her, minimized their social contact, including the summer vacations in Tannhausen, although the children's Saturday dinners at the Kauffmanns were still a must. With their cousins, Max and Käthe entertained their grandparents with plays and recitals. As they grew older, they took part in social events at the house and in the city. The great violinist Joseph Joachim sent Max a postcard as a memento of Max's piano accompaniment at one of his grandparents' musical soirees. Still, contention brewed beneath the surface of these visits. The Kauffmanns could not resist using Saturday dinners to criticize Gustav and Bertha's child rearing. In a sharp letter to Marie—the chief agitator—Gustav retorted that he was concerned for Max and Käthe's health and education, not for their introduction to Breslau's theaters, symphonies, and parties.

Ten-year-old Max called Bertha "Mama," appreciated her many kindnesses, and essentially enjoyed his new life. Only minutes from their new house lived one of his Kauffmann cousins and playmates, Franz Mugdan. After school, their favorite activity was creating colossal buildings out of toy bricks. Whether they followed the directions or not, they never liked the results—a reaction confirmed by their parents' raised eyebrows; so Max decided to try something more creative. Out of bricks no more than three inches long he built a vaulted bridge with a span of more than a yard, relying on no glue or external supports but solely on the principles of physics. Even Grandfather Kauffmann was impressed. Max never forgot the special moment

> when the last wedge was being fixed into the top of the polygonal arch resting temporarily on a pile of books, [and] the books were removed: would the bridge stand, would our purely intuitive calculation of the forces be correct? I think that this was the first time I had that kind of scientific adventure which consists in bringing a rational construction to the test by experiment, and it gave me the first idea of the meaning of definite rules of nature.

More often, Max and Franz built intricate cities of houses and castles, guarded by tin soldiers. With small spring-loaded guns firing shot or peas, they bombarded each other's creations. Houses and castles tumbled down, miniature bricks skittered about, tin soldiers fell to the rhythm of their loud "ka-booms." To achieve greater realism, Max even invented a method to fire shot from the little guns that resulted in a whiff of smoke along with a loud bang.

One night at supper, Max boasted to his father about the shattering defeat his army had inflicted on Franz's that afternoon. Gustav responded, "Shall I tell you about my own experiences in the war of 1870–71?" Max and Käthe listened, keen to hear of their father's exploits during Prussia's decimation of the French grand armies.

Gustav told them that at the outbreak of war, he was a medical student and had patriotically enlisted as an orderly, transporting the wounded and assisting physicians in the field. He saw no glamour, no romance, no glory—only the realities of war in the soldiers' broken bodies and spirit. And so he told his children that night

> about trains crowded with soldiers who had left their personalities behind and become numbers in a machine; about field ambulance stations where wounded and maimed men were brought in and operated on; about field hospitals full of suffering, death and horror; he spoke about the fate of the civilians in the battle zone, whose houses were burning, and about the famine in the beleaguered city of Paris.

With "suppressed passion," Gustav described "the degradation of the human personality to a cog in the wheel of the war machine, deprived of will and dignity." Max, who was about thirteen, did not fully comprehend this horror—but years later he would.

By this time, Max had entered the Kaiser Wilhelm Gymnasium, a school in the humanistic tradition, with Latin and Greek as the most important subjects, followed by mathematics. Max's school reports rated him, for the most part, *genügend*, or satisfactory, midway on the spectrum that started with *sehr gut* (very good) and ended with *nicht genügend* (not satisfactory). Occasionally, in physics he would get a *gut*, less often in mathematics, where his teachers' method relied on rote memory, a decided weakness on his part.

Only one teacher from that period remained memorable to him: Erich Maschke, who taught physics and mathematics in a manner that did not require Max to rely on memory. Instead, Dr. Maschke explained the larger concepts—"the meaning of the whole theory to which a theorem belonged, about its history and its practical applications." Max was excited by the world of ideas and the integration of mathematical concepts. The other subjects—Greek, Latin, German, history, geography—that required rote learning, were difficult and boring. "Content to be just average," he did not work hard, and his father did not push him.

Outside of school, Max had plenty of close friendships with his cousins, but at school, he was a solitary person, an outsider. He never made any real friends there—perhaps, he thought later, because of his shyness. As for girls, he met a few whom he liked, but felt too shy and awkward to pursue them—except for one, for whom he had developed a special longing. He expressed these feelings so outwardly when playing Chopin's E-minor prelude that his piano teacher, who was a bit of a ladies' man and had an ear for these things, exclaimed, "Max, you are in love. I hear it in your playing." After an innocent flirtation, however, the girl rejected him, closing even that emotional window.

A photo of fifteen-year-old Max shows a good-looking boy with an air of sadness and vulnerability. His early years without a mother had been difficult. Music was—and remained—an important release for him. Later, when trying to explain his emotional reserve, he thought that not having a mother "deprived us of a natural outlet for our feelings . . . So a kind of shell had developed around me, and nature had not given me the gift of expressing my emotions in words." Through music, he could at least express "a morsel" of his feelings. Under his teacher's tutelage, he developed a talent that allowed him a lifelong opportunity.

One evening in the winter of 1897–1898, Gustav arrived early for a dinner party at his sister Selma's and lay down to rest. Shortly afterward she heard him call for help. Hurrying in, she saw him covered with perspiration. Her husband, Joseph, who was also Gustav's physician, found his pulse and heart weak enough for him to die at that moment. He survived that night, and Bertha was able to take him home, but he suffered from a weak heart from then on. The next day, Uncle Joseph told Max about the serious state of his father's health—news that Max could plainly see was true. He saw his father as a "completely exhausted, broken man," as though the "worries of his life had worn him out." The "shadow of his suffering" descended over the family.

Gustav continued to work at his laboratory on a limited basis and in the evenings stayed home. In earlier days, Max would sometimes go by his father's lab to walk home with him, the two stopping on the way at a beer garden to sit under the Chestnut trees and eat scrambled eggs or Vienna sausages—peaceful evenings and a joy for Max. Now, the evenings were spent at home, but were still a joy. As Max recalled,

> We profited from that as he liked to chat with us in his unique way, mixing humour and earnestness, or to read to us aloud—if his breath allowed the exertion. During his last winter he read the whole of *Faust* to us; or rather he recited most of it by heart.

Sometimes a few of Gustav's colleagues would stop by to discuss their research along with politics and philosophy. Uncle Joseph visited, as well as Albert Neisser, the dermatologist who discovered the gonococcus. Occasionally, the anatomist Karl Weigert came from Frankfurt am Main, bringing his cousin, Paul Ehrlich, the future Nobel laureate who invented Salvarsan, a remedy for syphilis. These serious scientists welcomed Max into their circle. For a teenager who came alive in the world of ideas, it was inspiring—a final and lasting gift from his father.

Gustav's own research at the end of his life focused on determining the function of the corpus luteum, a small yellowish body in the ovaries of mammals. Its function was unknown at that time, along with much else about the gestational process. Gustav hypothesized that the corpus luteum was an endocrine gland whose secretions prepared the uterus to accept and protect an egg. He had found that its structure was close to that of other endocrine glands and that it was present only in mammals that formed true placentas. Although close to unraveling the mystery, he was just too sick to carry out the additional research necessary to test his hypothesis. On his deathbed, he called his former student Ludwig Fränkel to him and asked him to continue his work. Fränkel did so, and, with his partner Cohn, he later confirmed his teacher's theory. Within thirty years, other researchers had proven that the secretion from the corpus luteum was the hormone progesterone, critical for maintaining a pregnancy to the end of gestation.

Gustav Born, a scientist who was never appropriately recognized, knew none of that. In the early morning hours of July 6, 1900, he died from a heart attack. He was forty-nine years old. Max was not yet eighteen, and Käthe was just sixteen.

Gustav's family placed much of the blame for his early death on the poisoned atmosphere created by his supervisor, Professor Hasse. Even in death, Gustav was not free from his tormentor of twenty-six years. University officials asked Hasse, who was dean of the medical faculty, to deliver a speech at the ceremony following the funeral. It was a large and formal event with a performance by the university choir. Repeatedly in his speech, Hasse referred to Gustav as his *Untergebenen*—his subordinate, or inferior. Cousin Helene Schäfer, who was in the audience sitting just behind Max, saw Max become paler and paler as Hasse continued. At one point she feared he would spring up and throttle Hasse.

One obituary for Gustav began, "Professor Born is dead. This frightening news ran through our city in the early morning hours of July 6 and shattered everyone's heart." To his students and colleagues, it was the loss of "one of the best." Max himself simply described the days following his father's death as a "nightmare."

TWO

A HIGHER DESIRE

THE WEEKS FOLLOWING HIS FATHER'S DEATH SEEMED TO MAX AS THOUGH THEY WERE "covered by a veil." On the surface, life gradually returned to normal: classes at the Kaiser Wilhelm Gymnasium; sharing meals at home with his sister, half brother, and stepmother; visits with his Kauffmann grandparents and cousins. Underneath, however, was a void he could not bear to contemplate. At seventeen, Max retreated even further into shyness. Growing up without a mother was a loss that he felt and talked about throughout his life, but it took many years before Max could write about the loss of his father.

In the next year, Max finished at the Gymnasium and passed his leaving examination, the *Arbitur*, the credential needed for admission to a university. Since he had never been an outstanding student, his stepmother was so relieved that she gave him a party, even though she was still in mourning for Gustav.

Part of the credit for Max's success belonged to a family friend named Lachmann, a former medical student of Gustav's then in his late thirties. When a Kauffmann cousin had fallen ill some years earlier, Gustav had arranged for the newly accredited Dr. Lachmann to travel with him to Egypt as his medical advisor—and treat his own incipient tuberculosis. Back in Breslau, he became Max's tutor. His specialty was balneology, the therapeutic use of baths. During the busy spring and summer seasons, he supported his family by practicing in the spa town of Bad Landeck; in the winters, he came to Breslau to take courses at the university. Max recalled, "He was then deeply interested in astronomy and mathematics and it was much as a result of his influence that I took up these subjects. But I owe to him not only the deciding direction of my scientific interest, but also a general widening of my outlook on life."

For Dr. Lachmann was a socialist, an active member of the Social Democratic Party—at that time, synonymous in most people's eyes with communism. He offered up the troika of Marx, Kant, and Hegel to Max and asked probing questions about the contrast in living conditions between Max's family and the workers who streamed out of Grandfather Kauffmann's factories. "He was not afraid of proffering views which, in our bourgeois society, were still considered despicable." Grandfather Kauffmann, who provided well for his workers, had believed that the individual had to fight for his place. Max knew that he, like most other property owners, would have regarded Dr. Lachmann as an "enemy of society."

For Max, however, Lachmann offered a window onto a broader world. When he started at the University of Breslau in the fall, he changed his academic course from studying mechanical engineering to philosophy. Schoolboy memories of working with his fret saw and lathe—building little instruments for Dr. Maschke's Gymnasium physics class as well as an entire orchestra for his grandfather—could not compete with "the ancient metaphysical problems of space and time, matter and causality, the roots of ethics and the existence of God."

After a broad education in the Gymnasium, a university student was expected to focus immediately on courses related to his future. The German university was organized into separate faculties to facilitate this. Future lawyers studied in the Law Faculty, future doctors in the Medical Faculty, and ministers in the Theology Faculty. The Philosophical Faculty, which Max entered, comprised the sciences, mathematics, humanities, and languages. Students faced few academic requirements: They could move around to different universities, and were not obliged to attend specific courses but only to complete a certain number of terms at one university to be considered for a degree. They ended with only two options: either the state exam for a teaching certificate or the oral for a Ph.D.

Gustav Born, considering his own narrow specialization a major shortcoming, had adopted the dictum "what Hans doesn't learn, Hans never learns." Before his death, he counseled his son to explore a variety of intellectual areas. So Max filled his course list at the university with philosophy, the sciences, and mathematics—this last only because it was required for astronomy. His experience with the subject at the Gymnasium had left a bad taste.

Max soon saw that his choices did not stand the test of maintaining his interest. Aristotelian logic with syllogisms such as "All men are mortal; Socrates is a man, therefore Socrates is mortal" seemed to Max "the epitome of triviality." Zoology, physics, chemistry, and climatology bored him. Of the sciences, only astronomy excited him, especially cosmology and the relationship to infinity, but the professor did not focus on this aspect. His penchant was for endless measurements and calculations that Max found less congenial, as his own were full of errors. In front of his fellow students, the professor admonished Max that he needed to take a course in elementary arithmetic. Max's self-confidence and budding interest fizzled.

The lure of higher mathematics was different. Geometry and calculus introduced him to concepts that "mean nothing unless applied in a definite system of ideas to a definite problem where they can be made significant for that problem." At last, he was intrigued and challenged. Whereas the concept of infinity in philosophy had seemed "veiled in a mist of paradoxes," infinity in geometry became the symbolic point formed by the intersection of two parallel lines.

Here he found stimulating companionship. From these math classes grew a small circle of committed students who talked into the night. In Max's first twelve years in school, he never made a close friend. Those from this math circle would last his lifetime.

The social side of university life was dominated by *Burschenschaften*—male societies or clubs that engaged in traditional drinking and bravado but with a particular German twist—dueling—that harkened to bygone days. Members wore dueling facial scars as badges of honor. When a friend suggested that Max join a literary society true to this custom, he rejected the offer. Dueling, codes of honor, and the exclusively Jewish nature of this particular society were distasteful to him. But Max, who summarized himself as "sophisticated, shy and therefore lonely," did allow Dr. Lachmann to coax him into joining a mathematics society that did not require dueling.

The group's weekly meetings, held in the back room of a small beer hall that reeked of alcohol and tobacco, began with a dull mathematics lecture. Then the beer hall supplied the fuel for the rest of the evening—drinking games involving rousing renditions of German folk songs and strict ritual, regulated by a mixture of German and Latin, the rules of which required frequently downing a liter of beer. As their vigorous gatherings often led them into collisions with other societies, the group did encourage dueling lessons. Max eventually learned the peculiar art, which in this form was definitely not a sport: Rules called for the duelists to stand motionless except for the rapier arm that could only thrust straight forward, and the knuckles that acted as a pivot for the rapier, the paragon of the stiff Prussian. Making it through one year with no duels and unscarred cheeks was enough of this kind of socializing for Max.

Life in the Born family apartment was less appealing for Max without his father. Bertha was still a "hen who sat on a duck's egg"—kindly but socially awkward. Eighteen-year-old Käthe was surrounded by their many female cousins and numerous beaux—young men whom Max considered "priggish." Their penchant for reciting Nietzsche and Rilke in dimly lit rooms was too much for him. (Years later, looking back, he would see himself similarly, a "stupid little prig" . . . "primitive, uneducated, inexperienced, and narrow-minded.") But home served at least one useful purpose: In the parlor before an audience of Bertha, nine-year-old Wolfgang, and Käthe, Max assembled the first of his many trios—a violinist and cellist who joined him in performing Haydn and Mozart.

As a young adult, Max found the homes of the highly cultured Kauffmann women more alluring. Grandfather Kauffmann had died just after Gustav Born; his daughter-in-law, Luise Kauffmann, took over the family role of symphony benefactor, filling the Kleinburg villa with all the great artists who came to Breslau. On the two Bechstein grand pianos in her music room, Max had the pleasure of hearing Europe's finest pianists. In the old days, Johannes Brahms, Clara Schumann, and Joseph Joachim dazzled the audience. Max now listened to the artistry of violinist Pablo de Sarasate and pianists Artur Schnabel and Edwin Fischer. Schnabel's future wife, the singer Therese Behr was usually there. Gustav Mahler made an occasional appearance. At other times, Aunt Luise opened up the villa for sumptuous balls where Max—quite handsome and sophisticated in his tuxedo—danced the quadrille.

Across the city in another large villa, his mother's cousin Toni Neisser and her husband, Albert, a physician, maintained a grand salon. With dinner guests such as the poet Gerhard Hauptmann and the composer Richard Strauss, the Neissers listened to poetry and music within frescoed walls painted by the Art Nouveau master Fritz Erler; they moved among beautiful objects from Asia and India. In this rich company, Max was just another set of ears at the table, pleased to listen. "Villa von Neisser" was his refuge from home. In his childless "Aunt" Toni, the motherless Max found emotional sanctuary as well. She listened to his problems, shored him up, and introduced him with an almost parental pride to her influential friends. One was Albert's cousin Arnold Berliner, a physicist who worked in industry and who encouraged Max to pursue his scientific interests. Probably recognizing the young man's insecurity, Berliner playfully bet him several bottles of wine that in ten years he would be a professor. Berliner, who later became editor of an important scientific journal, often appeared at Max's elbow at small but important moments.

The tradition of the wandering medieval German scholars gave Max fairly frequent escapes from Breslau. Students changed universities to match the season, trading the winter term's big-city cultural events for the summer semester's outdoor sports and nature. In April 1902, Max and his cousin Hans Schäfer left Breslau for a summer in Heidelberg, nestled between the Neckar River and the hills of the Odenwald. On the hilly street leading up to Heidelberg's old Schloss, Max found a room in a large house, its vestibule guarded by an imposing painting of a knight in armor.

Having spent many summers together at Tannhausen (their mothers were sisters), Max and Hans also shared academic interests. Hans's was chemistry, and eventually he would take his interest in textiles and dyeing into the family firm. For now, their concerns were lighthearted. The first day of a mathematics class on determinants, when Max met the young geology student James Franck, began what Max described as a "happy-go-lucky" summer. As he said, "To my cousin Hans and to me it was very soon clear that Franck was our man."

The three mixed studying with a healthy appreciation for good company, often setting out to picnic on one of the Neckar's rocky islands. Testing their strength, Max and James rowed against the current; Hans, with one arm crippled from a childhood riding accident, helped with steering and navigation. Passersby on the riverbank encouraged the trio with comical cries. After a precarious landing, they might fill the evenings with talking, supping, and playing the three-way card game Skat.

On land, they ate and raised steins at local beer halls. At the end of such an evening, they would often commemorate it by sending a postcard to Hans's sister, Helene, in Breslau. One of these, drawn by the slightly unsteady hand of Mäckerle (as Max signed himself) portrayed dashing, dueling young men, all sporting mustaches, blood dripping from their epées. That was the beginning of May. By midsummer, when Helene came for a visit and they met her at the Heidelberg train station, James and Max had full beards. They had spent their razor money elsewhere.

Max good-naturedly called James "Noodle-head" because he could not decide what to study. Although similar in this respect, they were otherwise quite different; James would later describe Max as the classicist and himself as the romantic. To Max, James had a cleverness, a "peculiarly personal charm," and was "the best comrade in all activities." James praised Max's "goodness and intelligence." It was a friendship that would endure through extraordinary times.

Their happy-go-lucky summer ended with James deciding to study in Berlin and Hans and Max traveling to Italy, where Max suffered an asthma attack. He returned to Breslau to face the recurring dilemma of what career to choose. His blossoming interest in mathematics struck him as all too likely to lead to the discouraging life of a schoolteacher. The more exciting option of becoming a university professor revived his old insecurities.

Once again, Dr. Lachmann stepped in. He recognized the family wealth as an opportunity for Max to develop his interests rather than worry about his livelihood. If he failed—and Lachmann did not think he would—Max could fall back on these financial resources. So Max began the winter semester studying mathematics, physics, and astronomy. The following summer he spent in Zurich, dividing his time between hikes in the Alps and brilliant lectures on elliptical functions by the famed professor Adolf Hurwitz. Hurwitz must have recognized Max's promise, for he invited him and another student to his home for private lectures on the theory of functions of complex variables. Professor and student became friends. At Hurwitz's death, Max wrote to his widow, "The clarity of [your husband's] thoughts opened up for me wide areas of science."

Again Max returned to Breslau for the winter semester. A friend from the math circle, Otto Toeplitz, persuaded him to take a course on algebraic equations—a subject he found interesting, without suspecting how important it would turn out to be. He was more concerned about the Neissers' planned trip to the Dutch East Indies. What was the point of staying in Breslau without their sanctuary and stimulation? Once more, Max, now twenty-one, had to decide where to go and what to study.

Munich appealed to him, with mountain climbing in summer and cultural pursuits in winter. He now was leaning towards physics, having taken a course on Maxwell's theory of electromagnetism, which he would later describe as "a bad lecture, but . . . fascinating." But Otto Toeplitz—whom Max called his "self-appointed mentor"—argued for "pure" mathematics and more brilliant lectures at the University of Göttingen, where Professor Hurwitz's world-famous students taught. Max did not even know where the university was. Still, he took his friend's advice.

Göttingen's earth shook, at least figuratively, every Thursday afternoon when the "mandarins"—the high priests of mathematics—came together for their walk. Through the years, they had worked out the ritual. At 2:45, Hermann Minkowski left his house on Planckstrasse to walk two blocks to Wilhelm Weberstrasse. Perfectly timed, Felix Klein departed from the far end of Weberstrasse to stroll up the hill toward David Hilbert's front gate. As he walked past, Hilbert would come out so the two could meet Minkowski at the corner of Weberstrasse and Dahlmannstrasse at exactly 3:00. Then they strode up into the hills to have coffee at the Kehrhotel and to settle any raging issues. They were, after all, their generation's luminaries in the world of mathematics.

They followed closely in the footsteps of earlier mandarins—Carl Friedrich Gauss, Bernhard Riemann, and P. G. Dirichlet—who had made the University of Göttingen a world-famous center of mathematics in the nineteenth century. Now, those of the twentieth century had inaugurated another era of greatness. Felix Klein's reputation from his masterwork on geometric functions, David Hilbert's from number theory and the foundations of geometry, and Minkowski's from the geometrical theory of numbers had catapulted the university's standing in mathematics beyond that of the University of Berlin. And their personalities enhanced their mathematical renown: Klein, formidable with his leonine beard, supplied bureaucratic clout; Hilbert, exuding energy, offered eccentricity; and Minkowski, pince-nez delicately perched, provided sensitivity.

Born arrived in the small Hanoverian town for the beginning of the summer semester in April 1904. Göttingen possessed neither the magnificence of high mountains nor Breslau's culture. Instead, the highlights of the town were Bierfests in the sixteenth-century Black Bear Tavern, with sagging half-timbered façade, and in Die Mütze, a glorious fifteenth-century house converted into a wine bar, covered with rich biblical carvings and secular themes: two among scores of half-timbered buildings. Then there was the graceful statue of the *Gänsemädchen*, a young girl holding three geese under her arms and in a basket, with a fountain at her feet and a lovely rococo cupola over her head. The medieval Rathaus, still with massive crenellated walls, dominated the large square in which she gracefully, timelessly carried her wares.

The university, completely integrated into the town, brought it prestige. Classroom buildings dotted many narrow winding streets. Boarding rooms or homes of professors' widows housed students. Born recalls the town as "small and tame," but his first impression exiting the train must have bordered on shock. He had gone from balls and symphonies to a rural, medieval town that catered to a renowned university and little else. To his credit, Born adapted quickly and discovered other indulgences—namely, the maelstrom that was Göttingen mathematics. Within days of arriving, he became something new: an insider.

On the first day of Born's class with David Hilbert, the professor asked interested students to submit their class notes for consideration as his scribe. The lectures were a creation of the professor, not a recitation from a textbook; in a tradition unique to Göttingen, a student scribe wrote up each lecture for reference in the students' mathematics reading room. Born and about six others volunteered their notes. Thanks to earlier advice from Toeplitz, Born was already well practiced in the art of careful note taking. At the start of the second class, Hilbert announced, "There is one manuscript by far excelling all the others"—Born's. Born retrieved his notes from Hilbert, embarrassed by the attention but pleased to be chosen. He had taken his first step toward sharing the intellectual enthusiasm of the mandarins.

Being Hilbert's scribe meant making regular, invaluable visits to Hilbert's house. Whether he was clarifying his class notes or following as Hilbert worked out his theory of functions, Born witnessed the Hilbert Principle in action: strip any problem of nonessentials, simplify without losing the core, and solve. Properly specifying the problem built a bridge to the solution. As Hilbert said, "a perfect formulation of a problem is already half its solution." Of course, this was easier said than done.

Hilbert stressed searching for basic laws that provided the underpinning to scientific knowledge. Born's clear, neat hand duly recorded Hilbert's discussion of the axioms of mechanics, which, in his lecture, the professor listed grandly under the title "the axioms of natural science." In the first sentence, Hilbert acknowledged that the task of discovering these axioms was far from complete, but that such a drawback did not limit his conceptual dream. Without further ado, he addressed himself to mechanics, the area he evidently found most suited to his deductive approach. In his new scribe, Hilbert's concepts fell on a fertile mind. Just as important for Born, Hilbert liked him.

Born's introduction to Professor Hermann Minkowski came through family connections: Bertha Born had known the professor since their days together at dancing school in Königsberg. When Max delivered a letter from her to the Minkowski household, he promptly received an invitation to dinner on Sunday afternoon.

Any young student who felt reticent at his first meeting with an esteemed professor was likely to endure a painful visit with the equally reticent Minkowski. Someone as shy as Max Born could expect long awkward silences. Fortunately, the other guest, Dr. Constantin Carathéodory, a young Greek mathematician, charmed his fellow diners with tales of his adventures as a railroad engineer in Egypt. At the end of the dinner, Minkowski invited his guests to join him and his wife for a hike with the Hilberts to Die Burg Plesse, a crumbling medieval fortress a few miles away.

Hilbert and Minkowski were close friends, having grown up in Königsberg together. Through Hilbert's doggedness, they succeeded landing in the same faculty two years earlier. Now they spent Sundays together, enjoying picnics, hikes, or dancing at Maria Spring, a nearby beer garden, always discussing and dissecting as they went.

From his first days in Göttingen in 1895, Hilbert had operated outside the behavioral confines of German academia. An inviolate social barrier placed students on one side and professors on the other; they mingled in the lecture hall only. The irrepressible and unorthodox Hilbert invited his students for billiards, walks in the hills, lavish dinners, and afternoons in his splendid garden. While he filled their minds with mathematics, politics, and science, his quirkiness and egotism made him in equal parts challenging and supportive, blunt and dismissive. The stimulating atmosphere surrounding Professor Hilbert fostered the creativity that became Göttingen's hallmark.

Born quickly advanced from lecture scribe to assistant. Although one of many recipients of his mentor's intellectual largesse, he was the first to hold this unpaid, semi-official position. Most mornings, Born arrived at Hilbert's house to find Minkowski already there. Together, they framed the general principles of Hilbert's next lecture. Then Hilbert left, believing inspiration would fill in the details as he spoke. Minkowski and Born walked to Minkowski's house, the reticence of their first meeting long over, their conversation invigorated by the morning's mental exercises.

When Born had first arrived in Göttingen, he felt too timid to address students two years his senior. Now, as the valued young colleague of two renowned mathematicians, Born lived in a heady world and thoroughly enjoyed it. As Göttingen lured his mathematical friends from Breslau, he took pleasure in serving as their link, introducing the expanding Breslau Group to Hilbert, Minkowski, Carathéodory, and others in his new circle. His position in the mathematics hierarchy of Göttingen appeared firm.

Born had a different relationship with Felix Klein, the oldest of the three luminaries and the administrative hand behind Göttingen's success in mathematics. Felix the Great, as he was nicknamed, had strong ties with the minister of education, from whom he used to win extra resources for Göttingen mathematics. Whereas Hilbert approached each lecture with a few general principles to leave with his students, Klein dazzled with his breadth; as described in Born's words, "guiding [the students] in endless winding paths through apparently impenetrable ground and halting at each little hilltop to give a survey over the area covered."

Some students preferred Klein's courses to Hilbert's, but Born found them "too brilliant" (although later, his attitude shifted). Every time he signed up for one he found himself skipping. Unfortunately, as he would discover, even in a crowded hall, Klein noticed who was missing.

With the support of Hilbert and Minkowski, Born aimed for a Ph.D. in pure mathematics. In his second semester there, he took Hilbert's course, The Quadrature of the Circle, which examined unsolved problems in mathematics. Born gave an excellent presentation of Hilbert's proof of the transcendental character of the ratio of the circumference of a circle to its diameter—that is, that π cannot be constructed with a compass and ruler. For Born's dissertation topic, Hilbert suggested that he work out the proof of the transcendental character of the roots of the wavelike Bessel function. (Mathematically, *transcendental* has a clear meaning: in this case, that no algebraic equation with integral coefficients exists that has π as its root.) Weeks later, Born confessed to Hilbert that he could not solve the problem, that his skill at writing out mathematical presentations outpaced his ability to formulate this one. Hilbert's response—which Born later remembered simply as thoughtless and stinging—led Born to reexamine his mathematical ability. He decided that he had no genuine gift. He did not despair—his recent successes had been too encouraging for that—but decided to explore the possibilities of applying mathematics to science.

Particularly in Göttingen, where mathematics was the master discipline, the boundaries between pure mathematics, applied mathematics, and the natural sciences were a blur. The tentacles of the mathematicians (who were only "pure" some of the time) regularly reached into the realm of physics, if in distinctively different ways. These mathematicians were academic descendants of Carl Friedrich Gauss, who advocated expanding mathematical thought and applying advances to unsolved problems in physics. Hilbert, who once pronounced that physics was too important to be left to physicists, made "the mathematical treatment of the axioms of physics" number six out of twenty-three in his famous 1900 list of unsolved mathematical problems that could fundamentally contribute to the future development of the field. Minkowski had, throughout his career, investigated purely physical phenomena such as rigid bodies. A few years later, in his famous contribution to the development of relativity, he would write of "the pre-established harmony between pure mathematics and physics." Above all, Klein focused on the importance of mathematics to technological advances. For years he had worked to create an alliance for educating personnel who were mathematically sophisticated and could direct technological advancement in Germany.

Göttingen professors often offered interdisciplinary seminars to explore a specific physical problem. In the winter semester of 1904–1905, Klein and the professors of applied mechanics Carl Runge and Ludwig Prandtl held a seminar on elasticity, in keeping with Klein's philosophy of practical mathematical application. The following summer semester, Hilbert, Minkowski, and Emil Wiechert, professor of geophysics, explored the theory of electrons. Born participated in both, and, as he later wrote, "my science future was determined, [but] not by my deliberate choice."

The seminar on elasticity—the physical property that allows materials to return to their original configuration when stress is removed—did not in itself compel Born. He found it too applied, too practical, bordering on technical aspects with little importance to the deeper questions of science. Intending to remain in the background, he was thwarted by a requirement that all participants had to either present or be someone's deputy on a particular theme. He became the deputy for the topic of "the stability of the elastic line," and when the presenter, Arthur Haas, pleaded illness only a week before the presentation, Born had to replace him. Whether Haas was truly ill or simply afraid of Klein—the cause for much "illness"—Born could not be sure. Since Haas had little research to pass on, Born gave up on a literature review and in the short time left resorted to Hilbert's method: simplify the problem and solve. He began with the simple case of a wire with both ends fixed. His problem was to determine which curved configuration of the wire would minimize potential energy and thus be the most stable. For his analytical method, he drew on Hilbert's course on the calculus of variations—the mathematician's answer to complex minimization problems. Born determined the set of differential equations for the stability conditions of the curve and systematically solved for the specific case.

The presentation was a success, and Klein was pleased. When the term ended in March, Born left for Breslau. A week later came a letter from Göttingen. "Dear Herr Born! . . . We, the Philosophy Faculty, will probably arrange for an examination of the Stability of the Elastica in a Plane and Space as the prize exercise for the year 1905/1906." The thesis submitted for the prize could also double as a dissertation. Would Born be a student until Easter 1906—a year away—and thus be an eligible competitor? Not waiting for his reply, or perhaps anticipating it, the letter ended, "Anyway, it would probably be good if you familiarize yourself more thoroughly with this problem over the Easter holidays." Signed, "Your devoted, F. Klein."

The prize in question was a prestigious award that rotated among the university's many departments. As these included the humanities and sciences as well as mathematics, Klein's group rarely had the opportunity to determine the subject. By choosing the stability of the elastic line, Klein had chosen one dear to his heart, a topic that promoted broadening the attitudes of mathematicians.

With more promptness than judgment, Born sent Klein his regrets. He wrote back politely rejecting the professor's suggestion, asserting that his research interests lay elsewhere. Born and his circle of friends, who thought of themselves as mathematical purists, agreed that "Klein's predilection for applied mathematics" was a waste of his gifts, a product of his political ambitions. In their view, "students of minor mathematical gifts but greater adaptability were preferred and promoted by Klein if they flattered his partiality for applications." Even though Born had failed at his first dissertation topic on the Bessel function, he obviously did not consider his own talents minor. Yet his newfound confidence exposed tremendous naiveté and lack of intuition about people. At best, Klein would take Born's response to such a flattering invitation as cheeky and tactless. If Klein had already mentioned Born's name to his colleagues, he would feel blatantly insulted.

When Born returned to Göttingen a month later, he was pelted with questions. What had he done, his friends wanted to know, to perturb Felix the Great? Why was one of Göttingen's most powerful professors casting aspersions on Born's character? The laughter that greeted his explanation soon turned grim. In Germany, Klein's goodwill could propel careers, and his antagonisms could end them. Clearly, Born had tipped the balance in the wrong direction.

Born spent the summer semester trying to repair the damage. He visited Klein to make amends and offered to analyze the elastica problem, but his contrition was no balm for Klein's anger. The imperious professor treated Born like a naughty schoolboy, stating in icy tones that he was not interested in what Born did. He let his errant student struggle in misery.

Aside from the Klein problem, the summer semester had its rewards. Professors Hilbert, Minkowski, and Wiechert, and a young lecturer named Gustav Herglotz, held a seminar on electrodynamics to further the understanding of the electron and the new microscopic world that was just beginning to unfold.

The macroscopic world described by mechanics, electromagnetics, and thermodynamics had a new microscopic rival. Max was one of about fifteen participants to discuss the hottest topic in contemporary physics.

Cambridge physicist J. J. Thomson had discovered the electron, the first subatomic particle, in 1897. Eight years later, many scientists were not convinced of the existence of atoms, let alone electrons. The discoveries of other particles, such as protons and neutrons, and of the nucleus, which would help make the invisible atom comprehensible, remained years away. Yet some physicists had begun to see electrons as central to a view of the physical world that might provide a new foundation for physics.

The electron is a negatively charged particle, its motion giving rise to an electromagnetic field. Physicists firmly believed that this field spread out through and was supported by the ether, an indetectable medium assumed to fill space. They did not know if the electron had real physical mass or if its apparent mass was due to its motion, but experiments seemed consistent with the latter concept, raising the possibility that electromagnetism could explain and therefore, in effect, subsume mechanics—that is, instead of embracing an overarching mechanical worldview, these scientists championed an overarching electromagnetic worldview. It was a clash over the fundamentals for understanding nature, centered on the electron.

The small group in Göttingen—perhaps the largest concentration of theorists focusing on the theory of the electron and including some of the best mathematical brains in the world—had an opportunity to resolve some of these questions percolating through the scientific community.

For every theory about the electron, there seemed to be a countertheory, each with a different implication for how the universe and its myriad phenomena worked together—that for this group translated into how the electron's properties would be mathematically modeled. Some physicists played with its motion: Did it rotate, were they small or large rotations, did it vibrate? Some looked at its velocity: Could it travel faster than the speed of light, and if so, what were the ramifications? Was the speed of light a constant? The shape of the electron and the distribution of its charge mattered for the models. Some hypothesized that the electron was rigid, others that it was deformable—that is, that its shape would contract as its velocity increased.

In a famous theory offered by one of the most respected theoreticians, Dutch physicist Hendrik Lorentz suggested that matter could be deformed. This was his explanation for the unexpected results of experiments by Americans Albert Michelson and Edward Morley. They had designed an experiment to detect the ether that physicists were sure filled space and carried electromagnetic waves. Assuming that the Earth moved through the ether and created a wind, they constructed an instrument with equilength, perpendicular arms that could measure the time traveled by a ray of light as it went across the ether wind up and down one arm compared to the time light went through the ether back and forth on the second arm. The rays were like two men in rowboats, one going across the river's current and back and the other going the same distance and rowing up and down the river. The man rowing across the current always makes the trip faster. That was the expectation with the light rays, too—that resistance from the ether would be greater for the one moving "upstream"—but it did not hold. Both rays made the trip in the same amount of time. The instrument, known as an interferometer, floated on a bed of mercury to ensure a frictionless environment necessary to validate the experiment. Since there had to be an ether to propagate electromagnetic waves—just as sound waves need air and ocean waves need water—why did it offer no resistance, detectable as ether drift or wind?

Lorentz's answer was that motion through the stationary ether must cause a deformation of matter in the parallel arm that exactly offset the amount of resistance. This hypothesis caused him to rethink the whole concept of actions relative to one another in what he called "local time" (valid in a moving frame of reference). There was no doubt in his mind that if one could experiment outside of one's own frame of reference, an ether drift would exist in absolute time. Lorentz developed transformation equations to describe the physics of moving frames of reference, which would become central in the continuing investigation between space and time.

The Göttingen group saw itself as potentially in a position to resolve these issues and thereby establish the basis of physical nature. The excitement of the quest permeated their electron seminar. Minkowski began to bore into relationships between space and time, presenting some original ideas to the class. Meanwhile, a few hundred miles away in Bern, Switzerland, a twenty-six-year-old patent clerk named Albert Einstein was thinking about the speed of light and about objects moving at a constant speed relative to one another. In only a few months he would publish his own ideas on the relativity of time and space that would reframe the debate yet again.

Electron theory, the structure of matter, fascinated Born. He wanted to write his dissertation on it but was too shaken by the Klein fiasco to concentrate on work. Instead, he fled to Göttingen's beautiful parks with their summer orchestras and operettas, dancing in the beer garden at Maria Spring, and horseback riding in the hills. A phenomenal sight reader, he played transcriptions of operas by Bizet and Wagner and Schubert sonatas endlessly on the piano in his room. Summer soon turned to fall, and the winter term began with Born still aimless. He signed up for some courses, including one of Klein's on geometry, attended haphazardly as usual, when he realized that once again he was an outsider.

Among Göttingen's famous minor institutions was its Mathematics Society, cofounded by Felix Klein to discuss new ideas in mathematics. That year's members, ironically, included most of Born's friends. Every Tuesday night after the meeting, Born had to listen to their animated discussions. As much as he wanted to join the society, he did not dare. Otto Toeplitz (who had come to Göttingen a year after Born) agreed that Klein probably would veto him. Toeplitz offered a solution. "If you got the prize and your degree, I don't think Klein could object to your admission." Born took the matter too seriously to realize until much later that Toeplitz was manipulating him.

What he did see was that to be free to pursue what compelled him, he had to finish his dissertation. And if he tried for the faculty prize on the stability of the elastic line, perhaps he would appease Klein. The conclusion was easy: His dissertation topic had to be the stability of the elastic line. Only one obstacle remained. Since elasticity was Klein's interest—as opposed to Hilbert's or Minkowski's—Born would have to ask Klein to be his dissertation supervisor.

With his heart in his mouth, Born went to see Klein. "Why not?" Klein responded. "But I doubt if you know enough geometry to pass the oral [Ph.D.] exam. You have never attended my lectures." Born was thunderstruck: How could Klein have noticed his absence among the 100 students in the class? He countered defensively that he had read the books, to which Klein retorted, "How many inflexion points has the general curve of the third order?" Intimidated, unable to answer, Born left in despair.

Born convened his friends in a "council of war" to find a solution. Their reasoning was characteristically creative: Since Klein was in pure mathematics, Born could avoid him as a member of the oral examination committee, according to the rules, by asking Carl Runge, who was in applied mathematics, to be his supervisor. This Born did, and Runge agreed. Astronomy professor Karl Schwarzschild and physics professor Woldemar Voigt, from whom Born had taken a course on optics, became the other examiners. Born then set to work researching the elastic line. His first hurdle, the deadline for the prize, was April 1.

Throughout the winter he sweated through solving complex equations by applying once again the abstract methods of the calculus of variations, as he had done in Klein's seminar on elasticity six months earlier. He worked entirely alone. To keep submissions for the prize anonymous from the faculty judges meant advice from professors was forbidden.

Born built his analysis step by step, from the simplest case to the most difficult. He expanded on the two-dimensional stability conditions in his earlier report (where the wire was fixed at both ends) to complex analyses, where his hypothetical wire could loop, arch, and swoop, depending on the pressure exerted and restraints on the wire. After weeks of developing equations and churning through the calculus, he was able to predict the stability conditions that minimized potential energy.

A month before the deadline for the prize, his curiosity heightened. Born designed a simple apparatus to test his predictions and had a local firm construct it. One end of a thin metal strip was clamped onto a lever that, in turn, was anchored to a dial, and the other end was attached to a long rod with a weight on it. The lever could move to a different angle so that the wire would bend in different directions; the other end could attach to different points along the rod, thereby increasing or decreasing the weight; and the length of the metal strip could vary. To stress the wire almost to the breaking point became a balancing act. As Born spent nights calculating and days experimenting with his new contraption, he watched the metal wire gracefully bend and loop when he changed the length of the wire, the direction of the loop from the lever, and the weight at the other end. His predictions held: He had passed his own test, confirming his theory by experiment—"a moment of deepest satisfaction," as he later put it.

On the front of his prize article, Born quoted two lines from Goethe's *Faust*:

> *Irrtum verlässt uns nie; doch ziehet ein höher Bedürfnis*
> *Immer den strebenden Geist leise zur Wahrheit hinan.*
>
> *Error never leaves us; still, a higher desire gently draws*
> *the striving spirit always closer to the truth.*

Born probably learned the lines from his father.

Entries for the competition were judged during the following two months. On June 13, 1906, Born stood anxiously by the door of the anteroom to the Aula, the university's grand ceremonial hall. Inside, Göttingen's academic community gathered under the ornately coffered ceiling. He did not venture to sit among them. Listening to the rector's long-winded speech, waiting to learn who had won the Philosophy Faculty Prize, he finally heard his own name.

The prizewinner did not celebrate. In a few weeks he had to take the oral Ph.D. exam. On July 11, 1906, Born arrived at the Aula at 6:00 P.M. wearing the requisite formal dress suit, top hat, and white gloves. He entered the room and shook hands with the three professors sitting around the long table. With the first two—Runge and Voigt—questions went smoothly. After asking an hour's worth of questions on such topics as the roots of numbers, function theory, Cauchy's integrals, and the calculus of variations, Runge wrote on his evaluation sheet, "The candidate showed himself very well versed." After Voigt's half-hour exam concluded much the same way, Born was relaxed and confident. The astronomer Schwarzschild began by asking Born, "What would you do if you saw a falling star?" Born answered, "I'd make a wish." Though the others chortled, Schwarzschild was not amused and pressed his question. Born realized his impudence, collected himself, and answered satisfactorily. For his impressive work, the professors awarded him the Ph.D. in mathematics magna cum laude. Yet even with this degree, he knew he could not be a mathematician à la Hilbert or Minkowski. Just what he could be was not entirely clear to him.

This time, Born did celebrate. He stayed around long enough to toast his triumph with his friends at Andreasberg, the famous guesthouse north of Göttingen. Then he left, vowing never to return. As one friend put it, Klein had "broken his enthusiasm."

Born headed home to Breslau, where his compulsory one-year stint in the German army awaited him. Most of his friends had enlisted right after graduating from the Gymnasium, but when he had tried to, he was rejected because of asthma. Each year thereafter he requested a postponement to finish his studies. Now he was reporting for another physical exam, hoping the army would again reject him. Instead, he found himself a conscript in the Second Dragoon Guards Regiment.

The Dragoons were horsemen who drilled, rode, charged with lances, and cleaned stables. The disciplined life of a cavalry soldier came as something of a shock to a twenty-four-year-old scholar, but for his first four months, stationed in Berlin, Born enjoyed life in the city. He was among friends: His sister and her new husband, Georg Königsberger, various Kauffmann and Lipstein relatives, and his old comrade from Heidelberg, James Franck, also doing his military service, were all nearby. By this time, Franck had become a physicist, and he encouraged Born to go to Cambridge University sometime to see what real physicists do. It was advice for which Born thanked Franck some forty-five years later.

On January 27, 1907, the Second Dragoon Guards awoke early and arrayed themselves in spiked helmet and parade dress with all the flourishes. Born and the rest of the guard marched from their barracks at Tempelhof through miles of streets to the center of Berlin. They stood in formation on the terraces of the imperial palace of Kaiser Wilhelm II. In all, twenty thousand troops waited motionless for an hour or more, their hands, feet, and ears stinging from the extreme cold. Finally, with musical fanfare and the command "Present arms," the kaiser and his entourage, plumed and gilded, rode past on horseback. The kaiser slowly reviewed the long lines of his resplendent troops. What grander way for a king to celebrate his forty-eighth birthday?

Born had never questioned service to his country because military duty had little moral or practical implication. It was expected of him, just as it was accepted within his bourgeois world that European wars were events of the past. Standing there, watching the pomp, however, Born recalled his father's description of his wartime service—of being "a little cog in a little wheel in the big military machinery," and the resulting "degradation of human dignity." As he waited to pay homage to the kaiser along with thousands of other soldiers, his father's quiet but passionate disagreement with militarism became his own.

Born's service as a Dragoon was short-lived. The cold of January 27 led to an asthma attack, and he was discharged early for health reasons. He returned to Breslau, where he found his cousin Hans Schäfer planning a trip to England for the Kauffmann textile firm. It was just the opportunity Born needed to follow up on Franck's advice to study in England.

Just as Göttingen drew mathematicians from all over Europe and America to Germany, the Cavendish Laboratory drew physicists to England. The reclusive Henry Cavendish himself had discovered a number of important principles in chemistry and physics; in 1871, sixty years after his death, the eighth duke of Devonshire, a member of the family, endowed the Cavendish Laboratory at Cambridge University. Born did not go there immediately, but lingered in London with Schäfer. At first, they enjoyed the city from the comfort of an international hotel where they could "sit in the lounge with a glass of port and watch the elegant crowd." Yet thrift soon drove them to a shabby Bloomsbury boardinghouse. After a week, they separated—Schäfer to Bradford for business and Born to Cambridge for his first encounter with English physicists and universities.

Born arrived in Cambridge with no place to stay, no friends, and little English. He tried to apply for admission to one of the university's colleges, but in vain; no one could understand what he wanted, and besides, the term had already started. Discouraged and lonely, he was thinking of going home when, walking along the street, he heard some students speaking German. He described his predicament to them, and they were able to help him secure acceptance as an "advanced student" at Gonville and Caius College.

Born's studies at the Cavendish Laboratory went only slightly better than his admissions search. Professor Joseph Larmor's thick Irish accent made his course on electricity and magnetism almost unintelligible to Born. Professor J. J. Thomson, discoverer of the electron, pushed the young German toward the intricate demands of experimental work in the new realm of subatomic physics, but here Born was hampered by his narrow theoretical background. A basic laboratory course in electricity he found too simple. His most memorable classroom moment may well have been his exchange with Dr. George Searle when Born and his lab partner, a pretty young lady, had problems with an experiment. Born spoke first.

> "Dr. Searle, something is wrong here, what shall I do with this angel?" (meaning of course "angle"), whereupon Searle looked at both of us over his spectacles, shook his head, and said: "Kiss her." After that the shyness between us increased still more and was never overcome—a great pity, for she was a lovely creature.

More rewarding for Born were afternoons spent drifting in a punt on the River Cam, his head buried in a copy of J. Willard Gibbs's *The Elementary Principles in Statistical Mechanics* that he had found in a Cambridge shop. The age-old question, What is heat? had only recently been answered. Researchers such as Robert Mayer, James Joule, Hermann von Helmholtz, Rudolf Clausius, and William Thomson (Lord Kelvin) had established the two laws of thermodynamics, by the first of which, the energy of matter in motion, became the energy of heat and vice versa. It was the mechanics of this moving matter—that is, energetic atoms colliding at random—that James Clerk Maxwell and Ludwig Boltzmann had represented through the power of statistics, generalizing the movements of countless unseeable atoms. Gibbs's work presented Maxwell's and Boltzmann's statistical theories and created the field of statistical mechanics. The first time Born had tried to read Gibbs, his limited English held him back; this time, he found the theorems offered by Clausius and Kelvin "wonderful, like a miracle produced by a magician's wand."

Thermodynamics, it is often said, owes as much to the steam engine as the steam engine owes to thermodynamics. Born, who as always was less interested in application than underlying theory, became captured by a single question: What were the mathematical roots of the laws of thermodynamics, as opposed to their engineering foundations? This was a question in the grand Hilbert tradition, but one that Born could not answer. He had to leave it to his friend Constantin Carathéodory to derive an axiomatic proof of the second law of thermodynamics, the principle of entropy—the increasing disorder in a system as, for instance, mechanical energy is transformed into heat energy.

Born decided to tackle a narrower question. In his first article, "The Variational Principle of Heat," which he wrote with a friend from Göttingen, he determined the maximum value of the entropy of a system. The "engineering approach" to the second law involved studying a mechanical process, calculating the effects of cyclical changes in physical parameters such as pressure and temperature on a closed system (for instance, an engine). Born did not test any physical processes but hypothesized a closed system (the conduction of heat between two compressible liquids), generalized it to a system that developed through time, then using the calculus of variations went straight to the results. This approach would be characteristic of his future research—applying familiar mathematical techniques to new questions.

Back in Breslau at summer's end, Born had to come to terms with his future. Not having the mathematical creativity to be a professor in the Hilbert-Minkowski tradition—and not wanting to be second rate—he saw physics as a promising choice. Cambridge physics had been frustrating, but J. J. Thomson's lectures on the electron had sparked some interest, and the article he had coauthored on thermodynamics had proven that he could tackle questions in the field. If he could not invent new fields of mathematics, he could at least apply the existing forms creatively. So Born decided to become a "real physicist." It was a decision that drew on the delight of a young boy whose vaulted arch stood when the last brick was put into place, and on the satisfaction of a young doctoral student whose experiments with elasticity responded according to his theory.

Born's new direction seemed motivated less by any great sense of purpose or enthusiasm than by a decision not to become a mathematician. To some extent, his own vagueness of intent reflected the state of the discipline. The field of theoretical physics to which he aspired was then being created. Only nine years earlier, a prominent German physicist had declared that "physicists are almost exclusively interested in pure experiment and have scarcely any interest in theory." Likewise, mathematicians were turning away from theoretical physics. At the turn of the century, there were only seven senior chairs in the subject at German and Austrian universities. Junior theoreticians—hired mainly to teach physics—simply could not advance. Theoretical research required that professors had to do successful experimental work as well. How Born could develop his interests was unclear, but he knew what he had to do to begin.

The road to becoming a "real physicist" meant becoming a professor, and in the German academic system this was neither an easy nor a short process. To teach at a university, the young Ph.D. first had to finish another thesis, the Habilitation (a term derived from the Latin for "capable"). This was an apprenticeship: Rather than take courses, the aspirant worked as an assistant to a professor to demonstrate that he could do research. If the professor were pleased, he would recommend that the faculty invite the assistant to habilitate—that is, write an original research paper and give a lecture. A positive review by the faculty conferred the right to teach. The most difficult part in this formal and deliberative Habilitation process was the faculty's invitation. To obtain it, the assistant had to convince a professor that he was deserving of the honor.

Born had decided to try to habilitate at the University of Breslau. He applied to the professors in physics, the experimentalist Otto Lummer and the theoretician Ernst Pringsheim, for acceptance as a researcher in their lab. While waiting to hear, he resumed his old life of tennis in the Kleinburg gardens and horseback riding. He lived at home with his stepmother, who did her best to make him feel comfortable, hosting his friends and encouraging his music making.

This idyllic limbo was shattered by an order to report for military service. When he was at Cambridge, the German consulate in London had ordered Born to undergo another physical exam for the military. His casual student dress and informal greeting of *Guten Tag* had infuriated the stiff Prussian military doctor. Now that doctor had found him fit for duty.

A family friend came to his rescue with an unusual plan. A physician, he advised Born to volunteer for the most exclusive regiment, the Leib-Kürassier Regiment No. 1. Being assigned to the kaiser's bodyguard was an unlikely assignment for men not of the nobility—those without a *von* before their names—and unheard of for a Jew; but the chief military doctor for that regiment had been a student of Born's father. The family friend arranged with the military doctor to examine Born, admit him, and at a not too distant time, release him for health reasons. The plan worked. On October 1, 1907, Born was inducted into the Kürassier regiment and, as promised, discharged six weeks later. Greatly relieved, he took up his position at the university.

Born joined a small group of assistants who met daily to discuss their newest ideas. One member, Rudolf Ladenburg, elegant and athletic, was an experimental physicist from Breslau whom Born had known only superficially. They quickly became friends, sharing interests in music, art, science, and skiing. They also both needed a friend. Ladenburg had recently lost his mother and brother; and in a short time, Born found himself on the verge of being kicked out of the university.

Seven years earlier, the experiments of Lummer and Pringsheim had helped theorist Max Planck to find the theoretical law for the frequency distribution of energy blackbody radiation. A blackbody is an ideal body that absorbs and emits radiation perfectly; its closest approximation in the real world is an intensely heated porcelain pipe.

Born used a blackbody to follow up on Lummer and Pringsheim's experiments. One night, after everyone had left, a hose from a water-cooling apparatus connected to Born's blackbody came loose. For several hours, water spilled onto the floor, flooding the lab and seeping down into the basement and disintegrating the ceiling. The flood was not quite up to the standards of Noah but for Born it was, nevertheless, catastrophic. Lummer, who ran the lab, was furious and wanted to throw him out. "You'll never become a physicist." Whether this was a threat or a prediction, Lummer was clearly not going to allow him to habilitate; and without the offer, Born could not become a professor.

Frustrated with the experimental work from the start, Born had already begun playing with ideas from the Hilbert-Minkowski seminar on the electron. The mystery of the fundamental nature of matter still tugged at him. When he mentioned some of these to another assistant, Stanislaus Loria, he learned that research was being conducted on relativity by someone named Einstein, and that Max Planck supported it. Rudi Ladenburg wrote to Einstein—who was still at the patent office in Bern—to request three copies of the reprint, and Einstein sent them, expressing pleasure in their interest. Throughout the winter of Born's self-made flood, he, Loria, and Ladenburg studied Einstein's 1905 paper on the electrodynamics of moving bodies.

Einstein based his theory on two principles, the constancy of the speed of light and the principle of relativity: That the laws of motion are the same for two objects moving relative to each other at constant velocity. Combining these two principles and working through his argument, he came to startling results: Two observers in frames of reference that were moving at constant velocity with respect to each other would not observe the same lengths and times. If two observers had identical measuring rods and clocks when at rest, each would observe the other's measuring rod to decrease and clock rate to increase when moving at constant velocity with respect to each other.

Einstein's equations that described these relativistic transformations were identical to those developed by Lorentz to account for the null result of the Michelson-Morley experiment, which Born had studied in the electron seminar. Although the transformation equations were identical, there were fundamental differences between Einstein and Lorentz. Einstein derived them from his two general principles and found that the ether played no role in his theory. Lorentz derived them in an ad hoc way by considering a particular physical system—the Michelson-Morley experiment—and did *not* dispense with the ether as the absolute frame of reference of the universe.

Some physicists threw the ideas of Einstein and Lorentz together, referring to "their" theory as the "Lorentz-Einstein Principle of Relativity." Although some scientists recognized the conceptual differences between the two, they were widely regarded as being empirically equivalent. Born and his friends, however, appreciated Einstein's conceptual innovations and pursued these new ideas.

A particular spur for Born was Einstein's follow-up paper that focused on the relationship between energy and mass, for which he derived the equation

$E = mc^2$. After studying it closely, Born began developing a new, more direct method for calculating the electromagnetic mass of the electron. Excited to return to the ideas that had captivated him in Göttingen, he quickly outdistanced his friends. The flood, caused by his inept experimentation, receded in importance. Life for Born was brightening and seemed to have direction—until he heard about Minkowski's research.

In spring 1908, Hermann Minkowski published an article in the *Göttingen Nachrichten*, the journal of the Göttingen Academy of Science, entitled "The Basic Equations for the Electromagnetic Phenomena in Moving Bodies." During the summer days of the 1905 electron seminar, he had concluded that there was no absolute motion and no absolute time, as postulated in classical theory, because there was no absolute point of reference. His desire to work out the mathematical underpinnings of his theory had held back his publishing. He was surprised when he saw an article describing similar thoughts to his own, and he was even more surprised at the author—Albert Einstein, Minkowski's former student in Zurich, whom in those days he had deemed "a lazy dog."

Minkowski's 1908 article credited Lorentz with discovering the *relativity theorem*—the set of equations (the Lorentz transformations) that described the transition of an observer from one inertial frame of reference to another without changing Maxwell's equations. He also noted Einstein's contribution, what Minkowski referred to as the "relativity postulate" of expressing these equations as a new understanding of space and time, dictated by experimentally observed phenomena. Finally, Minkowski explained his own "relativity principle," in which he tried to develop a comprehensive relativistic electrodynamics and mechanics—starting from Einstein's new understanding of space and time.

Among Minkowski's ideas, Born recognized some that resembled his own. His brilliant new extension was not so original after all.

Finding himself eclipsed by his former teacher, Born was devastated. Even his friends could see how nervous and distressed he was, although characteristically, he kept his concerns to himself. His stepmother grew so worried about him that she finally asked Loria if he knew what was going on. Yes, he did, he told her, and advised her to persuade Born to write to Minkowski and to share the results of his research. "You will see that Minkowski will have Max come to Göttingen and propose Habilitation to him," he predicted.

Bertha Born did as Loria advised. Born took the suggestions; and Minkowski invited him to Göttingen to work together, with the possibility of doing his Habilitation. Despite his vow never to go back there, Born accepted with delight.

Before heading to Göttingen, Born attended the September meeting of the Society of German Scientists and Physicians in Cologne. On September 21, he heard Minkowski's renowned lecture on space and time. "Henceforth, space by itself and time by itself, are doomed to fade away into mere shadows, and only a kind of union of the two will preserve an independent reality. . . . Three-dimensional geometry becomes a chapter in four-dimensional physics." Minkowski's lecture heightened the interest in relativity and helped to define the nature of the theory.

During the next month in Breslau, Born studied Minkowski's ideas. He arrived in Göttingen on November 10, 1908, moving back to his former rooming house where Toeplitz still resided. Born spent part of every day going over ideas with Minkowski and the rest, struggling to understand inertial mass in the context of electrodynamics and relativity. This topic was to be the first article in his larger plan to make electrodynamics consistent with relativity. Toeplitz helped Born brush up on his matrix algebra so that Born could apply Minkowski's four-dimensional space-time notation. In the evenings, Born often visited the Minkowski household, listening to his professor quote *Faust* by heart, much as his own father had done in the months before he died.

At Christmas, Born went home for a few weeks' holiday. He felt he was entering "a happy time, full of scientific excitement, but also rich in experience of a personal sort." Before returning to Göttingen, he sent the article on electrodynamics and relativity for publication.

Returning to Göttingen from the holidays, Born learned that Minkowski had become seriously ill with appendicitis and was about to undergo an emergency operation, which was unsuccessful. Shortly after, Born stood at Minkowski's bedside, both aware that the professor had only a short time to live. Born listened as Minkowski rationalized his death, saying that perhaps it was best for the propagation of his ideas that he was stepping aside. Born took this admission as a testament "to the modesty and generosity of this rare spirit." Minkowski died on January 12, 1909, at the age of forty-four.

The next day, the mathematics students asked Born to speak on their behalf at Minkowski's funeral. After a day and night of writing and memorizing, Born stood before the mourners. His words extolled Minkowski, the man and teacher who shared his mind and self with his students, giving them gifts "out of a source of beautiful thoughts that never dried up." The tribute touched many in the audience, winning him, as he later discovered, "friendly feelings in many hearts, even in that of Klein." In a long obituary written roughly a week later, Born focused on Minkowski's mathematical contributions, calling him a pioneer of new ideas and describing his space-time framework. Not until fifty years later, in a tribute to Minkowski on the anniversary of his death, did Born set down his more personal feelings; he wrote, "With him also sank my hopes and prospects. However, that was not what moved me most deeply. I mourned over the loss of a mentor and friend, a great thinker and scientist."

THREE

MATTERS PHYSICAL

A FEW WEEKS AFTER MINKOWSKI'S DEATH, THE GÖTTINGEN MATHEMATICS SOCIETY HELD ITS regular meeting. As always, it was a daunting gathering. In front of a long blackboard, Felix Klein and the other mathematical and scientific notables sat imposingly at the main table. Filling two long tables perpendicular to it were the younger and less established faculty members eager for a chance to distinguish themselves. Klein opened the meeting by discussing the recently published scientific books piled on the table. Then he disbursed them among the younger members, who looked through them, affecting an air of indifference as the evening's speaker rose. In actuality, they waited to pounce.

Max Born prepared to begin the presentation of his new research. He was anxious, deeply aware of how much was riding on this particular evening. Not having Minkowski's support, he needed to persuade the faculty to accept his Habilitation thesis. In his path loomed the specter of Klein, the spoiler.

Since Minkowski's death, Born had extended his research to the effect of acceleration on the shape and structure of the electron. It was a subject surrounded by controversy. Scientists disagreed over whether and how an electron changed shape when in constant motion, even without the burden of acceleration. On one side was Lorentz, who believed that bodies moving at a constant velocity contract in the direction of their motion, the closer the velocity is to the speed of light. In opposition was the strident antirelativist Max Abraham, who believed that electrons stayed rigid even at high speeds. Physicists, including Lorentz and Abraham, also disagreed over whether and how much the motion of an electron changed its mass—that is, its motion generating an electromagnetic field that interacts with the electron, creating a "self energy" and thereby increasing its mass. Based on a contracting electron, Lorentz derived a simple equation for calculating this self-generated mass. Abraham, rejecting Lorentz's formulation, developed a more complex formula for his rigid electron. Abraham, who was well known to Born and well known for his biting sarcasm, sat at one of the long tables that evening.

The aim of Born's lecture was to expand Lorentz's ideas from the effect of motion on an electron with a constant velocity to the effect of acceleration on a relativistic rigid electron. By analyzing the forces exerted on a rigid electron, he could gain a description of its motion. But first he had to resolve the problem that relativity causes deformation, and rigidity means no deformation. This he tried to do by developing a new definition of rigidity that linked space and time through a rotation in four-dimensional Minkowski space. His new definition allowed someone within the frame of reference of the body in motion to see the body as rigid, whereas someone in another reference frame would see deformation. This was truly a new concept of *rigid body*.

Standing before the blackboard, Born began his lecture with his redefinition—and that was as far as he got. Klein stopped him cold, accusing him of misunderstanding the concept. Either because he did not grasp the redefinition or chose to disregard it, Klein told Born "to study the mathematical literature before addressing the society."

When Born replied to the criticism, an attack came from one of the long tables. It was Abraham, asserting that Born's "knowledge of physics seemed to be just as scanty as that of mathematics."

Further affronts threw Born completely off balance. Klein called the presentation to a halt and proclaimed that it was the worst lecture he had ever had to sit through. Born returned to his seat utterly crushed.

As Born tried to retreat inconspicuously after the meeting, Carl Runge, his doctoral supervisor, stopped him. "I am interested in your work," Runge told him. "Klein seems to have missed the point. Call on me tomorrow morning and explain it to me."

This consolation, in the most "wretched time" of his life, helped Born survive the night. Still, taking his failed lecture as a sign of "inadequacy and presumption," he resolved to leave for Breslau and pursue a career in engineering. Talks with Runge and Hilbert, followed by an invitation to make another presentation, did not change his mind. Finally, Runge accused Born of being afraid to try. Stung by the rebuke of cowardice, Born agreed to give the lecture. His future plans would depend on the outcome.

After a few weeks of meticulous preparation, Born again stood nervously before the society. This time Runge, rather than Klein, made the introduction; he told the group that Klein had misunderstood and asked that Born repeat his lecture. Neither criticisms nor comments interrupted Born's discussion of electron theory. When he concluded, he looked up from his notes to a room of friendly faces. Woldemar Voigt, professor of theoretical physics, sought him out after the meeting to offer to sponsor him if he wished to write his Habilitation thesis. This was a night to celebrate. At the wine cellar Mütze's, Born drowned all thoughts of Breslau and engineering.

Born began the biographical sketch required for his Habilitation application with the words "I, Max Born, of the Jewish religion. . . ." Once habilitated, he still faced a long hill to climb from lowly lecturer to professor. In 1909 Germany, only about twenty full professors out of twelve hundred were Jewish. At many universities, an unwritten rule allowed no more than one Jewish professor in any specific faculty. Göttingen had only broken the rule in 1902, at Hilbert's insistence and with Klein's help, to bring in Minkowski, despite the presence of Jewish professor Karl Schwarzschild in the astronomy department. There was a waiting line for professorships, and Jews were at the back. The situation had changed little since Gustav Born's time.

Around this time another young physicist, Paul Ehrenfest, whom Born knew briefly from the early days in Göttingen, wrote to request a copy of Born's first paper on relativity. Through their correspondence, they discovered that they shared the belief that Einstein's theory of relativity did not provide a theory of the electron. In a fifteen-page letter, Born tried to enlist Ehrenfest as a "valiant ally" in sorting out the confusion surrounding relativity, agreeing with him that

> for Einstein, Planck, and Minkowski, 1) the mass is not electromagnetic, 2) Electricity is completely structureless. As a consequence of this, one does not understand what an "electron" should be essentially, why it, if it exists, does not explode with an audible crack, etc., etc.

So, he asked, how was one to explain electron theory since relativity did not? Born gave Ehrenfest the answer contained in his earlier speech before the Mathematics Society and since expanded into a soon-to-be published article: To define relativity as a generalized kinematics (the branch of physics that describes motion) that incorporates ordinary mechanics as a special boundary case, then to place into this view of relativity an accelerated, rigid electron.

Born's article on relativity, "The Theory of Rigid Bodies in the Kinematics of the Relativity Principle," appeared in the August edition of the *Annalen der Physik*. His approach has been considered a "mathematical *tour de force*" as he worked out the hyperbolic trajectory of a rigid body in Minkowski's four-dimensional space-time. One of his conclusions was that the charge of a rigid electron had to be distributed in concentric circles. Another was that the greater the acceleration, the smaller the electron would appear. Born saw these results as an important confirmation of an atomistic approach to electrodynamics.

This article was only the beginning. His personal reserve and modesty enveloped a strong intellectual ambition. He wrote to Ehrenfest,

> Before all, I need time. In 10 years I hope to be able to lay before you a dynamics which encompasses electricity and mechanics, which leaves nothing wanting in regard to rigor, and which will deliver a good foundation for consideration of nature, as long as new discoveries do not turn us to wholly new paths.

Emotion may have played a part in these aspirations as well. Just six months later, while discussing articles relating to relativity and rigidity, he told Hilbert, "I place such things close to my heart in the hope that there is a new experimental confirmation of the 'World Postulate.'" This was Minkowski's name for his concept of relativity.

Initially, Ehrenfest responded positively to Born's article, writing to a friend, "I began to read it last night—it is clear and elegant throughout—Minkowski + Born represent an enormous advance—hopefully the whole devil's business will fall apart much more quickly than I had hoped!" But then Ehrenfest began to reconsider Born's new definition of rigidity, and in an article proposed an alternative but equivalent one. His article provoked a controversy that quickly overshadowed Born's general findings. Ehrenfest showed that definitions of rigidity led to a paradox, which he illustrated with a disk rapidly rotating like a frisbee. Since deformation occurs only in the direction of travel, a ruler placed between the end points of two radii should contract, thus causing the disk to contract. Yet, since there is no motion in the radial direction, there is no change in the radius. Ehrenfest's paradoxical conclusion was that on the one hand, the disk contracts; on the other hand, it does not.

Many physicists—including Einstein—weighed in on the discussion during the next two years, unintentionally fulfilling one of Born's goals: sorting out the confusion concerning the relations between relativity and electron theory, showing that relativity could not explain the latter. Born's ten-year plan had to wait.

Born's article earned him an invitation to present his research on rigid electrons at the September meeting of the Society of German Scientists and Physicians in Salzburg. This was his first invitation to speak at a major gathering of scientists. He was not alone: Four years after publication of journal articles that had turned physics upside down, Albert Einstein, too, had finally been invited to make a formal address at a major physics conference. Ironically, Born—whose presentation immediately preceded Einstein's—was the primary voice of relativity there, while Einstein spoke on the quantum nature of light, arguing for the first time for wave-particle duality and suggesting a future fusion of wave and particle views. In later years, amused by this twist, Born noted, "Einstein had already proceeded beyond special relativity which he left to minor prophets, while he himself pondered about the new riddles arising from the quantum structure of light."

This first meeting between Einstein and Born did not launch a friendship, although they did have a private chat. A few months afterward, Einstein implicitly criticized Born by remarking to Ehrenfest, "You are one of the few theoreticians who has not been robbed of his common sense by the mathematical contagion!" But Einstein's consideration of Ehrenfest's rotating disk paradox may have helped him to conclude later that he needed non-Euclidian geometry—a geometry for curved space—to treat accelerated motion in explaining gravitation in his general theory of relativity. When he realized this, he remarked to a friend, "I have become imbued with great respect for mathematics, the subtler part of which I had in my simple-mindedness regarded as pure luxury until now."

Once Born was back in Göttingen, professors in the philosophical faculty came to the Aula on Saturday, October 23, by invitation of the dean. At noon, they heard Born's Habilitation lecture on Thomson's "plum pudding" atomic model (so called because the electrons were distributed throughout the atom like plums in a pudding). Born poetically introduced Thomson's model as the effort of a musician who, upon hearing a new symphony performed by an orchestra he could not see, attempts to capture the structure of the score and succeeds in providing a piano arrangement. He used this metaphor to exemplify the limited understanding of the structure of matter, and for the next fifteen minutes, summarized the ideas of Thomson and Lorentz.

The lecture was largely a formality, the more important requirement being an original piece of research—in Born's case, his work on the rigid electron, which he had already submitted to the faculty committee. Still, it was a relief when two days later the dean's letter arrived congratulating him on receiving the *venia legendi*, the official permission to teach at the university level. He was finally a lecturer in the Philosophical Faculty of the University of Göttingen.

Like his father, Born escaped to nature to ground and rejuvenate himself. If he lived frugally on his funds from the Kauffmann inheritance, he could afford winter ski trips with Rudi Ladenburg or Hans Schäfer and spring excursions to the Italian Alps or Swiss Engadine, one of his father's favorites. During life's trials, the majesty of the Alps always lent Born strength.

Summer holidays were devoted to family and music. Early on he and his cousin Fritz Jacobi regularly went to the Wagner Festival in August and the

Mozart Festival in September. Later, Born accompanied the Neissers who, as committed Wagnerians, spent several weeks every summer in Bayreuth, attending all the opera performances at the famed opera house, the Festspielhaus. The music was sublime; the atmosphere sometimes less so. On the day when Wagner's widow, Cosima, a strident anti-Semite, opened the doors of her estate, Haus Wahnfried, for an elegant garden party, she barred Jews. The Neissers' large contributions to the Festspielhaus and decades of attendance did not overcome their birth—they were Jews. On the excursion they always planned for that day, no one mentioned the party they could not attend.

There was great irony in this snub. Years before the glory days of Bayreuth, the Festspielhaus, and Cosima, Richard Wagner had been a friend of Max's great-uncle, Stephan Born. The revolution that swept across Europe between March 1848 and June 1849 had brought them together.

During that period, German political reformers created a national assembly in Frankfurt and wrote a constitution that called for a constitutional monarchy and a loose federal system. All the states in the loose German confederation became involved. By April 1849, twenty-nine had approved the constitution, but as change seemed more certain, reactionary forces took action. The king of Saxony, for one, dissolved his state's parliament in Dresden and then fled. Reformers established a government and set the stage for resistance. The king's cabinet requested troops from Saxony's sister state, Prussia.

Richard Wagner, the Kapellmeister of Dresden and for a few weeks in April and May a "mouthpiece of revolution," prepared for the revolt by organizing the distribution of arms. Stephan Born, a printer by trade and one of Germany's first recognized labor leaders, came from Leipzig to help defend Dresden. When fighting broke out on May 3, Born, one of the main military commanders, manned a pivotal barricade—the "Dardanelle Pass" as he called it—in the heart of town for six days. As the resistance collapsed on May 9, both men fled Dresden—Wagner at 4:00 A.M., escaping to Freiberg, and Born a couple of hours later, abandoning his barricade only minutes before royalist troops breached it. On arriving at a friend's house in Freiberg, Born was greeted by "an enthusiastic man" who, as he later wrote, "rushed to me with open arms, kissed me, and said in glowing words, 'Nothing is lost! The young, the young, the young will make everything good again, everything saved.' It was Richard Wagner." It was the first time the two had met. Ending up in Zurich after outpacing the Saxon guards, they became friends.

Max learned of this incongruous piece of family history only later from a stash of old letters saved by Bertha Born. Riding through the countryside to avoid Haus Wahnfried, the Neissers' circle had no idea that Max's great-uncle and Wagner had been comrades-in-arms.

In fall 1909, Born settled down to the life of a young Göttingen academic: lecturing, researching, and enjoying evenings with colleagues. In the German university system, lecturers were like nonvoting members of a club: They reaped the benefits of association but held no influence over decisions. More significantly, lecturers received no salary from the university, being paid directly by students who chose to attend. Having outside income was essential, given that many first-time lecturers did not attract large classes. (Professors, who received a salary as well as fees from students, reserved introductory classes—with their large audiences—for themselves.)

Born's zeal to enlighten through the logic of mathematics limited his audience as well as his income. One of the four or five attendees in his course on electron theory in the spring later described Born's lecturing technique. "He said a few words, then turned around, wrote something on the blackboard, and said, 'So that's how it comes out.'" Hours of long calculations with little explanation. In later years, when he questioned these students on points of science, they blamed their ignorance on his abominable early lectures. Fortunately, he improved with time.

In summer 1911, Albert Michelson, of the Michelson-Morley team—whose experiments Born had studied in the electron seminar—visited from the University of Chicago. Ironically, despite the role Michelson's experiment came to play in the advance (although not the origin) of relativity, he never accepted the theory, and Born had little academic contact with him. The friendship between the two grew, instead, on the tennis court, where, for all of Michelson's fifty-nine years, he regularly beat Born.

A few months after Michelson left, he invited Max to give a course on relativity at the University of Chicago during the 1912 summer semester. In those early years of the twentieth century, with most breakthroughs in physics originating in Europe, few young European physicists traveled to the United States. Born looked forward to an adventure.

At the beginning of May, a fast German steamer carried him from Bremen to New York. He stayed for a few days, impressed by the grandeur of Manhattan's skyscrapers and disgusted by the contrast between rich and poor. Through contacts of his father's, Max gained entrance to New York's wealthy Jewish families. At one dinner party, he watched in dismay as each guest was served a whole chicken, which—being one of multiple courses—they barely nibbled. Earlier in the day, he had seen Jews in the East River neighborhoods searching for food in the trash.

Chicago was a combination of the orderly world of academe, a hellishly hot dormitory room, and, on weekends, bits of pleasant landscape thrown in, such as the Michelsons' summer house on Washington Island in the northern part of

Lake Michigan. During the week, Born gave lectures on relativity and worked in Michelson's lab photographing spectra, the lines of light emitted by atoms. Physicists had only recently developed sophisticated instruments to record these lines, each of a definite wavelength, and Michelson was one of the best in making photographic plates for this purpose. Born practiced the technique. Learning to produce some decent photos, he became intrigued by what the lines, unique to each chemical element, might reveal about the internal workings of an atom—and thus, the structure of matter. This interest, started by the electron seminar in Göttingen, had been a focus of much of his short career.

Halfway through the summer, Born became caught up in the U.S. brand of democracy. Former president Theodore Roosevelt had bolted from the Republican convention in June; now—feeling "fit as a bull moose"—he launched a new party to win his old job back. The Progressive Party—or Bull Moose Party, as it was nicknamed—held its convention in Chicago. At the convention hall, filled with cacophonous bands, roaring crowds, and posters of an "elk," Born heard Roosevelt proclaim his new party's fervor: "We stand at Armageddon, and we battle for the Lord." The carefully orchestrated political world of Wilhelmian Germany had not prepared Born for the planned chaos of an American political convention, especially one featuring a bull moose. He was not sure that he was at ease with the "pandemonium" of American democracy or, as he had witnessed it in New York, capitalism.

With the summer semester over, Born took the Canadian Pacific transcontinental on its northern route, rolling past Canadian wheat fields to Glacier Lake and the Engadine-like Lake Louise. He changed direction to enjoy a week's stay in Yosemite Valley, hiking and touring there by horse-drawn coach, went further south to visit the observatory at Mount Wilson, then took the southerly route back to New York. He was awestruck by the Grand Canyon, where he rode a mule 4,000 feet down to the riverbed.

Once back in New York, Born received a cable from a Göttingen friend, Theodore von Kármán, that read something like, "Will you share a house with me and three other lunatics?" Born and his group of bachelor friends had been looking for a comfortable house to rent with a good housekeeper and maid; the house they could find—the housekeeper they could not. Born assumed that von Kármán's telegram meant that he had found a housekeeper, and indeed he had.

The stout Sister Annie had planned to open a home for the mentally ill. To persuade her instead to keep house for them, a member of the bachelor group told her that other than for the house being near the university, there was "very little difference" between their plan and her own—hence "the lunatics."

In Born's cumulative six years in Göttingen, he had rented rooms in sundry boarding houses that ranged from modest dwellings to large villas whose owners had financial problems. The houses often mixed students, young professors, and visiting scholars under one roof, where they shared amenities, including the usual German daily fare: a big breakfast, hot dinner served at noon, and a supper consisting of cold cuts, cheese, and bread. Born augmented the plainly furnished rooms with his own piano.

Since 1909, Born had lived at 49 Nicholausberger Weg, a massive yellow brick boarding house, its rectangular turret topped by whimsical Victorian filigree. Most days, the Hungarian applied mathematician von Kármán—whom Born found a curious blend of cleverness, kindness, and cynicism—and fellow countryman and mathematician Alfred Haar, joined Born in his room for dinner and lively conversation. Sister Annie changed that.

Born arrived back in Göttingen at the end of October to find his three housemates already moved into Dahlmannstrasse 17. Joining in, he had rolled up his sleeves to help with the painting, furniture arranging, and picture hanging, when there was a knock on the door. First making their appearance presentable by putting on their suit jackets, the men opened the door and found, to Born's amazement, a pretty black-haired woman who was interested in the room advertised in the newspaper. The men had wanted a female roomer to "neutralize" the intrusive Sister Annie at the dinner table, where she corrected the men's manners and restricted their conversation. Ella Philipson, a medical student, trooped up the stairs to inspect the available attic room and took it at once; but Ella was no help with Sister Annie. She simply laughed away her interference and instead focused her attention on Paul Ewald, one of their visiting physics friends.

The entrance gate to Dahlmannstrasse 17 soon posted a sign inscribed *El Bokarebo*—*El* for Ella, *Ka* for Kármán, *Re* for Albrecht Renner (another medical student), and *Bo* for Born and Hans Bolza (a physics student). Richard Courant, a mathematics student from Breslau, labeled it the "in group," with von Kármán as leader. At breakfast they discussed mathematics; during the noon dinner a courtship progressed between Ella and Paul, who soon became engaged; and in the evening they held parties. They were a creative and eclectic bunch, acting out original poems at supper parties, where personae such as God (Renner), Archangel Gabriel (Ewald), and Lucifer (von Kármán) handed out the last judgment to the assembled guests. At Hilbert's festive fiftieth birthday party, they celebrated the toils of his assistants in a medley of international verses and spoofed his many flirtations in an alphabet love poem, with *A* for Ada, and *B* for Bertha, all the way to *X*, *Y*, and *Z* for those unknown.

Despite their popularity, these gentlemen were acutely aware of their place on the bottom rung of the academic ladder. As lecturers, they needed to make their mark to receive a "call" to a university position. In the German university, a teacher could receive two different kinds of lifetime calls: the junior Extraordinarius professorship or the senior Ordinarius professorship. At most universities, individual fields of study had one Ordinarius professor who headed the institute, with a few at the Extraordinarius level below him. The numerous lecturers (called *Privatdozenten*) waited with more or less patience for a vacant position—the reason why Hilbert, as a lecturer, had begun his day by reading the obituaries.

In fall 1911, just before *El Bokarebo*, Born was lecturing on the kinetic theory of gases and still "drowning in the formalisms of relativity theory," as he would describe it ten years later. Von Kármán, an applied mathematician, was researching fluid motion. His examination of the flow of liquid around an object was leading him to understand the pattern of vortices—as in the wake behind a boat or in the air currents behind an object moving in space—later known as the "Kármán vortex street." Both men also sat in on Hilbert's seminar, the "Mechanics of the Continuum." Conversation afterward brought them around to the internal structure of a solid, and more specifically, the effect of heat on the temperature of a solid. With von Kármán nudging Born out of his preoccupation with relativity, the two decided to coauthor an article.

The effect of an increase in heat on the temperature of a solid had been under examination throughout much of the nineteenth century. Thermodynamics predicted that one unit of heat would increase the temperature of a solid by the same amount. The specific heats of solids was constant, independent of temperature; but with improved experimentation, researchers found that certain solids (such as diamond) did not follow this pattern: The specific heat decreased below this constant value as the temperature decreased to zero. In 1906, Einstein came up with an explanation by merging the statistical mechanics of James Clerk Maxwell and Ludwig Boltzmann with Max Planck's new quantum hypothesis. A year earlier, in contrast to Planck, Einstein had hypothesized that radiation, under certain circumstances—as in blackbody radiation (high frequency)—behaves as if it consists of discrete units: Light quanta were real.

For his theory of specific heat, Einstein now applied the concept of quanta to matter, thus generalizing the quantum theory to encompass both radiation and matter. He found that this theory of specific heats accounted for the decrease in the specific heat of diamond (carbon) and other substances below the classical thermodynamic constant value as the temperature went to zero.

No one paid much attention to Einstein's work. Then, a few years later, two developments happened that caught the interest of Born and von Kármán. The well-known physical chemist Walther Nernst noted that his measurements of specific heat at low temperatures were compatible with Einstein's theory. At the same time, in Göttingen, the lecturer Erwin Madelung refined Einstein's idea by regarding the motion of the atoms in solids not as independent but as dependent on one another—that is, atoms exchanged energy back and forth between them—an atomic version of pinball. Von Kármán recognized the importance of Madelung's refinement as well as his simplistic mathematical treatment.

Von Kármán and Born decided to build a theoretical model of a solid—whose real inner structure was as yet unknown—then explore its thermal properties as well as its optical and electrical ones. The question they addressed was, What is vibrating? To answer it, they developed a mathematical formulation for a repeating three-dimensional solid structure—a lattice. It was as though they had a concept of how the heating system of a house worked, and in order to investigate it, they first designed a prototypical house.

Born, with Hilbert's support, had initially thought that if the theoretical description of the lattice were more exact, he could avoid using Einstein's hypothesis about the quantization of energy. Von Kármán disagreed, arguing that the deviations between theory and experiment were too large to explain without drawing on this concept. Born finally gave in, worked out the mathematical model, and found theoretical results on specific heat that fit experimental ones, even in the low temperature range where Einstein's theory had departed slightly from experiment. In the last paragraph of the paper, Born wrote that he could not "suppress certain doubts" about the assumptions underlying the quantum theory of radiation. Yet he also realized that there was something fundamental to the idea—exactly what was not clear to him.

The *Physikalische Zeitschrift* received Born and von Kármán's paper on March 20, 1912, for publication the following month. Shortly after publication, Arnold Sommerfeld, professor at Munich and former assistant to Felix Klein, wrote to tell them that he liked their theory of specific heat and that he thought their model of the solid—which to Born and von Kármán was a secondary product of the research—had to be correct. A visit from Sommerfeld to Göttingen a few months later, however, blunted part of this accolade. Born and von Kármán learned from him that the Dutchman Peter Debye, then professor in Zurich, had presented a paper on specific heat in Geneva and published it in the *Archive de Genève* shortly before Born and von Kármán's submission. Priority for the improved theory of specific heat went to Debye by just a few days.

Debye, they discovered, had proposed a solid with no lattice substructure (a "house" with no walls) in which the vibrations of its quantized oscillators were independent. He found that the specific heat approached zero as the temperature approached zero in a manner similar—but not identical—to Einstein's. Born had, instead, labored over a parabolic curve of energy as a function of frequency. (Debye had explored a similar complex three-dimensional cube, but after consulting with Born and von Kármán's old friend Alfred Haar, decided it was too complicated.) Born's mathematical solution was more elegant and complex than the average physicist wanted, especially experimentalists, who preferred Debye's theory.

Shortly, an experiment proposed by the Berlin physicist Max Laue, then in Sommerfeld's institute in Munich, showed that X rays can be refracted like electromagnetic waves by a crystal, thereby confirming the lattice structure mathematically modeled by Born and von Kármán. Sommerfeld had been correct in his original assessment of their work. In their next paper on specific heat, which again used the lattice model, Born and von Kármán did not even reference Laue's work. They treated their theoretical lattice model as an established fact—"it could not be otherwise"—rather than as a basis for what was to become a major field of research: solid-state physics.

In depicting the mathematical description of a lattice, the model offered the ability to answer questions not yet asked. The approach was quintessential Born—understanding the macroscopic properties of matter from a minimum set of assumptions about the microscopic structure and doing so in the most general framework possible. Only through time—as the complexities of physics required more accurate predictions—was the power of Born and von Kármán's method fully appreciated. For the next ten years, Born and his later collaborators extended and refined this treatment, while von Kármán turned his attention to aerodynamics.

Born later credited von Kármán with teaching him "the essentials of mathematical physics," a transition point, as he saw it, in his journey to becoming a physicist. From von Kármán, he learned the importance of such research tools as knowing all the facts, estimating the order of magnitude of the results he expected before calculating, and using approximations appropriate to the precision required. Born melded Hilbert's stress on stripping a problem to its essence with von Kármán's approach, forming his own method to provide the most general rule to address a specific physical problem. Born perceived his mathematical knowledge as the key to his creativity, to "penetrate the mystery of nature, and discover a secret of creation."

By now Born had matured from a youthfully good-looking student, somewhat serious and sensitive, into a handsome man with a dash of the cosmopolitan. At twenty-nine, he was personable, athletic, financially secure, intellectually and culturally well-educated, and socially graceful. Yet romance was still missing—a lack he felt acutely. His rapport with his sister and many female cousins did not translate into ease with women. Timid and awkward in their presence, he had been spurned enough that his friends felt his dejection. Otto Toeplitz went so far as to despair of the female sex, because "it was so slow to turn to the person [Max] who seemed to me the most worthy."

Born's best hope lay in a grand old Göttingen tradition. The professors—especially those with daughters—always invited lecturers such as Born to elegant dinners and lively parties. The genealogies of German academic families were a bit like those of Europe's royal dynasties; there was much cross-fertilization.

Born became good friends with the musically talented daughters of his mentor Carl Runge; they introduced him to their friends. One of these, an old friend of Iris Runge's, arrived at the university in fall 1912 to take various lecture courses. Twenty-year-old Fräulein Hedwig Ehrenberg—Hedi to her friends—came from a family Born already knew. Her father, Victor, was a professor of insurance law at the University of Leipzig, who earlier had been on the Göttingen faculty. As Victor's wife, Helene, was a social butterfly, Max had enjoyed parties at their home. Rudi, one of Hedi's older brothers, was a professor on the medical faculty and had attended one of Max's disastrous lecture seminars. Despite the seminar, they had remained friends.

Fräulein Ehrenberg was a pretty, graceful young woman, who wore her thick brown hair fashionably plaited and wound at the nape of her neck. Her clever intellect and bright conversation bespoke a social ease that complemented Born's more reserved demeanor. He had found a young woman who shared his interests and appreciated his gifts. Among his first attempts to win her over was an invitation to tea at *El Bokarebo*. "Today we have another large tea, but this time *really* with women. Come across. The tea begins at 4; if you come earlier, we can go for a stroll."

The courtship prospered. On May 8, Max and Hedi became engaged. Their friends received a thick, engraved invitation:

We are honored to announce the engagement of our daughter Hedwig
to Herr Dr. Max Born,
Privatdozent in Physics at the University of Göttingen.
Professor Dr. Victor Ehrenberg and Frau, Leipzig,
Whitsun, 1913, Bismarckstrasse 8,
Reception day, Sunday the first of June.

Just a week after this reception, Max was in Breslau sending Hedi "morning greetings." The stationery he chose for his note had belonged to his mother; on the upper left corner was a hand-drawn, elaborately decorated easel upon which sat a small, square photo of sweet, curly-headed, one-year-old Max. Explaining that his mother had used this "quite special paper" when he was 'ein kleiner Bub'—a little lad—he asked Hedi if she would do so with their children. After assuring her of his love, he closed, "Your Bub."

That summer Hedi came to Breslau to meet the dozens of Born and Kauffmann relatives. (It was the last time that Max saw Toni Neisser, who died of cancer a couple of weeks after the visit.) From June to the beginning of August, Max was a regular on the train, first between Göttingen and Leipzig and then between Göttingen and the festive spa town of Bad Pyrmont, about forty-five miles to the northwest. Hedi's parents had sent her there with two girlfriends to "strengthen her for matrimony." She had become run-down—a frequent complaint through the coming years. Hedi was a hothouse flower who, in keeping with the German tradition of spas and sanatoriums, sought almost annual periods of rest to keep herself in bloom. In Pyrmont, she and her friends could luxuriate in mineral springs and rest under palms in an enclosed tropical garden.

While Hedi was there, a packet with a shipping slip estimate of 10,000 rubles arrived for her. Accompanied by a footman for safety, she collected it from the post office. Inside was a sixteenth-century necklace that had been in the family of the sender, Prince Leon Galitzin, for generations. His friendship with Hedi's grandfather, Rudolf von Ihering, the most famous German scholar of jurisprudence, had prompted him to send it once he heard from Helene Ehrenberg about the wedding—because "the granddaughter of Ihering was to wear it!" Twenty-two gold-colored disks, each containing a semiprecious stone, formed a chain that suspended an ornate, stone-encrusted cross.

The cross was a comfortable symbol for Hedi, less so for Max. Indeed, religion was the only discordant note in their wedding plans. Hedi's father was a Jew who had become a Lutheran to please his wife, Helene, at the time of their marriage. Helene wanted Max do the same so that he and Hedi could have a proper church wedding in Göttingen, but Max would not agree either to baptism or to marriage in a formal Christian ceremony. This perturbed Helene Ehrenberg, a willful German matron in the mold of Max's Grandmother Kauffmann, particularly owing to some Göttingen gossip she heard: Two years earlier, Max had desperately wanted to marry a young woman named Paula Cramers and had *volunteered* to convert in order to make his proposal more acceptable to her parents (who remained unswayed). The Ehrenbergs had not made conversion a condition for the marriage, but Helene certainly expected it. Max finally agreed to limited instruction—but not conversion—in order to be married by a Lutheran minister; but that did not end the to-do. Max's comment to Hedi, that the experience was a "*Schweinerei*"—a dirty mess—was one she would recall with irritation even forty years later.

Max later traced his reluctance to his father, who had taught him not to believe in a God who punished, rewarded, or performed miracles. Like his father, he based his morality on his "own conscience and on an understanding of human life within a framework of natural law." And he had his memories—of his father resisting conversion at the expense of a needed job, despite his distance from Judaism, and of schoolboys laughing at his own recitation of the Lord's Prayer. He did not mention his willingness to convert for Paula Cramers.

Helene had to give up her dream of a fancy church wedding for Hedi, though she must have wondered what Max had to lose by converting, given that he never went to synagogue or celebrated even the holiest of holy days. After all, his sister Käthe had converted to Christianity soon after her marriage; and Helene's husband, Victor, had consented, even though his family had strong ties to Jewish religious values. At the turn of the nineteenth century, Victor's grandfather, Samuel Meyer Ehrenberg, had served as director of a poor Talmudic school, the Samsonschule in the German town of Wolfenbüttel, which he and his son Philip gradually transformed into an "enlightened" teaching institution, leaving Talmudic study—although not Judaism—behind. They reshaped the curriculum to make it easier for graduates such as the Jewish reformist and scholar Leopold Zunz, famous for his emphasis on the nature and analysis of Jewish knowledge, to share in the non-Jewish world around them.

This "enlightened" teaching had actually led twentieth-century German Jews to become more involved in German society, lose their cohesion, and develop a range of religious expression—assimilation, conversion, reform, and intermarriage—as Helene had witnessed. Daughter Hedi, who had been baptized at birth, considered herself completely Lutheran. Still, their religious heritage gave some Jews a deep sense of belonging that Helene did not appreciate. Samuel Meyer's great-grandson, Franz Rosenzweig (Hedi's cousin), who was raised in an assimilated household, later became an important twentieth-century philosopher, returning to the Orthodox traditions that the family had set aside and devoting his life to Judaism—its study, teaching, and practice. Judaism was certainly not like the homogeneous religion that the Lutheran Helene Ehrenberg, a direct descendant of Martin Luther, identified with.

Helene Ehrenberg and Max Born resolved their dispute by changing the venue of the wedding, instead of Max's religion. On August 2, 1913, family and friends gathered in the Berlin suburb of Grünau, where Käthe Born Königsberger had moved with architect husband Georg shortly after their marriage in 1906. Their large stucco house on Regatta Weg—already the noisy home of six-year-old Susi, five-year-old Otto, and two-year-old Janne—was an idyllic setting, with gardens and lawns that swept down to the Dahme River.

On that sunny Saturday afternoon, in a striped tent outlined with masses of flower garlands, Hedi and Max stood before Pastor Luther. Hedi wore a long veil, a white dress with a train, and her bejeweled Russian cross around her neck. The ceremony was "semi-religious," according to the compromise Max and Hedi's mother had reached, and was followed by a sumptuous dinner in lavish Ehrenberg style. Echoing Max and Käthe's childhood activities at Tannhausen, the Königsberger children entertained the guests with an account of Max's 1912 trip in America. Five-year-old Otto starred as his Uncle Max, attired in a miniature version of his typical long blue trousers, sport coat, and straw boater. Susi played a reporter, and little Janne had a silent role as a newsboy. Max called it "a wedding in the grand old style."

Hedi and Max had known each other for a year—but a year spent within rigid social conventions that inhibited couples from real emotional intimacy until after marriage. Did Max realize that Hedi was high-strung and overindulged? Did Hedi see how intensely Max was involved in his work? The course of their two-month honeymoon might have seemed, in retrospect, a preview of their future. First came the warmth and glow of the pastel facades on Copenhagen's waterfront, followed by a storm-plagued steamer trip through the Swedish canals and rivers connecting Göteborg with Stockholm, and then—in a reversion to a different era—a trip to Tannhausen, where Max introduced Hedi to the paradise of his youth.

From there, they made a southwesterly detour to a conference in Frauenfeld near Zurich to hear Einstein speak on the foundations of gravitational theory. (Born had brought copies of Einstein's first two papers on general relativity and studied them along the way, much to Hedi's displeasure.) They stayed long enough for Born to participate in the discussion afterward; then the honeymooners were off to their last stop, Vienna, to attend the eighty-fifth meeting of the Society of German Scientists and Physicians.

Vienna itself added the pomp and circumstance of the Hapsburg court, fabulous palaces, museums, and opera—pleasures for them both. At the meeting, Max presented a paper titled "The Theory of Eötvös' Law, an Application of Quantum Theory," while Hedi met Max's friends and colleagues and attended at least one of the sessions. A photograph of members of the physics section captured them together in the eighth row of the great lecture hall, Hedi turning to speak to Max.

Discussion at the meeting focused on the recent work of twenty-eight-year-old Danish physicist Niels Bohr on the quantum theory of the atom. In three articles in five months, Bohr combined Planck's quantum concept with the atomic theory of Ernest Rutherford, which, based on experiment, showed that atoms have a positively charged nucleus. Bohr, with a model resembling planets (electrons) circling the sun (the nucleus) on different orbital planes (stationary states or discrete energy levels), hypothesized that electrons can only occupy certain stationary states and not a continuum of possible energies, as previously thought. In the Bohr model, electrons make transitions from one energy level to another by absorbing or emitting a fixed amount of energy. The frequency of the electron is proportional to the difference in energy between the two levels. Bohr's model was shown to be consistent with observed optical spectra, the discrete lines now understood as corresponding to an electron emitting energy as it makes a transition from a higher to a lower energy level.

Like many others at the meeting, Born was fascinated but skeptical. Later that fall, Harald Bohr wrote from Göttingen to his brother Niels that Born was one of a number of scientists who found "the assumptions too 'bold' and 'fantastic,'" and "do not dare to believe that they can be objectively right."

Returning to Göttingen after their trip, Hedi and Max set up house in a first-floor apartment that backed up to the town's military barracks and its adjacent parade grounds. Marching drills often drowned out Max's Bach partitas and reminded of the diplomatic tensions hovering over Europe. For some time, they had frequently seen "Kaiser Wilhelm's spike-helmeted soldiers, goose-stepping through the streets" of the medieval town.

Born's research interests now diverged, mimicking the state of physics in general—a constant tension between puzzling but engrossing new quantum principles with familiar classical theory. Reverting to his earlier interests with Minkowski, he wrote a short paper as a follow-up to articles by Gustav Mie on the classical approach to generalizing electrodynamics to explain the existence of the self energy of electrons. Then, collaborating with Richard Courant, a mathematics student from Breslau, he applied Planck's quantum theory to surface vibrations in an approach similar to his work on the specific heat of solids, a continuation of the paper he had presented at the meeting in Vienna.

Almost five years since becoming a lecturer, Born continued to wait for an offer of a university chair. One by one, his friends had been called to different universities and technical institutes: By spring 1914, von Kármán was in Aachen, Ladenburg in Breslau, Toeplitz in Kiel, and Ehrenfest in Leiden, with Hellinger on his way to the new university in Frankfurt a.M. Only one of these, Ehrenfest, was in theoretical physics. The field still had growing pains and its limited opportunities may have hampered Born. Anti-Semitism, another possible factor, was unlikely here; these friends were all Jewish.

Born, in fact, was no longer Jewish. His mother-in-law had worn him down. In March 1914, after a few religion lessons in Berlin, he was baptized a Lutheran by the pastor who had married him to Hedi. As he later explained,

> there were . . . forces pulling in the opposite direction [to my own feelings]. The strongest of these was the necessity of defending my position again and again, and the feeling of futility produced by these discussions. In the end I made up my mind that a rational being as I wished to be, ought to regard

> religious professions and churches as a matter of no importance. . . . It has not changed me, yet I never regretted it. I did not want to live in a Jewish world, and one cannot live in a Christian world as an outsider. However, I made up my mind never to conceal my Jewish origin.

As for Born's research, by the end of 1913, he had published an impressive twenty-seven articles. His work on relativity and on specific heat and crystal lattices had brought him favorable attention. Yet, at the same time, it might have inhibited job options. Einstein's response to a colleague's request for an opinion of Born stated that he was "a good mathematician, but so far he has not demonstrated much acumen for matters physical." Sommerfeld—a Felix Klein protégé who had a similarly strong focus on mathematics—offered him a lecturer's position in Munich, similar to the one in Göttingen. Sommerfeld also recommended Born for a chair in Heidelberg, which eventually went to someone else.

May 1914, however, was a time for other thoughts. Sitting in bed at Neu-Bethlehem Hospital in Göttingen, Hedi recorded hers in a poem entitled "Days of Waiting" that began,

> *A wreath of days, that with deep peace*
> *Separates past and future being—*
> *To a quieter entrance it would content us,*
> *Before new happiness calls us to new obligations.*

Six days later, on May 25, a daughter arrived, and they named her Irene, after the goddess of peace. She was a beautiful baby, her bright eyes and chubby cheeks crowned with masses of dark hair. Like her mother, she was baptized Lutheran. For the most part, in that Göttingen summer of sunny skies, Irene's enthusiastic coos and gurgles helped to create a serenity that shut out the ominous clamor over an assassination of a Hapsburg archduke in Sarajevo.

This spell was broken on August 2, the Borns' first wedding anniversary, when Göttingen came alive with cheering crowds, parading troops, and flags fluttering from half-timbered houses. With the continent's efforts at peace a failure, Germany had just declared war on Russia and was mounting an invasion of France. Göttingen residents, like the rest of the country, proudly and patriotically supported their fatherland's decision to defend itself against attack, as German newspapers vigorously proclaimed. Flaxen-haired maidens strewed flowers on cobblestone streets as troops marched to the station to board trains headed west. Within days, the Borns had said good-byes to several friends and relatives heading to the front.

In the 1870 Franco-Prussian War, Germany had crushed France in mere months. Like little Maxel Born and his cousin Hans Schäfer, every German schoolboy learned and acted out the exploits of General Helmut von Moltke, war hero. Since those victories, the kaiser and another general, Moltke's nephew and namesake, had expanded the German fleet and army and prepared every resource that might be needed in case of another conflict. What they considered a brilliant military strategy, the Schlieffen Plan, had lain in a drawer, waiting to be executed; and in August, they launched it—cutting a devastating swath through the small towns of neutral Luxembourg and Belgium to attack and once again overwhelm their western neighbor, France. A swift conquest to the west would allow the German army to redirect its might at the East and Russia.

In the midst of this nationwide fervor, Max Planck, the esteemed dean of German theoretical physics, wrote to Born. Planck explained that the Education Ministry (which ran and funded the university system) had just created an "Extraordinarius" chair in theoretical physics at the University of Berlin to take over much of his teaching load. Originally, the university had offered the position to Max von Laue, but von Laue accepted an offer of a more prestigious Ordinarius professorship from the University of Frankfurt. The department—and most importantly, Planck—had selected Born in von Laue's stead. Planck expressed the faculty's confidence in him. "I know, and my colleagues, too, that no one is more suitable than you in every respect." Planck and Born shared a theoretical basis—a commitment to relativity and an appreciation of and cautious attitude toward quantum theory.

"Joyous surprise" was Born's quick response to the honor of working with Planck. Born was privately pleased that Planck's interest must have arisen from his publications, since they did not know each other well. Planck answered that he would contact Born in November, when the faculty reconvened.

The country hoped for—and for the most part counted on—the same military result as in 1870: a victory within months. Although the letters between Planck and Born highlighted their concerns about the war, they also looked to the future. The war, Planck said, did not release him from the obligation of thinking ahead. For his part, Born hoped that there would be so quick a victory "upon which everyone here counts with confidence, that the winter can be devoted again to the usual work of peace." Why should they expect otherwise, when their everyday routine remained unaltered—shopping, socializing, reading, even mail delivery? Although the country's trains were occupied rushing soldiers to its western borders, the exchange of Planck and Born's three letters between Berlin and Göttingen—a distance of about 200 miles—had taken only a total of six days. Even eight months into the war, Born wrote a friend, "all internal life in Germany goes on as if there were perfect peace and not 10 million armed men on both borders."

So for the next three months, the Borns anticipated their move to Berlin and the letter from Planck as they anxiously kept up with war reports. By the beginning of September, Hedi's brother Kurt had taken part in three battles against the British. In October, a French shell almost hit Born's friend Richard Courant and destroyed binoculars Born had given him that had once belonged to Grandfather Kauffmann. That same month, Rudi Ladenburg, now a captain in a cavalry unit at the front, predicted confidently in a letter to Max that they would see progress at Verdun. There was pain too—like the death of young Captain Lodemann, whose wife and child lived in the apartment above them. Hedi and Max struggled with the contradictions of war and finally decided it was fruitless to torment themselves with what they could not really understand.

Hedi cared for infant Irene and found comfort in singing the songs of Schumann and Brahms. With his asthma holding him back from volunteering for the army, Max looked for other ways to support the war effort. He and Alfred Landé—one of Born's first students four years earlier—volunteered to help with the harvest. Around Göttingen, farmland stretched as far as the eye could see. With the shortage of men, harvesting these vast fields in the late summer was more crucial than ever.

Each morning for a week, Born and Landé rode off on their bicycles to a farm near Northeim, twelve miles to the north. Working dawn to dusk under a hot sun in thick dust was no work for a scholar—nor an asthmatic. After a week, Born had to leave Landé to carry on alone.

By October, the growing casualties at the front provoked Born to write to Paul Ehrenfest in Leiden, describing his bitterness at Russian brutality against the Germans in the east. Later, he recalled these emotions. "We were told every day about the abominable atrocities committed by the Cossacks in East Prussia. The idea of these 'Asiatic hordes' destroying the nice tidy German villages, torturing women and children infuriated me." Ehrenfest, whose wife, Tatyana, was Russian and who lived in the Netherlands, was not sympathetic. In reply, he pointed out, as Born recalled, "what the Dutch thought about the 'German hordes' destroying the nice tidy Belgian villages and torturing women and children."

Indeed, from the start, the battles fought from the pressroom were almost as heated as the physical combat. The German press regularly assured its citizens that their country had been attacked and had to defend itself; that German troops would not have gone into Belgium unless the French did first; and that the Russians were savaging the defenseless Prussians in the eastern states. Meanwhile, the English and French papers just as emphatically decried German aggression and brutality. Both sides played to their home audiences to maintain war fever, and also to neutral countries such as the Netherlands, Denmark, and the United States, to win their sympathy.

On October 4, 1914, ninety-three eminent German scientists and intellectuals (including Max Planck and Paul Ehrlich) put themselves on record in the Manifesto of the 93. (Hilbert refused to sign; Born was too junior to be involved.) They disavowed the litany of charges thrown at Germany. They wrote, "It is not true that Germany bears the responsibility for this war. . . . It is not true that we violated the neutrality of Belgium. . . . It is not true that the life and property of even a single Belgian citizen was touched by our soldiers." Such was the intensity of patriotic feelings that some supporters had not read the document before signing it. The aim of these distinguished men in their "Appeal to the Cultured World" was to sway the "civilized" (or neutral) countries to the German side. A friend sent Born a newspaper article entitled "Eine Neutrale Stimme" (A Neutral Voice) that reproduced a letter from a Dutch academic to a German friend. The Hollander complained about being "overwhelmed with literature saying 'it is not true.' What does it mean if they [the Germans] keep repeating 'It is not true?' Maybe it isn't true but that is only clear to anyone who accepts in advance that everything done by the Germans must be absolutely good and right."

The strategy of the declaration backfired. Instead of persuading, it enraged. English and American scientists responded publicly to the Manifesto of the 93. Six eminent British physicists, among them William Henry Bragg and J. J. Thomson, answered with their own declaration that pledged England and Belgium's fight to the end for freedom and peace. Individual physicists then took up the charges with pointed letters to one another. Mutual enmity between scientists continued for years, well after the war's end. Born would not escape the repercussions.

Between summer and winter semesters, as professors and students volunteered or were called up, the number of users in Göttingen's mathematics reading room fell from 350 to 98. Born's courses listed for the winter semester (in mechanics and the accompanying exercises) were dropped. He was restless as he waited to hear from either the army or Planck and divided his time between writing and sitting in on two series of lectures by Hilbert—one on crystal optics and the other on the Bohr atomic model. General interest in the latter had partly sprung from Niels Bohr's first visit to Göttingen the previous summer.

The previous July, Bohr had lectured in halting German on his new atomic theory. The eminent professors who had sat in the front row shook their heads, muttering, "If it's not nonsense, still it doesn't make sense." Born was less skeptical. Calculations based on Bohr's quantized atom matched spectroscopic evidence for the simple hydrogen atom. Born told Landé that, incredible as it seemed, Bohr, who looked so like "an original genius," must be on to something. In the fall, Born decided to work through Bohr's theory for Hilbert's seminar and wrote a ten-page report on the "Stability of the Bohr Atomic Model," which examined Bohr's model as well as new work by Ludwig Föppl.

This effort only took up part of his time. Born's major effort was a long article that quickly mushroomed into a book and incorporated all of his research on the dynamics of crystal lattices, including the early work on specific heat. His goal was "to derive all crystal properties from the assumption of a lattice whose particles could be displaced under the action of internal forces," including the integration of "infrared and optical vibrations of a lattice." The publication of *The Dynamics of Crystal Lattices* in the next year—though overshadowed by the war—met with admiration. Scientists welcomed it as a masterful presentation and expansion of the field as well as strong evidence of Born's intellectual potential.

Planck's promised letter arrived on November 27. The faculty had reconvened, Planck had sent Born's name forward as first choice, and was awaiting the reply. War aside, the Borns had much to rejoice at year's end: the excitement of Berlin, the birthdays of Hedi and Max, and Irene's first Christmas. In the lighthearted spirit of the season, six-month-old Irene "wrote" her first birthday poem to her father. For her "Da da da's" and "Ho ho ho's," Hedi translated: "I love you tenderly, Papa." New Year 1915 started with great promise, until they received Max Planck's letter of January 6.

With deep regret, Planck explained "that the matter of your call has taken a new direction." Von Laue had paid Planck a visit on New Year's Day and told him that, having been found unfit for military service, he was better able to take the Berlin call offered to him in the summer and wished to do so. Von Laue really wanted to be in Berlin to work with Planck. Planck told him of the offer to Born and advised him to resolve the issue with a direct request to the Ministry. Planck shared this with Born so that he did not "wait in vain for the call from the Ministry."

The Borns together labored over a reply that concealed their great disappointment and instead conveyed "a natural regret but no trace of resentment, acknowledging Laue's superiority and expressing the feeling that in these days of war, personal affairs did not matter anyhow."

Born's admission of von Laue's superiority was not false modesty. Three years older than Born and a friend from the Göttingen days, von Laue had just won the Nobel Prize for his work on X ray diffraction two years earlier that had confirmed the Born-Kármán theoretical structure of a crystal.

Born sent the letter to Planck right away. Before he heard back, however, he received a letter from an official of the ministry. In a short but friendly note dated January 8, the official stated his support for Born's candidacy and asked him to sign the enclosed employment contract immediately. Born took no chances, signing the contract on January 9, the day he received it and returned it at once. Planck's letter a day later described in greater depth his pain at having put Born through this incident and addressed the von Laue issue by merely confirming von Laue's plan to stay in Frankfurt a.M. As Born later learned, the government official whom von Laue had visited evidently found his request "an interference" and promptly sent off the offer to Born.

Among papers kept by the Born family is an old newspaper clipping with the heading "Local," Göttingen, 12 January, from the university. "As we have heard," it starts, "Privatdozent Dr. Born in Göttingen has received a call as Extraordinarius professor for theoretical physics at the University of Berlin." Arnold Berliner had won, by a few months, the bet he had made almost ten years before with a diffident young scholar who did not think he would ever receive such a call. Born gladly paid off the bet with the agreed-upon bottles of wine. By March 11, 1915, Max, Hedi, and Irene were on their way to prepare for the spring semester at the University of Berlin.

FOUR

A BITTER PILL TO SWALLOW

BORN RECEIVED AN UNEXPECTED OFFER WHEN HE MET FRITZ HABER ON THE STREET IN Berlin. It was one that he never forgot. Haber asked if he would like to work in his lab on chemical war instruments such as gas masks.

Haber, a native of Breslau and part of the Neisser social circle, was a brilliant chemist who would win the 1918 Nobel Prize. In 1915, he was director of the Kaiser Wilhelm Institute for Physical Chemistry in Berlin and primary developer of Germany's gas warfare.

With a gleaming shaved head and piercing eyes set behind pince-nez glasses, Haber could have posed for a caricature of the German patriot that he was. His name was prominent among the ninety-three who had signed the manifesto. When the fighting quickly became bogged down in trench warfare, he worried about ways to break the stalemate. Inspired by a demonstration of tear gas released from a shell, he proposed lethal gas, specifically chlorine gas, heavier than air so it would settle into the trenches. Some of Born's friends, including James Franck, were among the first recruits to Haber's program.

When Haber and Born met, April 22 had already passed. On that day, French soldiers in trenches near Ypres, Belgium, curiously watched as a yellow-green cloud spread out toward them. A northeasterly breeze carried it along, together with a strange smell that they said was like pineapples and pepper. As it enveloped them, their breathing became labored, each gasp sucking in more of the chlorine gas that would destroy their lungs and start their slow asphyxiation. Five thousand soldiers died at Ypres that day and another ten thousand suffered with corroded lungs for the rest of their lives.

Born did not reject Haber's offer immediately, but ultimately he did. If he recalled correctly, the letter with his response was very pointed, similar in spirit to later writings when he described poison as a cowardly instrument of murder and not sanctioned as a war weapon (an opinion shared by many Prussian officers). "Without a limitation of the allowable soon everything might be allowed," he said. His letter angered Haber who answered that one is no less dead when killed by a bomb or bullet than when killed by gas. The men discontinued their friendship.

His reaction to Haber aside, Born openly professed his patriotism. He described slain physicist soldiers as "those who have died a hero's death in the fight for the fatherland." He closed one letter to the physicist Jacob Laub, "Germany's strength is great and its task good. We are glad to be her sons." Still, his genuine excitement over battles won by General von Hindenburg translated into dreams of peace, not of victory. In private, he felt less fervid. Convictions inherited from his father left him with a tremendous internal struggle. With friends and colleagues dying, he and Hedi had long, intense talks over the futility of war versus duty to the fatherland.

Born's public embrace of flag and country was a common stance. All around him, Germans swore devotion to the fatherland, not least among them his own family. His brother, Wolfgang, now twenty-one, immediately enlisted in the army and went off to the western front, one of more than 10,000 Jews who did so during the first few weeks of the war. Käthe's husband, Georg, enlisted in October, bringing his own motorboat with him to help patrol the Vistula on the Russian border. Many of the Kauffmanns, as well as Bertha Born, put their considerable wealth into bonds to help finance the war. Jewish organizations and newspapers issued statements such as, "We shall as German citizens gladly fulfill all demands on our possessions, on life and blood." They urged "unlimited sacrifices for the great aims of the war." They appealed to Jews "to devote your resources to the fatherland beyond the call of duty." Assimilated Jews, especially, saw the war as their opportunity to display loyalty to the fatherland and, thus, finally earn complete acceptance as Germans. Forty-one-year-old Karl Schwarzschild—Born's former astronomy professor in Göttingen—enlisted for these reasons. (He died of a rare skin disease contracted during the war.) Academics, Jewish or not, responded to the patriotic call. The ninety-three who signed the infamous manifesto embodied the spirit of Germany's academic world.

With war casualties mounting, Born realized that the army would soon call him up, even with his asthmatic medical history. On March 15, just after arriving in Berlin and consulting with Max Planck, he enlisted in a communications unit of physicists and technicians assigned to operate "wireless" systems such as telegraphs. Born's enlistment, however, did not mean that he had to forfeit becoming an Extraordinarius professor. The military deferred his reporting date until the end of the university's summer semester. Beginning in April, he boarded the tram each day to ride the several miles from his apartment in the southwest corner of Berlin to the university in the heart of the city.

In their year and a half of marriage, Max and Hedi had had a child, witnessed the start of a war, changed jobs, and moved to a new city. In Berlin, they settled in the suburb of Grunewald on Teplitzer Strasse, the dividing line between the elegant villas there and the very middle-class district of Schmargendorf. Their fellow apartment dwellers were a quirky lot: a singer who accompanied herself on a lute (Hedi took singing lessons from her); a dancer with striking, exotic students who lounged on the stairs; and a former nurse who cared for a raucous old prince with a penchant for coming home at dawn. Their eccentricities enlivened an atmosphere that was uncertain, but at the time, not much more.

As she grew, Irene spent her days going to the nearby park with the maid, Lina, or playing with little Siegfried Wachtel, the boy next door. (Hedi—who blamed Irene's incipient temper tantrums on Siegfried's influence—was not so enthusiastic about the friendship.) The early stages of the war had little effect on her; Grandmother Born visited and brought her expensive frilly dolls with fluttering eyelids and coiffed hair. They were too delicate for her to play with, but she adored the presents. One thing made her curious about her grandmother: They slept in the same room, and one night, Irene, who was peeking from under the covers, saw her grandmother "take her hair off." Much as she tried the next day, her own would not come off. Irene enjoyed her days; the ugly and dangerous parts of the war were hundreds of miles removed from the parks of Berlin.

Born lost no time in joining in the life of the city's robust physics community. Berlin's well-established colloquia and Physical Society meetings continued as usual. Within weeks, he presented research on the dynamics of crystal lattices to the German Physical Society, his first of many lectures throughout the war.

At the end of June, Born's deferment ended abruptly; university classrooms had emptied as more students were called up. He packed a suitcase, said goodbye to Hedi (who was four months pregnant), and went off to Döberitz, a large military camp near Spandau, twelve miles west of Berlin. For four weeks, he practiced listening to and writing Morse code, the medium for telegraph equipment; drilled and marched; and held private seminars on wireless technology with the other eight physicists in the group. The training prepared Born to become a radio operator. He was not encouraged by the prospect of being an ordinary field soldier. In the off hours, he read through galley proofs of his first book, *The Dynamics of Crystal Lattices*, and on weekends, he often took the train home. Once, he returned home with a cold and asthma, the result of marching in a downpour. Arnold Berliner found a high-ranking military doctor to certify that Born was unable to report back to camp, and he stayed in bed.

There, Rudi Ladenburg found him convalescing a few days later. Ladenburg was now stationed in Berlin, heading up an army research group. Since his mistaken prediction of progress at Verdun the previous fall, his cavalry unit had been unhorsed and stranded in the trenches. Listening to the thundering discharge of artillery all day, he thought about ways to determine the position of the enemy's artillery by measuring the arrival times of the discharge noise. He persuaded the army to send him back to Berlin to assemble a research team on artillery. Now he asked for Born to be transferred into his unit on a temporary basis.

Ladenburg needed a mathematician to measure the distance and effectiveness of new weapons. At first, Born worked mostly at home, giving him an opportunity to clear up his asthma. Making frequent trips to the shooting range for a month, he developed a measurement method and drawings, from which a Berlin draftsman produced diagrams. Over mulled cider at Josty's—a café near Potsdamerplatz—Born reviewed the details with Ladenburg so that they could schedule trials. A week later, they had successful results.

Born's achievement brought him a permanent niche for the duration of the war. October found him as a sergeant in Ladenburg's unit of the Artillerie-Prüfungs-Kommission (APK), the army's technical division dedicated to research on artillery. Their offices were in a residential apartment building on Spichern Strasse—a relatively tranquil backwater a few miles away from the mighty Prussian military storm in the main headquarters on Kaiser-Allee—where the staff could do their research without much interference and keep nine-to-four work hours. Head of the division was Captain Fritz von Jagwitz—a prototype of the lean, tall, fair Prussian officer whose goal was to avoid the front. He gathered as many experts as he could and used their successes to make himself invaluable to the commander of the APK.

The activities at the APK kept Born from any meaningful scientific research, but the challenge of combining mathematics, physics, artillery, probability, and psychology to explore Ladenburg's original idea about sound-ranging interested him. He had to figure out the basics of determining the position of the enemy's artillery: How to collect the arrival time of the discharge sound from a piece of enemy artillery at different points and with these data, to calculate back to where the sound waves intersected—the enemy's gun. Born worked out charts and graphs for use at the front, and with the help of Captain von Jagwitz, he transformed the material into appropriate military guidelines.

While Born's days were routine and comfortable, the nights let dark thoughts seep in. The war's first-year anniversary in August forced recognition of a new reality: This conflict would not emulate the quick and decisive German victory over France in 1871. Sitting one evening in their apartment, with Irene sleeping peacefully, Born wrote to his wife who was visiting her parents. He weighed death and war against love of country and duty, with doubt shaking any fleeting balance. In the quiet, his thoughts turned to the losses of "the terrible times of last August and September."

> Then, I see the misery of Frau Lodemann before me, and the lovely face of Hans Werner appears and all the other poor, tormented people. And I ask myself: Why all this? And should it go on?
>
> If one could pile all the sorrow into a mountain, it would be higher than all the mountain ranges, and later generations would take it as a warning. There must be a great poet who is capable of erecting such a spiritual mountain. But he will probably not come and, in 20 years, everything will be forgotten. One will read about it in the history books. The French boys will be enthusiastic about the glorious Marne battle; the German boys about Tannenberg; and the Americans will be filled with enthusiasm about nothing. The sorrow of a mother for her only son, the sorrow of a young wife for her husband often appears to me greater than the great fight of the people. But one ought not to think so and if these thoughts come, one must simply cut them off and repress them, although one feels that the dear God would prefer these ideas to the great courage of sacrifice. But the devil plays his part in the world, and if he is not to swallow you up, you have to make a deal with him. One has at least to pretend that one crawls to the cross before him, as if what is happening in the world is as it should be, all this killing from *noble* motives and burning patriotism. But between ourselves we can admit that all of this is a bitter pill to swallow: love your enemy, but kill him. The best thing to do is to one's duty and not to think but to enjoy life while one has it and the little wife and the "Schnörkleins" [babies].

He and Hedi had gone over similar ground a year earlier and found it futile. He apologized to her for his relapse, which took one "not a hair's breadth further in understanding." In spite of doubts, however, he was still comforted by love of country, ending his letter, "The only consolation is that the security and future of the dear German fatherland is an important step further."

The times provided few sanctuaries for the uncertain. Born had only Hedi, and she did not want to rehash the same moral dilemma. Born found two men who did. The first lived alone in an apartment on Wittelsbacherstrasse, only a short tram ride away. He was Albert Einstein.

The thirty-seven-year-old Einstein was a pacifist, an internationalist, and since 1901 a Swiss citizen. He had been in Berlin only since 1914, when Max Planck and Walther Nernst recruited him as the director of the Kaiser Wilhelm Institute for Physics, one of several new science research facilities. Einstein was no admirer of the Prussian military spirit, and being the citizen of a neutral country buffered him from German nationalistic fervor. Rather than sign the Manifesto of the 93, he had the freedom to contribute to a counter declaration, "Appeal to Europeans." Where the Manifesto of the 93 justified German aggression, this declaration emphasized the need for "international bonds" and "a common world culture." The author was G. F. Nicolai, and he and Einstein were two of the four signers.

Unlike Born, Einstein was certain of the immorality of the Prussian military. He told friends in neutral countries that he hoped "for a victory by the Allies, which will ruin the power of Prussia and of the dynasty." After seeing atrocities on both sides, he later wrote to a colleague that he had grown more tolerant, but he had not changed his views in the slightest.

Born and Einstein's brief meetings at conferences over these years had not led to friendship; but once in Berlin, Born sought him out, and by fall 1915, they had become friends. Einstein too had little outlet for his feelings about the war, so they had much to talk about; besides, they shared a passion for physics and music.

A gift for music is not uncommon among theoretical physicists and mathematicians. Like mathematics—in representing extraordinary creations with its symbolic language—music seems to produce a sense of the familiar in the physicist's mind. Einstein, a violinist, thought that his music and work were both "born of the same source and complement each other through the satisfaction they bestow."

Einstein's first visit to the Borns' home in early 1916 was to play duets with Born. Hedi described meeting him when he entered the apartment. "He shook my hand and said, 'I hear that you have had a youngster.' And then he put down his violin, took off his 'Rollchen'—the detached cuffs of a thrifty man—and threw them in a corner. They played Haydn, which he especially liked at that time." It was the first of many evenings the three spent in music and conversation.

Einstein's antimilitaristic views strengthened the thirty-four-year-old Born's convictions about the "abhorrence of the slaughter," and before long, had won over Born to Einstein's conclusion that Germany must lose the war. Keeping faith with an optimistic view, they envisioned a new democratic Germany rising from the destruction. For anyone in Germany, views such as these had to be private—but especially for Born, who wore an army uniform every day.

Born's first real taste of war reinforced these beliefs. In summer 1916, he and Captain von Jagwitz traveled to the Somme Valley to inspect sound-ranging stations. When they arrived behind the lines, enemy artillery was shelling all the villages on the way to the first station. They traveled remote country lanes to bypass the danger and get to the station. After checking the site, they learned that other communication units had come under fire, and they had to remain in place for the day.

Uncomfortable with the stuffiness in the dugout, Born went outside to lie in a flat ditch on the hilltop. From there, he looked out over miles of the front and saw "a terrible wall of smoke and fire, and [heard] a hellish noise." Aerial dog-fights careened over his head. His military discipline initially kept the horrors of the Battle of the Somme at bay, while he absorbed the "spectacle." Later, his reflections on the "abominable mass slaughter" sobered him, and he never forgot what he had seen.

Born and Einstein balanced their evening philosophical discussions with afternoons steeped in physics. Einstein had just published the general theory of relativity, which Born later wrote was "the greatest feat of human thinking about nature." Yet at the time, Einstein's important and revolutionary field equations had little empirical basis. Although he encouraged confirmation and did as much as he could to promote wider public dissemination of the theory, Einstein's mainstay was his confidence in the structure of the theory and the opinion of what he called "convinced adherents" such as Born, Hilbert, and Planck. When Born wrote "Einstein's Theory of Gravitation and General Relativity"—a nine-page summary for physicists uncomfortable with the differential calculus—Einstein was pleased to be "completely understood and acknowledged by one of the best of my colleagues." Born, for his part, was fascinated to see firsthand what he called Einstein's "uncanny insight into the working of nature."

The friendship of Born and Einstein would endure for forty years through scientific disagreements, across oceans, and in spite of a deep moral divide that developed in later years over the general nature and culpability of the German people. When Born later reflected on the years in Berlin, he said that even in "the dark, depressing time . . . , with much hunger and anxieties, . . . it was one of the happiest periods of our life because we were near to Einstein."

Hedi expressed similar views. "His utter independence and objectivity, and his serene outlook," she wrote, "enabled me to ride up over the awful darkness of those days and to look far beyond the desperate day-to-day conditions." She too was a regular visitor of Einstein's, writing silly poetry and signing herself "Hedi Regina" to cheer up Einstein, who was sometimes bedridden with stomach ulcers. For his thirty-ninth birthday, she baked him a cake in the shape of an enormous castle and key and wrote a long poem called the "The Born Insurance Institute." The verses teased Einstein about his protecting his apartment from burglaries—his constant fear—with every possible means.

Born discovered his other philosophical refuge around the corner. During summer 1916, baby Irene developed a serious skin infection, and Arnold Berliner, ever ready with a medical referral, suggested Dr. Alfred Blaschko, a well-known dermatologist. Born had heard of Blaschko and had seen him around the neighborhood but had been too reticent to speak to him. Blaschko answered their call right away, and after treating Irene, he and the Borns delved into a lively political discussion. The friendship had begun.

Like Max Born's father, Blaschko was both a physician and the son of a physician whose life had connected with the working poor. Besides, he was friends with Arnold Berliner and Albert Neisser. It was easy for Born to trust and admire him.

The fifty-eight-year-old Blaschko was also an internationalist and an activist. He had refused to sign the Manifesto of the 93, and like Einstein, had goaded the nationalists by publicizing his views and calling for nations to come together. Activism came naturally to him. Recognizing that occupations bred certain diseases—the skin lesions of silver platers resulted from silver particles; the dermatoses of chemical workers arose from aniline dyes and organic solvents—he had turned to the social politics of disease prevention early in his career. His later examination of the social impact of syphilis and gonorrhea was an outgrowth of this involvement. (German dermatologists have traditionally been venereologists, undoubtedly because of the skin lesions associated with the diseases.) He made his most important public contribution by drafting Germany's first "Law for Combating Venereal Diseases," although he did not live long enough to see the Reichstag pass it in 1927.

Hedi, as often as Max, visited the Blaschkos, who took a parental interest in them. One night when Hedi was away, Max wrote to her about the Blaschkos' compliments on her clarity of thought and maturity—she was still in her twenties. "It was good you were not there because you would become vain. I made no objections and enjoyed in silence every word of the good people."

On his regular visits, Born often talked with Blaschko's good friend Eduard Bernstein, the leader of Germany's moderate socialists. He, like Blaschko, believed in a socialism of ethics and compassion—not one of Marxist economic doctrine. Especially after laboring through the "extremely dull and boring" *Das Kapital*, Born acquired similar beliefs. He later summed up his own attitude on socialism as "skeptical with regard to economic beliefs" and "not based on rational arguments but, like Blaschko's, on ethical principles."

The similar ethical and religious values on which Born and Blaschko had built their lives led to a sense of familiarity and comfort that Born later wrote about.

> He shared my skepticism of revealed religion. I loved listening to him when he talked to me about these things in his quiet, clear way. In less serious matters there was often a twinkle in his eyes and a joke on his lips. Since my father's death I have loved no man more than him.

As Max resolved his conflicts about the war, Hedi plunged into a serious depression. In November 1915, she had given birth to Margarethe (named after

Max's mother). The stresses of war interfered with her physical and emotional recovery. The food and supplies available for running the household when they had first arrived had become increasingly limited after the British blockade of Germany's coast began earlier in the year. With soldiers getting priority, the blockade hit hardest on the cities. Hedi struggled to feed her two small daughters. During the period when Max was going to the front to inspect sound-ranging equipment, her health failed so seriously that he took her to a sanatorium at Bad Nassau. She stayed there for two months.

Hedi was hardly back on her feet when an early frost killed off most of Germany's crops. The country turned to turnips—two to six pounds a week for adults in Berlin. To feed the girls something other than turnips, she rode the train out to the countryside and walked around villages asking farmers to sell her vegetables. At the end of a fruitless day scavenging in fields, she was forced, like the rest of the country, to rely on domestic ingenuity to turn turnips into a variety of edibles—everything from jam to flour. The "turnip winter" of 1916–1917, Germany's most severe food shortage, was marked by starvation, disease, and food riots, but throughout it all, the war effort continued nonstop.

At the APK, Born still needed to refine the accuracy of the timing of the artillery discharges. Effective sound ranging requires precise measurement of very small time intervals. Since oscillographs—that electrically record the time of sound intervals—were not manufactured during the war, Born and Ladenburg brought in two psychologists, Max Wertheimer and Erich van Hornbostel, and purchased several expensive Swiss stopwatches. The job of the psychologists was to train personnel to have a consistent reaction in timing the sound of the artillery discharges with a stopwatch.

Born quickly developed a close bond with Wertheimer, a provocative thinker who became one of his best friends in Berlin. Born described Wertheimer as "a person with a thousand shaded sides, interested in everything and philosophically deep but without real clarity. Discussions with him are difficult for me because his vague concepts make me nervous." The two engaged in hours of intellectual wrestling.

Together with Hornbostel and others, Wertheimer had introduced revolutionary ideas to create the field of Gestalt psychology, in which they argued that the actual form or shape of an event, when examined closely, was found to differ from sensory perception. Wertheimer talked about a perception of motion, as with movies, where a rapid sequence of still pictures "creates" motion. Extremely skeptical, he regarded "any observation as a deception of the senses" until proved by experimentation.

Through long discussions, Born came to agree with Wertheimer that the shapes, or *Gestalten*, were "the raw material of all observations." Yet, when Wertheimer's skeptical nature questioned logic and points of arithmetic, Born could not agree. Wertheimer asserted that primitive peoples who could not count past five had a different kind of arithmetic, and he tried to construct a metalogic and a meta-arithmetic. Years later, Born reconsidered and thought that these arguments might have some merit, but at the time, his debates with Wertheimer mostly helped to ease war anxieties.

Numerous research initiatives at the APK—overcoming the effect of wind, background noise, and the shells' bow waves on the propagation of sound, improving searchlights, and developing sonar—created a need for more scientists. Born began a recruiting effort, especially among his old Breslau and Göttingen associates, quieting his despair at the loss of the country's youth. He viewed it as a crucial opportunity to shelter young scientists from the front and save them for Germany's future.

The first recruit was his former student Alfred Landé, who after leaving Göttingen had volunteered for the Red Cross. When the APK transfer papers caught up with him, Landé had just been drafted into the army. Erich Wätzmann and Fritz Reiche—part of Born's short-lived Habilitation group in Breslau—Göttingen alumnus Erwin Madelung, and Rudi Minkowski, nephew of Born's former mentor Hermann, were added to the roster. Born did not have a place for Einstein's former assistant, Otto Stern, but with a sense of urgency he helped Stern transfer to a research division at the University of Berlin.

Born desperately wanted to bring the young mathematician Herbert Herkner to the APK but was repeatedly frustrated by bureaucratic delays. During Herkner's one-year stay in Göttingen before the war, Born and the other lecturers considered him "a phenomenon," who would bring "new blood to mathematics." He stood out as an "inspired genius" among the exceptionally talented. Herkner had already seen war service in Flanders, Soisson, the Donau Crossing, and Verdun before the APK transfer papers finally reached him on November 19, 1917. He was at Cambrai, a small French town safely behind Germany's massive defensive structure, the Hindenburg Line. At 6:20 on the morning after his papers arrived, the British began a surprise attack, spearheaded by a massive tank formation. During the next two days, British troops pushed over the trenches and the Hindenburg Line, crossed the St. Quentin Canal, and captured the Bourlon Woods north of the town. Herkner was killed in the onslaught. Two weeks later, a German counteroffensive retook the territory.

In the wake of Herkner's death, Born blamed himself for not moving the transfer papers quickly enough. He convinced Arnold Berliner, editor of *Naturwissenschaften*, that Herkner, though only a student, deserved an obituary in the science journal. In a long essay, as laudatory as any for an esteemed colleague, Born wrote "of this noble youth with whom a part of the spiritual future of Germany has been destroyed." It was a paean to a brilliant mind and a heroic person, but it also expressed Born's vision of the horrors of war and his own cloaked antimilitarism.

Gradually, the research in Born's unit was completed, and on quiet Spichern Strasse, under the unobservant eye of Captain Fritz von Jagwitz, Born began redirecting the work of his small but stellar staff. Walking into the office one day, James Franck and fellow physicist Walther Gerlach saw large rolls of paper filled with graphs spread out on the desks of Born and Madelung. As soon as the two workers recognized their guests, they quickly removed the paper rolls to reveal pads filled with calculations of the energy of ionic crystal lattices. Born, Madelung, and Landé had each emptied a desk drawer of its military items and substituted research books and notes. "Physics does not have to work for the war," they would say, "but the war has to work for physics."

For the first two years at the APK, Born had to take advantage of precious minutes to explore the many ideas stirring in his mind. Evenings often found him working at his desk at home, simultaneously rocking the baby while Irene played with her doll under his desk. He was frustrated with his slow progress and managed to keep up with the field mainly through lectures given at the various physics societies and colloquia. In 1917, he wrote to a colleague, "I am but a soldier and do not have time to publish," and in 1918, in a thank-you note to Niels Bohr for sending him a "beautiful paper on quantum theory," he wrote, "[I] almost never get to scientific activity. I envy you your freedom."

Freedom was on Born's mind as much as his work was. When a few of his investigatory field trips aggravated his asthma, von Jagwitz confined him to Berlin, eliminating even private trips to such places as Göttingen. The best he could do to keep in touch with friends was to ask Hilbert to encourage people to write to him. When responses were slow, he even suggested that Hilbert organize a special "card time" for friends to gather and write. He felt like a bird in a cage.

The article that Bohr had sent Born was the recently published "On the Quantum Theory of the Line Spectra (Pt. 1)." Born promptly mined Bohr's new information on the stability of the atom as the starting point for his continuing research on the structure of solids. Using the simple hydrogen atom, with its nucleus and single electron, Bohr had conceptualized it as a ring-model in a plane, much like that of the planetary system. Arnold Sommerfeld in Munich had elaborated this concept to atoms with several electrons, and Born decided to investigate the simple ionic crystal—rock salt (sodium chloride)—using a ring model. He asked Landé to collaborate with him in calculating the forces between the lattice points that would determine the structure and stability of the crystal.

Try as they might, the mathematical expression that Born and Landé derived contained a summation of terms that would not converge. Sitting across from Born and watching his frustration, Madelung offered a solution. His interest in the problem stemmed from his own research in Göttingen on lattice energies that, six years earlier, had been a catalyst for Born and von Kármán's articles on specific heat. The new mathematical method he provided for convergence allowed Born and Landé to calculate the electrostatic energy between neighboring atoms (a value now known as the Madelung constant). Their results for lattice constants of ionic solids made up of light, metal halides (such as sodium or potassium chloride), and the compressibility of these crystals agreed with experimental results.

Born and Landé sent their paper to Einstein for publication by the Berlin Academy, and a few weeks later received the page proofs. One morning Born arrived at the APK to find Landé "pale and desperate." He had discovered an error in the tables that ruined the compressibility results; and it was not just any error—it was the error of a novice. In calculating the mutual energy of the rings, they had neglected to divide by two. Their theoretical crystals were twice as soft as they should have been—more like rubber balls.

Born hurried off to Einstein's house to withdraw the paper. After Born explained the situation, Einstein responded with a hearty laugh. He thought it was "marvelous" that Born had made a mistake. "I thought you would never make one," he said. He suggested publishing the first half of the paper on the results of the lattice constants, which stood on its own, and reworking the second part on compressibility for a later article. Born complied, concluding the paper with a note saying that the theoretical values of the compressibility were wrong and would be reexamined.

Given that dividing the compressibility results by two would destroy their model's agreement with the experimental results, Born and Landé knew they had made a major conceptual error; but Born did not try to fix it by reworking the model. He realized an important relationship: compressibility, for which there were experimental values, was a function of two variables. For one of those (the lattice constant) they had observations; for the other (a variable related to the repulsive forces in the crystal) they had the value calculated from the Bohr-Sommerfeld ring model, which, Born realized, was evidently incorrect. The second Born-Landé paper concluded, "The planar electron orbits are insufficient, atoms are seemingly spatial objects." He needed a space-filling, three-dimensional model, like a cube.

Later, Born would recall that that result was the "second hint given by lattice dynamics that in atomic dimensions we ought to face the fact that quite a new mechanics was needed." (The first hint had come in his earlier work with von Kármán.) "From that moment on," he said, "my endeavor was not to find confirmation of the orbital [planar] model but to provide arguments for its insufficiency."

The distance from the A.P.K. to Einstein's apartment was only about a ten-minute walk. In fall 1917, Einstein had moved to 5 Haberlandstrasse in the Bavarian Quarter, next door to his cousin Elsa. This arrangement made it easier for her to nurse him and his stomach ailment. (Two years later, she became his second wife, Einstein having left his first, Mileva, and their two sons in Switzerland.) Once recovered, he and Born met frequently for lunch, where they often lost themselves in new physical theories. When James Franck joined them, he told them that their discussions about relativity sounded like "gibberish." They replied diplomatically that "one could do physics in different ways."

Other days, Einstein met up with Born and Wertheimer. In summer 1918, the Borns depicted this trio in a postcard containing a photomontage (a popular craft at the time). They cut the men's three individual pictures in the shape of ovals and arranged them in a line with the ends overlapping. Einstein, the recipient, whose picture was in the middle, referred to "the clover leaf" as "three incorrigible hobby horse riders in brotherly unity; two are introspective, one" (himself) "stares unconcernedly into space."

As the Borns composed their montage, Wilhelmian Germany was coming to an end. The country was falling apart. The war had strained the resolve of the population, and the British blockade had created austere conditions. In Berlin, the average food consumption of the 4 million inhabitants was about 1,000 calories per day; for meat about three-tenths of a pound per week. By the end of October, the Borns saw a serious risk of a Bolshevik revolution and asked Hilbert's advise on Göttingen serving as an escape from the possible turmoil and on the likelihood of getting there, should the time come. As civil disorder increased through the next week, the Borns knew the old government would end. The question was how, especially with rumors circulating in Berlin of the revolutionaries bringing in machine guns and canons to fight the military.

On November 4, sailors in Kiel demonstrated against rumors of a last-ditch attack on the English fleet. This small revolt quickly spread into a broader revolution, and socialist and communist councils organized in its wake throughout the country. Chancellor Prince Max of Baden, fearful that the monarchy would fall to a communist takeover, resigned and persuaded Kaiser Wilhelm to abdicate. (The kaiser fled to Holland.) A new government formed under Friedrich Ebert, leader of the moderate Social Democratic Party. On November 9, everything that Born and Einstein had talked about for three years had become reality—a defeated Germany, an exiled kaiser, and the establishment of a socialist government.

The streets in Berlin were not peaceful, but they were largely bloodless. The military, including the special guard regiments, threw their ammunition into the canal, and Spartakus, a radical communist group, did not get the opposition it wanted. Revolutionaries manned machine guns, and officers feared that the crowd would exact revenge for Germany's misery and defeat, but only brief skirmishes resulted. The Borns and their friends spent a few sleepless nights worrying that Bolsheviks would come to wealthy neighborhoods such as Grunewald to plunder houses. Military officers, such as Captain von Jagwitz and other officers at the APK, quickly changed into civilian clothes and disappeared into the chaos, clearing the way for marauding soldiers to loot their offices. These gangs stole everything, from Born's scientific notes and books to the Swiss stopwatches used for sound ranging. Born had dearly wanted one.

Born followed the change of state from his sickbed, where he was suffering through another attack of bronchitis. Two days after the changeover, he received a call from Einstein saying that a council of students had taken over the University of Berlin and arrested the rector and other officials. Einstein worried that their erstwhile colleagues could be in serious danger and wanted Born to help him rescue them. Born agreed, and despite his illness, walked the two miles to Einstein's apartment. (The phones still worked, but the buses and trams did not.) Warily navigating "through streets full of wild looking and shouting youths with red badges," he met Einstein and Wertheimer, whom Einstein had also called, and the hobby-horse trio set off for the Reichstag.

An enthusiastic crowd and heavily armed soldiers sporting red ribbons swarmed around the Reichstag. The three men forced their way through the throng to the front, only to have "red" soldiers bar them from entering the building. There they stood, absorbing the excitement and cheers for socialist notables who were arriving, unable to penetrate the cordon. Being stymied, they changed their plans and made the short walk to the Chancellor's Palace, which housed the new government.

At the Palace, footmen clad in imperial livery waited on the magnificent staircase as though ready to usher in generals and royalty to the kaiser. Representatives of the new government and social order—just ordinary people—raced up and down the stairs. The three scholars milled about in the reception hall with about twenty others, watching events unfold and waiting for a cabinet meeting to end so that they could have an audience with Ebert, the new president. When they finally saw him, Ebert quickly gave Einstein a card with instructions for the release of the officials and apologized for his haste. He had just received the terms of the Armistice from the Entente.

Born returned home exhausted but pleased by the events. The next day he, Einstein, and Wertheimer went back to the Reichstag where they were able to negotiate with the students for the release of the university officials. In the process, Born saw the students' commendable mind-set, an asset that he felt had helped to thwart a Bolshevik takeover. With the kaiser in exile and a new government pledged to become a republic, Born, Einstein, and Wertheimer felt that they had witnessed the birth of "a free, democratic, socialist Germany."

Born's friends and family had all come through the war. Besides the close Breslau-Göttingen coterie of Ladenburg, Reiche, Madelung, Landé, and Minkowski at the APK, others such as Franck, von Kármán, Toeplitz, Hellinger, and Courant were healthy. Brother-in-law Georg Königsberger returned safely to Born's sister, Käthe, and their children in Grünau, although without his boat, which had been lost with most of its crew. Early in the war, brother Wolfgang had had a close call when his entire regiment was wiped out in the battle of Verdun. The night before, he had been hospitalized with typhus. He spent much of the war assigned to a medical division, drawing cranial operations for the scientific record.

One serious casualty of the war was the Jewish community's hopes to gain true equality through their show of patriotism. Most Germans were not as phlegmatic as Born and friends about the fatherland's humiliating defeat. Some, unwilling to blame themselves or their leaders, blamed the Jews for the war, the defeat, and everything that followed. Funneling frustration and shame into anti-Semitism, they questioned Jewish participation in the fighting by publishing "official" reports claiming that only 35,000 had served at the front rather than the actual 80,000. Meanwhile, the value of war bonds fell to nothing, destroying the wealth of such families as the Kauffmanns and Borns, who had invested most of their resources in them. Max and Hedi's comfortable financial cushion had largely disappeared.

When a summons came from Max Planck, Born was as glad as APK director von Jagwitz to change clothes and slip away. Eight days after the start of the Armistice, Planck requested Born's immediate transfer back to the university, "in the interest of a speedy restoration of . . . instruction." His request had nothing to do with teaching (there were no students); rather, he needed Born back in an official academic position because Born had the possibility of a promotion.

Since taking the chair in Frankfurt, Max von Laue had never abandoned interest in returning to Berlin to be near Planck. In summer 1917, he had thought up a somewhat far-fetched plan that he proposed to Einstein when he visited Frankfurt.

> I have a request: Do commend our colleague Born as much as possible in Frankfurt. This can be done with a good conscience, of course. For if all else fails, I should like to try to swap positions with him; and the only problem that could possibly arise in this, as far as I can see, lies in Frankfurt.

Von Laue was unperturbed by the Ministry's strict regulations on appointments.

Born became aware of the plan in the beginning of 1918 and straightaway received Einstein's advice: "Accept unconditionally." Einstein wrote to Hedi,

> I have no need to assure you how fond I am of you both and how glad I am to have you as friends and kindred spirits in this . . . desert. But one should not refuse such an ideal post, where one is completely independent. There is a wider and freer sphere of activity than here, and it gives your husband a better chance to display his powers.

Von Laue was extremely pleased to hear from Einstein that Born agreed.

The Berlin faculty approved Planck's draft letter to the Ministry agreeing to the exchange, and the faculty at the University of Frankfurt concurred. Owing to the unusual circumstances, the University of Frankfurt did not make a formal application and as a consequence, it did not receive recommendations or personal information on Born. On November 25, the University of Frankfurt wrote to the Ministry that it needed to announce the lectures for the spring semester. Who would give them—von Laue or Born?

Born was anxious about the situation. He wanted to leave the "stinking air" and the chaos of Berlin, where extreme left-wing and right-wing elements still skirmished in the streets. He wrote to Hedi's brother, Rudi, "I have received nothing in writing about my call to Frankfurt. If a new government should come in tomorrow, perhaps nothing will come of this. I will not go away from Berlin unwillingly, in spite of many good friends."

Released from his duties at the APK and with no students to teach at the university, Born divided his time between research and organizing a special colloquium for demobilized physicist soldiers. These physicists, who clamored to catch up on what they had missed, first approached Heinrich Rubens, a professor at the university who for years had held the famous Wednesday physics colloquium. Yet he declined—he was too exhausted from the stresses of the war—and turned to Born. Born was not an obvious choice, since Rubens was an experimental physicist and Born a theoretician, but they were friends and shared family ties. (Rubens was godfather to the son of Alfred Lipstein, Bertha Born's youngest brother and one of Max's traveling companions in his younger days.) Besides, Born was also an active contributor to both the Wednesday colloquium and Friday lectures at the German Physical Society.

For the new colloquium, Born chose the topics, undertook the scheduling, and, as he later would say, headed the most distinguished collection of physicists of any time in his career. One memorable presentation was by James Franck and Gustav Hertz on their 1913 experiment on atomic excitation in which they had unwittingly verified for the first time an aspect of Bohr's quantum theory. Sending electrons through mercury vapor in a tube, they found that electrons with low energy passed through. When they gradually increased the energy of the electrons, the mercury atoms were able to absorb the electron energy at a critical value and make the transition to a higher energy level. As Bohr had postulated, this demonstrated that atomic energy levels existed in discrete, quantized states.

Whether Born lectured on his own research is not recorded, but throughout these winter months, both he and Landé presented their work at the meetings of the Physical Society, including Born's new work on ionic states. Born had calculated the lattice energies of ionic crystals—the amount of energy released when ions in a gas phase come together to form a solid—but could not verify his calculation because lattice energies cannot be measured directly. Born thought he could solve this by comparing his theoretical lattice energy calculations to measured data on their chemical companion, heats of formation: The amount of energy emitted during a chemical reaction, in which metals and nonmetals react to form ionic solids.

In late November, with trams stationary due to street fighting, Born walked from his apartment in Grunewald across open fields and meadows south to Dahlem to visit Franck—his frequent source of answers to experimental questions. Born found him in his office at the Kaiser Wilhelm Institute for Physical Chemistry, where he had worked through much of the war. The two began discussing Born's ideas, when who should come in but institute director Fritz Haber, with whom Born had not been on friendly terms since their exchanges concerning gas warfare.

Born, uncomfortable with face-to-face conflict, could not hold a grudge against Haber; as the three talked, Born found Haber "so charming and kind that I forgot my aversion to the gas war and he his resentment about my aversion." Not only was Haber someone whom Born genuinely liked—apart from his war research—but he was also a part of Born's scientific world.

With his fast, clear mind open to new ideas, Haber immediately grasped Born's problem. He said that no one had calculated chemical energy successfully from physical data; he liked Born's idea of using information from his investigation of lattice energies. Through the next few days, the three continued the discussion.

Together they dissected the process by which metals (such as solid sodium) react with nonmetals (such as chlorine gas) to form an ionic compound (such as sodium chloride). They envisioned the solid metal absorbing energy as it vaporizes to a gas, then requiring more energy (the ionization potential) to ionize it to a positive ion. The nonmetal—the chlorine molecule—first absorbs energy as it dissociates into atoms (dissociation energy), then releases energy as it acquires an electron and becomes a negative ion. The positive and negative ions finally coalesce into an ionic solid, releasing great quantities of energy (lattice energy) as they form a crystalline lattice. Together, these energy changes add up to the heat of formation, a quantity that can be experimentally determined. With the exception of the lattice energy, the energies of all the steps were measurable or could be calculated, and since lattice energy was the only unknown variable, they solved for it algebraically. (The value for lattice energy that Born had derived from theory was so accurate that it is now often used to calculate electron affinities.) Haber drew a diagram that organized the entire process into a series of steps. Known as the Born-Haber cycle, it would soon be a fixture in chemistry texts.

On January 4, 1919, the Prussian minister of education made Born's call to Frankfurt official. He was to become an Ordinarius professor in the science faculty at the University of Frankfurt beginning in April with the summer semester.

The rest of the winter was a whirlwind. The Borns bought their first house after intense negotiations caused by Frankfurt's tight housing market. Max held classes in Berlin and traveled to Frankfurt every weekend to lecture to thirty students or so on electricity theory. (It was the *Zwischensemester*, the interim semester for war participants.) When they were preparing to leave Berlin at the beginning of April, little Margarethe (nicknamed Gritli) came down with scarlet fever. The quarantine was six weeks; they had to get out of the apartment in three. Somehow they made it.

Earlier, Max had written to Hedi's brother with a vision of promise. "In Frankfurt I hope to find a much greater effectiveness and independence. I have so many work plans, and if these succeed only as reasonably as my last works, then I can occupy an entire Institute."

FIVE

THERE IS NO OTHER BORN IN GERMANY

"IT IS IDEAL HERE," HEDI WROTE TO FRIENDS IN BERLIN. "THE PEOPLE FEEL VERY MUCH LIKE you and the Schurs." For most Germans, including the Borns, *ideal* was a relative word—Frankfurt was safer than Berlin. Hedi shopped without worrying about street fighting blocking her path. The little girls climbed up the trees in their backyard. Max organized his new institute. Yet, in every small town and large city in Germany, the consequences of war threatened people's existence.

Cold and hunger dominated Germany's present, and many feared the same for the future. Four years of war had left the country and people ravaged, and the terms of the Armistice stripped them of the resources to regroup. On the night of the Armistice, November 11, 1918, the British commander Douglas Haig had recorded the reaction of the French in his diary.

> The Germans pointed out that if the rolling stock (which have been handed over in the terms of the Armistice) are given up, then Germans east of the Rhine will starve. Report says Foch was rather brutal to the German delegates, and replied that that was their affair!

The Allies wanted revenge—especially the French, who lost 1.5 million soldiers in battle, with another 4 million wounded.

The British were no more merciful. To make sure that the Germans came to the peace table ready to agree to the Allies' demands, the British blockaded the German coast for another eight months after the Armistice. With a devastated country and no imports, children starved, babies went without milk, and suicides floated down Berlin's Spree River. Everyone was cold for lack of heating fuel. Whatever relief the end of the war might have brought to Germany's people the continuation of the blockade destroyed.

The Born family was merely undernourished and chronically hungry. They had friends such as Carl Oseen, a physicist living in neutral Sweden, who would occasionally send packages of food appropriately called *Lebensmittel* ("means of life"). Born's worry was for the hungry people around him who, he feared, would lose "all concept of honesty and responsibility" in their bid to survive. The starvation conditions caused by the blockade became his preoccupation in every letter. Even so, he anticipated a fair treaty. He believed U.S. President Woodrow Wilson who, one year earlier, had enunciated his Fourteen Points for a program of world peace including: "the principle of justice to all peoples and nationalities, and their right to live on equal terms of liberty and safety with one another, whether they be strong or weak."

Born knew that the animosity of the Allies toward Germany was not confined to the military. The Manifesto of the 93 had injected acrimony into the science community that persisted. With resources for research in Germany extremely limited, however, Born wanted to ask the Harvard chemist Theodore W. Richards to measure the compressibility of a specific list of salts, which would help Born continue to investigate Bohr's planar concept of the structure of matter. Wary of approaching him directly, Born turned to Carl Oseen, who agreed to write to Richards.

An inflammatory article that appeared in *Science*, "National Prestige in Scientific Achievement," by the physicist P. G. Nutting about German scientists soon proved that Born's wariness was justified.

> Plagiarism and piracy were common practices, and from personal knowledge I doubt whether a third of even the more eminent German scientists were free from this taint. Further, work of foreigners was taught as the work of Germans in both literature and science. Neither fairy tale nor scientific discovery, if in an obscure publication, was safe from adoption as their own while the misleading of the young student was easy and common.

Repetition of the claims by the British journal *Nature* increased the insult. Arnold Berliner, editor of *Naturwissenschaften*, wanted to reply, but Born and Einstein cautioned him not to bring more attention to the matter. Postwar animosity was not one-sided. Fiercely nationalistic Germany felt humiliated by the Allies' detention of German war prisoners, continuation of the blockade, and threats to cut off German borderlands. Highlighting the article would have further increased tension in the scientific community.

A few weeks later, Born received Theodore Richards's reply via Oseen and read that Richards did not foresee a renewal of ties between scientists until "far in the future." Richards could not forgive the ninety-three German scientists who had signed the manifesto declaring Germany's actions noble and intentions good when Prussian troops had overrun Belgium and instituted a "campaign of terrorism." "Until the German scientific world acknowledges its error," Richards wrote, "it will be very hard for the civilized world to accept their friendship." He wanted repentance; he did not think that he could shake the hand of a German scientist. With regard to Born, Richards would try to support his research request since Born had not signed the manifesto.

In a letter to Oseen, Born responded with fury to Richards's remarks.

> We German pacifists who fervently believed in Wilson and his principles have been completely and entirely refuted. After the publication of the peace conditions of the Entente there is no more doubt that we have fallen victim to an outrageous deception as we gave up the fight in trusting Wilson's promise of a peaceful reconciliation. Whether there will now be more or less mitigation at the negotiations in Versailles, the matter of the deception remains and will never leave our memories. . . . All hopes of reconciliation and forgiveness are now buried. . . . You write him that *we* will *not* forgive as long as the body politic exists which holds, both as a party and a judge, for its right to impose starvation on an entire people.

The British lifted the blockade two weeks after the Germans signed the Treaty of Versailles on June 28, 1919. The treaty gave Germany's northeastern borderlands to Poland, blamed Germany for the war, imposed large reparation payments, and deeded the Saar coal mines to France.

Born had been a tangle of conflicting beliefs at the beginning of the war, but he felt no confusion about the peace terms. He was devastated. He believed that the Treaty of Versailles would "greatly weaken if not destroy the germs" of democracy and socialism he had hoped the German people would embrace. In one particularly low moment, he wrote a letter to Einstein, a phrase from which Einstein repeated back to him in reply. Einstein wondered if Born should be allowed to say "with tears in his eyes that he has lost his faith in humanity?" Einstein counseled optimism. "Conditions [of the treaty] are hard, but they will never be enforced. They are more to satisfy the enemy's eye than his stomach." He believed in the "slovenliness" of the French and eventual disunity among the Allies to undermine the intent of the treaty. His reaction was unusual: Most Germans were furious, not only at the Allies but also at their new government that signed this treaty—a five-month-old democratic republic led by Friedrich Ebert.

For all of the hardship in Germany, Hedi's enthusiasm for Frankfurt did not vanish. Frankfurt had a small pocket of the "ideal," and the Borns enjoyed it, even when cold and hungry. They settled into a spacious house at Cronstetten Strasse 9 that was surrounded by a large garden of peach and pear trees and gooseberry bushes—the luscious fruit a delight to their daughters, Irene and Gritli. The house was located on the northern edge of the city, and from their back window, they could see the Taunus Mountains, about twenty miles away. In front of them lay a bustling, cultured city.

Frankfurt am Main was both an old town and a new city. The center next to the Main River was a medieval mélange, anchored by a Gothic town hall called the Römer, fountains, a Gothic cathedral, and half-timbered houses with painted farming scenes and window boxes adorning the facades. Beyond was the modern city, a hub for the banking and chemical industries that were the foundation of the city's great wealth.

Five hundred ninety-nine millionaires lived in Frankfurt in the prewar era. Jewish businessmen, who had flourished there for the hundred years since Napoleon had literally unlocked the gates to their narrow, crowded ghetto, were a large part of this wealth. Their fortunes furnished 80 percent of the 8 million marks needed to endow the University of Frankfurt, which opened in 1914. The number of postwar millionaires in Frankfurt had shrunk by 90 percent, but the donor who supported Born's chair of theoretical physics, one Herr Oppenheim, a dealer in precious stones, was still among the remaining 10 percent.

Artur von Weinberg, co-owner of the chemical company Casella, was another generous benefactor to Born's institute. The Weinbergs "lived in a princely style," with footmen, personal waiters, works of art, polo, and stables for dozens of horses. Born was friendly with them but found their extravagance uncomfortable, preferring the plain elegance of the Oppenheims. The Oppenheims opened up Frankfurt's thriving life of operas and concerts to the Borns and entertained them at parties, where Born and Frau Oppenheim sometimes played four-handed pieces for the guests. At home, Born assembled a quartet of the most superb musicians with whom he would ever play. One guest, listening to Brahms, Reger, and Bach fill the room, rated the results "extraordinarily beautiful."

Born's research ambitions may have been, as he had said, to keep "an entire institute" busy, but he first had to tackle an institute that wartime and the unhappiness of his predecessor, Max von Laue, had left "little established." Born organized the institute's lecture series for the different areas of physics and drew up a weighty list for his own first-semester courses: analytic mechanics, quantum theory, introductory theoretical physics for war participants, and an evening colloquium in conjunction with the head of the department, the experimentalist Richard Wachsmuth. He hired a young assistant with good experimental skills, Fräulein Elisabeth Bormann, and a highly capable mechanic, Meister Schmidt, to construct the equipment needed for experiments. (The staff encouraged Schmidt's industry with that scarcest of goods in postwar Germany—cigarettes.)

Born filled out the small professional staff just as he had at the Artillerie-Prüfungs-Kommission—with the Breslau Göttingen circle. Otto Stern, an acquaintance whom Born had helped find safe haven in Berlin, was the lecturer, a position he had held at the beginning of the war. Alfred Landé, who completed his Habilitation in the summer, became his assistant. After the Armistice, Landé "had thrown away" his uniform, staying in Berlin only long enough to lecture with Born on December 13 at the Physical Society on their lattice research. Landé then took a job teaching music at the Odenwaldschule, a private school thirty-five miles from Frankfurt. He came to Frankfurt to lecture once a week, keeping his job at the Odenwaldschule since he had no regular income from his lecturer duties.

Born and Landé had worked together on and off for almost ten years. In the postscript of his first letter after Landé left Berlin, Born instructed him to stop addressing letters to "Herr Professor" but instead, "lieber B. or some such." It was a rare relaxation of the formalities between German student and professor.

Ernst Hellinger returned to the chair in mathematics he held before the war, completing the circle of old friends. Owing to a tight housing market, the Borns rented their attic apartment to him and his sister. Gritli and Irene adored "Uncle Ernst," the patter of little feet on the stairs announcing their frequent visits.

Unlike many mathematicians, Hellinger was interested in application, a trait that led Born to think of him affectionately as "a 'tame mathematician' whom I could consult whenever I had a problem beyond my own resources." Born, who attributed his creativity in physics to mathematics, did not like to stray too far from his muse.

Born's institute was tucked into two rooms, a large and neat workroom for experiments and a small office for thinking. Within "this tiny, restful department," as Born called it, ideas were stirring. A walk through the institute would find Otto Stern in the lab fiddling with equipment to measure the properties of atoms and molecules; Landé in the small office working on one of nine papers he wrote in eighteen months on the cubic-atom research begun in Berlin; and Born across the desk from Landé focusing on the mobility of ions during electrolysis, a topic he pursued as a result of gentle nagging from faculty member Richard Lorenz. A few graduate students might be working on experiments.

Using beams of silver atoms, Stern wanted to verify that the distribution and the mean of the velocities of atoms followed the Maxwell-Boltzmann distribution. Except for being a good glassblower, Stern had no more manual ability than Born did to assemble the needed equipment, but he had great ingenuity in designing the experiment. Herr Schmidt followed Stern's instructions for constructing a special cylindrical vacuum tube through which atoms could travel from one end to the other. Stern invented this molecular beam method, which would prove crucial for his future examination of properties of atoms and molecules.

In the summer, the bustle began to slow as Born helplessly watched creeping inflation eat up research funds, undermining Stern's expensive experimental work. Born wrote to Einstein listing the institute's five experimental research tasks, beginning with Stern's. Einstein, who was director of the Kaiser Wilhelm Institute for Physics in Berlin, said that he would do his best "to squeeze some funds out," but nothing came. As November arrived, Born's 3,000-mark budget ran out. The university had kept it at the same level, even though the mark was at that time one-tenth to one-twentieth of its former value. Born wrote to a colleague, "My institute is completely frozen; temporarily I can see no prospects whatsoever for experimental work."

Two days later, Einstein had something to offer besides money—his newfound celebrity. On November 6, the news went out that British astronomer Arthur Eddington had confirmed Einstein's general theory of relativity. Standing on Principe Island in the Gulf of Guinea on May 29, 1919, he had measured the deflection of light through the gravitational field of the sun during an eclipse. The world became enthralled with Albert Einstein and his new universe. On November 23, Born celebrated Einstein's triumph in a lengthy article in the *Frankfurter Zeitung*. He began with drama and exultation.

> On May 29th of this year, a solar eclipse took place that, although invisible in Europe, darkened a narrow strip in the southern half of the globe for a few minutes. To this unspectacular event is linked one of the greatest victories of the human spirit over nature, not a triumph of roaring technology but of pure insight: the confirmation of Einstein's theory of gravitation and general relativity.

Threading the reader deftly through Einstein's initial insights into time and space and his thought experiments, Born ended by debunking Kant. Kant had proved why and how, based on a priori reasoning, "the human spirit" invented such ideas as Euclidian geometry, he explained, but Einstein's measurement of time and space reduced these ideas to "ordinary empiricism." For Einstein, he said, "things are only real if they can be measured."

Not everyone in the physics community was ready to accept Einstein's theory. A couple of weeks later, Munich physicist William Wien, unsure about the theory even with Eddington's evidence, wrote to his colleague Arnold Sommerfeld that Born's article was "unfortunate and damaging."

In January 1920, for three consecutive Tuesday evenings, Born held lectures on general relativity in the large auditorium at the university. The admission he charged the general public raised almost 7,000 marks, enough "to get the institute off the ground" and allow Stern to finish his experiment. Born became so enthused by Stern's work that he and Fräulein Bormann began a related project—to measure the free path of silver atoms in order to look at molecular collision cross sections. It was Born's first and only successful experiment.

Einstein was impressed with Born's industry and teased admiringly,

> And you, Max, are giving lectures on relativity to save the institute from penury, and writing papers as if you were a single young man living in splendid isolation in his own specially heated apartment, with none of the worries of a *paterfamilias*. How do you do it?

In fact, Born at times felt as though he were barely surviving. He struggled with asthma, as he did most winters. Hedi and the children were constantly ill with flu, measles, and colds. In between conducting his research and running the department, he cared for them, often alone because their maids Lina and Minna left to be married and those newly hired seemed to run off on a regular basis. One of the few bright moments was Hedi's birthday in December, when he wrote her a poem that began

> *You have wrestled yourself from the heavens*
> *Have broken yourself from the pious*
> *You have overcome the heart's thirst*
> *And have come down to me.*

As an early spring beckoned, however, things looked up. Hedi was strong enough to start gardening; the girls were well; and Max signed a contract for two books with Springer Publishers—the first was based on his relativity lectures, and the second on the structure of matter. On the same day—approximately nine months after receiving Richards's letter—he wrote to an American scientist for the first time.

His correspondent was Gilbert N. Lewis, a professor of chemistry at the University of California at Berkeley. Born had first met Lewis in Pontresina with the Neissers one summer before the war. Lewis had been in Europe looking for Fritz Haber, who was also with the Neissers. Born always remembered the day when, relaxing in a mountaineering outfit, he saw Lewis walking up a path in the Alps wearing "a black suit, gray striped pants, shiny black shoes and a bowler." Lewis and Born's friendship quickly developed from discussions on the theory of special relativity and was renewed when Born visited Berkeley in 1912.

Born asked Lewis to help rescue Paul Epstein, a young physicist whom Born knew only slightly, from the effects of postwar nationalism. His letter bristled with defensive sarcasm, but beneath it was the same anger and despair over the conditions in Europe and the Allies' attitudes toward German scientists.

> The unfortunate rupture with the States makes it difficult for us Germans to write to an American. In my own case, I have not done that. But I am forced to because a member of one of the states who belongs to the "Allies," a Pole, asked me to make a connection for him in America. . . . The splendid war of the Entente [Allies] for "justice and civilization" which has been won through American help has transformed all of Europe into a field of ruins in which hunger and misery are at home. The Poles . . . suffer the same need as we, if not worse. I hold it as my duty to help each person whatever nationality he is. That's why I am fulfilling the request of this man and write to you on the assumption that you also place humaneness above nationality.

The outcome of the war had placed Epstein in the unusual situation of a person with too many national identities and no place to go. Born explained that Epstein had been born and raised in Russia. He had studied physics with Sommerfeld in Munich. With the recreation of a Polish state after the war, his home district in Russia had become part of Poland. So now, for passport purposes, he was Polish. Epstein could not teach in Poland because the country's nationalism was too strong to accept a German-educated professor on the faculty. He could not teach in Germany because the Germans considered him a Pole and, therefore, a citizen of a country that had just stripped Germany of its northeastern border. Russia was dangerous because of its disorder: He had already lost almost all his money through the Russian Revolution. He had tried for a position in Switzerland with no success. Since French, Belgian, and English scientists had no official contact with Germany, they could not sponsor him.

Having been stung deeply by Richards's rejection of German scientists, Born's conclusion to the letter girded him for more of the same.

> Should you not take the view of so many Americans that we Germans are Huns, I would be pleased by a sign of life from you. Unfortunately a highly esteemed man such as your colleague Richards appears to have fallen apart from war psychosis. I possess a letter from him to my friend Oseen in Sweden which I keep as a document of human foolishness.

This letter to Lewis was cathartic for Born. Not waiting for a reply, he wrote to another American, the chemist Irving Langmuir, a friend from the early Göttingen years. Again he described the hunger and inflation, but this time he shared the frustrations of not knowing what the Americans were working on and of not being able to collaborate with them. "Several days ago, Niels Bohr was in Berlin," he wrote to Langmuir. "I discussed much with him about quantum theory and atomic structure. It is a pity that American scholars cannot participate in the discussion; otherwise one would reach agreement quickly."

Born then received Lewis's reply. Born's return letter changed tone dramatically.

> I hurry to answer your letter because it really caused me great joy. In the misfortune that haunts our land, it is the greatest bitterness that the bands of art and science that should connect all men are now in tatters. It pleases me so much that you have kept your personal friendship towards me and are prepared to take part in the reproduction of intellectual collaboration of researchers of all countries. I myself see in each person the good will to love instead of hate, the friend and companion indifferent to which nation he belongs. To assign right and wrong is not our affair; but to alleviate the misfortune that human passion has produced can be helpful for everybody. Your letter is the first from the other side of the sad separation, it gives me hope for a better future for mankind for which I thank you heartily and assure you of my heartfelt friendship.

From Lewis and Langmuir, he also pleaded for scientific reprints. With the inflation, the Germans could not afford American publications. Both men complied.

The political situation in Germany was worsening. Right-wing extremist Wolfgang Kapp attempted a coup to unseat the new government under Ebert. The military, sympathetic to Kapp, refused to intercede. In turn, workers supportive of Ebert called for a general strike that paralyzed Berlin to such a degree that the rebels' attempt failed. The strike grew throughout the country. The Borns watched and waited as the strike unfolded in Frankfurt. They regarded the little "shoot-outs" that resulted from the strike as rather minor after the turmoil they had lived through in Berlin. As that threat receded, French troops marched into Frankfurt and the surrounding area in response to German troops going into Ruhr mining towns to quash left-wing agitation. Once, when Max was taking the two girls to a party, they just missed an incident in the center of the city where French Algerian troops shot into the crowd, killing or wounding nine onlookers.

Germany was vulnerable and demoralized. The results of the national elections in June reflected the population's insecurity and its increased radicalism: as people became more polarized, the middle lost out, with increasing power going to advocates of force and violence.

In the midst of the turmoil, Hedi's mother visited and came down with the flu. It was a particular virulent strain that killed 20 million people worldwide. Max and Hedi worried about the girls getting it and kept Helene isolated. This quick and lethal strain favored the elderly, and Helene became one of its victims. Hedi, twenty-eight years old, collapsed. She wrote to Einstein that "the further the hour of her death lies behind us, the stronger is my longing for the departed; the darker and more incomprehensible seems the enigma of death." Max too felt the loss deeply. Helene Ehrenberg had been "a real mother" to him.

Another stress was added to their environment of loss and political instability. Göttingen's Philosophy Faculty had been looking for a replacement for physicist Peter Debye, who had taken a chair in Zurich. Hilbert, working behind the scenes for Born, wrote to Einstein for a recommendation. Einstein's reply called Born "the most important (with Debye) of theoretical physicists" and praised his scientific contributions and "his noble and courageous character." Even so, the faculty's initial list was Sommerfeld first, Born second, and Gustav Mie third. Sommerfeld refused the offer.

Born was aware of the developments. On the same day that the Göttingen faculty received Sommerfeld's refusal, Einstein was answering Born's request for advice.

"It is difficult to know what to advise you," he wrote. "Theoretical physics will flourish wherever *you* happen to be; there is no other Born to be found in Germany today. Therefore the question really is: where do you find it more pleasant?" Einstein's personal recommendation, though, was not to take the offer. He viewed Göttingen's academic personae as "self-important," "unfeeling," and "narrow-minded." They were the people, he reminded Born, who had made life difficult for Hilbert.

Born's name headed the next list that the Göttingen faculty sent to the Ministry. Born was in a quandary: Did he really want to exchange the cultural pleasures of Frankfurt for the backwaters of Göttingen, or sacrifice a real home for whatever he could find in a town with no available housing? The mere idea of the physical move daunted him, considering the difficulties with disrupted service and overfilled cars simply from taking a train from point A to point B.

Yet he also knew that the position in Göttingen was "a real honor"—and Göttingen, one of the major mathematical centers of the world, was where he had been academically nurtured and disciplined. Now he was being invited back to be the head of the family, certainly an emotionally powerful offer.

Born was sufficiently drawn to the offer to go to Berlin to negotiate with Herr Geheimrat Wende, Ministerial Councillor for Art, Science, and Education. And for all of his ambivalence about whether to accept, Born knew exactly what he wanted, as he explained to Wende. For him to accept the call to the theoretical chair, the Ministry had to call a second physicist—an experimentalist to work closely with him. He could not and would not teach experimental physics as the position required. He proposed that a chair available in an associated department be redefined for this purpose. He also asked for an Extraordinarius professor of theoretical physics who would carry some of his teaching load.

Wende agreed to create an experimental chair, unusual as it was. With regard to Born's second request, he handed him the Ministry's budget estimates so Born could see for himself the insurmountable financial constraints to creating another position. Born was not good at this kind of detail; yet there, amidst all the budgetary columns, he discovered a clerical error: the termination conditions of a certain position had been assigned to the wrong one. In a game of musical chairs that exploited this error, Born proposed a way to get each of his requests, and Wende's superior, who was called in, agreed, invoking the spirit of the revolution.

In his head, Born had already filled the chairs—his old friend James Franck as the experimentalist and Otto Stern as the theoretician. Franck was still the director of the physics division of the Kaiser Wilhelm Institute for Physical Chemistry in Berlin, and before departing for negotiations in Göttingen, Born got his assurance of accepting the offer. Franck liked the prospects both of living in a small town to raise his children and of having his own laboratory.

After visiting Göttingen, Born wrote Wende a thirteen-page letter telling of his successful discussions with the Göttingen faculty, including his agreement with the other experimental physics professor, Robert Pohl, and his elaborate plan to reorganize the department. Yet, throughout June, Born's apprehension increased in spite of these accomplishments. At the end of the month, he wrote to Elsa Einstein, "The question 'Göttingen, yes or no?' worries us a great deal. We are still undecided."

Pohl, a well-respected experimentalist in solid-state physics, had been an Extraordinarius professor on the Göttingen physics faculty since 1916. He was opinionated, feisty, and concerned that the arrival of Born and Franck would negatively affect his status. Who, Pohl wanted to know, would teach the beginning course, who would be Ordinarius professors, who would use which lecture rooms? Peter Debye had warned Born about the constant squabbles with Pohl over the use of the large lecture hall. Born's anxiety about dealing with Pohl was one of the main reasons for his insistence on recruiting Franck.

In June, Franck's call began to unravel. The Göttingen faculty had decided that the chair Born planned to appropriate for Franck could not be transferred to Born's department. Born started a new round of negotiations in Berlin and Göttingen. Ultimately, the Ministry and faculty agreed to provide a chair for Franck if Born gave up bringing Stern and being head of the entire department. Born agreed. More important to him was gaining "a collaborator to whom I feel connected because we have a shared view on basic scientific concepts. From this I expect great advantages for teaching and research." David Hilbert, extremely pleased with the outcome, wrote to Richard Courant, now head of the mathematics faculty, "Franck + Born are the best imaginable replacement for Debye! . . . We have Born's energy to thank for it!"

As Born waited for written confirmation from the Ministry—which arrived in the beginning of August—the University of Frankfurt tried to woo him to stay. Benefactor Artur von Weinberg made it possible for the university to match Göttingen's financial offer. Money, however, had not been Born's main condition for staying: He had asked the university—and, therefore, the Ministry—to promote Stern from lecturer to Extraordinarius professor. The Ministry rejected that condition.

Born began working on Stern's future. He wrote to Einstein,

> Now the question of my successor becomes urgent. [Arthur] Schönflies wanted to write to you and to ask for your expert opinion. I would, of course, like to have Stern. But Wachsmuth does not; he said to me, "I think very highly of Stern, but he has such an analytical Jewish intellect." At least it is open anti-Semitism. But Schönflies and Lorenz want to help me. . . . Stern has raised the standard of our little Institute and really deserves recognition. . . . I have asked Laue to give his opinion; perhaps it would be best if you were to talk to him about it, so that your verdicts do not clash.

A couple of weeks later, Hedi wrote to Einstein, "Wachsmuth is agitating against Stern on anti-Semitic grounds." Even at the "Jewish University," with bylaws forbidding religious discrimination in faculty appointments, anti-Semitism could not be eradicated.

Max and Hedi decided it was time to escape from six years of fear and stress. For the month of August, they headed to Bolzano in the South Tirol, a place special to Max since he had stayed there with Ladenburg in Easter 1909, and one, not incidentally, where the exchange rate was favorable.

At 6 A.M. on August 6, the arched concourse of Frankfurt's Hauptbahnhof receded behind Max and Hedi as they traveled off, the children staying with the nanny. They changed in Munich to the Brenner train, stood in a line at 3 A.M. for two hours for passport checks and luggage registration in Kufstein, Austria, and waited at the "clear and bitterly cold" Brenner Pass for reregistration at Italian customs. At noon, thirty hours later, they climbed down from the train into the warmth and beauty of Bolzano, the gateway to the magnificent Dolomite Mountains and South Tirol's romantic blend of Italy, Germany, and Austria. With promenades and covered arcades, well-stocked fruit and vegetable markets, and serene mountains, it rivaled Tannhausen. Max was saddened to see the Italian flag flying over a former German city—a result of the Versailles peace treaty—but pleased that the people did not have "a trace of the spiteful and presumptuous attitude of the French."

Hedi was amazed at the graciousness of life outside of Germany and how "unbelievably pampered" she felt at the hotel in Bolzano. "God, this is beautiful," she exclaimed. There was plenty of white bread and even whipped cream at dinner. She could place clothing and shoes for cleaning at the door, leave jewelry in the room, and hang washing overnight in the meadow without fear of theft. "Poor, corrupt Deutschland," she wrote. "All confidence, all honesty broken."

After a day and a half in Bolzano, the Borns traveled up through Etschtal, Merano, Spondinig—where they ran into the Plancks, who were hiking the range even though Planck was sixty-two—and finally to the Suldenhotel. It too was beautiful, "2000 meters above the sea, a high alpine summer station . . . an exit point for lovely parties by wagon and by horse" nestled at the bottom of the glaciers of the Ortler, Königspitze, and Cevedale, their snow-capped peaks jutting into the clear sky. Here was Hedi's "*Schlaraffen* existence," a fantasy of luxury and ease with "fabulous" food: soup, appetizers, vegetables, roasts, desserts, and "so much butter that we have to cover the butter with bread and not the reverse." To Max, the butter and whipped cream were "fairy tale things."

Reminders of the war were scattered about—shells, steel helmets, and cannon placements, even on the sides of the highest peaks—but for the Borns it was "heavenly restful" as they soaked in "the splendor of the glacial world . . . through every pore." The only unpleasantness occurred at their next stay in Venice, where Americans at their boarding house treated them as outcasts, neither speaking nor dining with them.

Their Italian episode did not compare with the insults being heaped on Einstein back in Germany. On August 24, 1920, the Study Group of German Natural Philosophers, as they called themselves, held a meeting in Berlin's Philharmonic Hall. Most of the scientists comprising the group were unknowns, but one, the Nobel physicist Philipp Lenard, gave them visibility. Speaking before a sold-out crowd, these scientists denounced Einstein and his general theory of relativity. They cloaked their rampant anti-Semitism, which his international fame had stirred up, under a guise of scientific disagreement. Einstein himself attended part of the meeting, listening to the harangue with a seemingly bemused manner. He saved his real reaction for a letter to the *Berliner Tageblatt* a few days later, in which he criticized the detractors as frauds, defended his science, and stated, "Were I a member of a nationalistic German party with or without a swastika instead of a Jew of liberal, national conviction, then. . . ." Einstein did not finish the sentence, preferring that his readers fill it in.

The Borns did not learn of the attack until they reached Munich on their trip home on September 5. Hedi wrote Einstein a couple of days after arriving in Frankfurt.

> You must have suffered very much from them, for otherwise you would not have allowed yourself to be goaded into that rather unfortunate reply in the newspapers. Those who know you are sad and suffer with you, because they can see that you have taken this infamous mischief-making very much to heart. Those who do not know you get a false picture of you. That hurts too.

Distinguished German scientists did not bicker about important scientific issues in the press but debated calmly in scientific assemblies, at least theoretically. A month later, the meeting of the Society of German Scientists and Physicians in Bad Nauheim, a spa town twenty miles outside of Frankfurt, was to be such an event, with a special joint session of the mathematics and physics sections dedicated to a discussion of relativity between Einstein and his critics.

Upon arriving home, Born had to focus on a related issue, the publication of his book *Einstein's Theory of Relativity and Its Physical Foundations*, which was based on his Frankfurt lectures. Using nothing but elementary mathematics, the book traced the development of physics from classical mechanics, through optics and electrodynamics, and then to relativity, in each area presenting unsolved problems and explaining in the following section how those theories gave solutions that then posed further unsolved problems. Born had created a powerful teaching tool and written an exciting science book—the philosopher Karl Popper later modeled his geometry thesis on Born's method.

This book, which Born hoped to have out before the Bad Nauheim meeting at the end of September, contained more than substance; highly unusual for the time, it included a biography and photograph of Einstein. When Born returned home, a letter from Max von Laue, urging him to omit Einstein's picture, awaited him. From the start, Born had been sensitive about the image of "burning incense before the idol." After receiving von Laue's plea, he wrote of his concerns to the publisher Ferdinand Springer to see if there was time to leave the picture out. Backed by Arnold Berliner, Springer decided to leave it in. He thought it would be easy to deal with anyone who took exception. No one did. The first printing of 2,000 copies sold out quickly without incident and was followed by a second, third, and fourth edition, none of which, however, contained the Einstein photo.

For the meeting of the Society of German Scientists and Physicians, Einstein stayed with the Borns. (Elsa accompanied him to Frankfurt but stayed with the Oppenheims in posher surroundings.) Each morning of the six-day meeting, the two men took the hour-long train ride to join the other 2,500 physicians and scientists at Bad Nauheim. This health spa, nestled in the foothills of the Taunus Mountains, is surrounded by gentle hills—a serene backdrop to complement the expected civility of which the opening remarks of the president of the society reminded the audience.

> Scientific questions of such difficulty and of such great importance as the theory of relativity cannot be voted on in popular meetings with demagogic slogans and in the political press with venomous personal attacks. In contrast, within the tight circle of true experts they receive an objective evaluation that does justice to the importance of their ingenious creator.

On September 25, the day of the debate on relativity, Born and Einstein walked out of the train station to face a changed scene: guards armed with fixed bayonets. Not trusting the scientists' sangfroid, the government prepared for trouble, unnecessarily, as it turned out.

In Bathhouse 8, five to six hundred eager listeners "squeezed together on seats, stood along the walls, filled the balcony." First came hours of invited papers, until session chair Max Planck finally opened the floor to discussion. Lenard spoke first. When Einstein followed, Planck was forced to silence heckling, perhaps orchestrated. Lenard and Einstein rebutted each other's comments, as others in the audience asked questions and offered opinions, including Born. Then Planck, who had maintained a more dignified proceeding than many had considered possible, observed that relativity theory still had not made it possible to extend the time for the meeting and ended the discussion.

Einstein was disappointed in his performance. "I will . . . not allow myself to get excited again, as in Nauheim," he wrote Born. "It is quite inconceivable to me how I could have lost my sense of humour to such an extent through being in bad company." Born knew that Einstein suffered under the attacks and worried that he might leave Germany. This was all the more reason for the Borns to react strongly to the swirl of publicity that surrounded Einstein and, in their opinion, made him more vulnerable to attack—publicity that the Borns and other friends such as Max Wertheimer attributed to Einstein's good nature and Elsa's enjoyment of the attention.

After Nauheim, Hedi took matters into her own hands. Speaking for herself and Max, she wrote a long letter to Einstein, imploring him not to allow publication of a book that the Borns and many other friends thought would be exploitative, demeaning, and construed as "self-advertising" by his enemies. Einstein attempted to stop publication of *Conversations with Einstein*, which was written by a slightly disreputable journalist who happened to be Jewish, but was unsuccessful. Contrary to expectations, the Earth did not shake, and the to-do faded.

One piece of it continued, however. Einstein wrote to Born, "Today I write principally because I want solemnly to bury the hatchet. I have had a tiff with your wife for the sake of mine, mainly because of a rather exaggerated letter that she wrote to her." Einstein relayed his displeasure in an obvious manner. He referred to Born in the formal *Sie* rather than the intimate *Du* reserved for one's family and close friends that he had begun using about a year earlier. Born had felt honored to be in that group.

When Max talked to Hedi, he discovered there was much to apologize for. Hedi's six-page letter to Elsa had been scathing, accusing her of being open to all kinds of flattery, of dulling Albert's judgment, of taking advantage of his great heart, and of selling his freedom when he was too sick to object. Born replied to Einstein that he had taken the matter "very much to heart, more so than anything else I can remember." The friendship continued, as did Einstein's flirtations with the press, and after a couple of months, his use of *Du*.

Freed from the conservative cultural and political regime of the kaiser, Jews were beginning to thrive in an open German society. They were part of the government for the first time and part of the highly visible explosion in avant-garde music, art, architecture, literature, and drama; but underneath was an undercurrent of anti-Semitism. With Germany roiled by defeat and desperate economic conditions, the public was foaming, and Jews instantly were made the scapegoat. The presence of some notable Jews, such as Walther Rathenau, in the same government that had accepted the harsh terms of the Treaty of Versailles created a convenient target.

Historically, anti-Semitism had infected the atmosphere in the theoretical physics community less so than in other sciences. As the neglected stepchild of experimental physics, its lower prestige made it easier for Jews to get appointments. Einstein's ascendancy changed the hierarchical place of theoretical physics both in Germany and internationally. The academic establishment that had so carefully controlled the employment of Jews now found the most famous scientist in the world, a Jew, in its midst.

The attacks on Einstein were from a small, organized group that wanted to reestablish influence on theoretical physics. Born had watched the efforts at Bad Nauheim. In a long letter to Felix Klein, he warned of the importance of the "correct choice of members" on scientific committees, emphasizing that "after the Nauheim proceedings, one cannot gloss over that in physics a 'South-German particularism' [a special subgroup] exists for which the spokesmen are Wien and Stark." In another two years, Nobel Prize–winner Johannes Stark would write the anti-Semitic tract, "The Present Crisis in German Physics" and later ridicule relativity and quantum theory as Jewish science in his lecture, "Jewish and German Physics."

Despite divisiveness in the physics world, within Born's institute, harmony reigned. He wrote to Gilbert Lewis, "The physics is so beautiful that one would like to wish for days with 48 hours in which one could create more." He was busy working on lattice theory and on his experiment with Fräulein Bormann to measure the free path of silver atoms. Landé still explored the cubic atom, and Stern wanted to expand his research methods to test Sommerfeld's theory about the spatial orientation of electrons in atoms. Earlier, Sommerfeld, working on Bohr's theory of the atom, had hypothesized that when a magnetic force is applied to an atom, it could orient itself in only certain discrete directions—a phenomenon known as *space quantization*—while conventional wisdom called for the atoms to form a continuous distribution. Using the molecular beam method, Stern planned to explore what would happen to the orientation of a silver atom as it interacted with a magnetic field. Born, like other theorists, considered Sommerfeld's formula to be largely symbolic and cautioned Stern. Stern was convinced, however, and Born agreed to find financial support for his work.

Somewhat flippantly, Born told a German friend who was on his way to New York that should he meet a rich German American who still had a place in his heart for the old country, he should ask him to support Born's institute. A few weeks later, the friend sent a postcard saying, "Your man is Henry Goldman," and gave instructions to write to Goldman at the St. Regis Hotel. Born did so and received a check from Goldman for a few hundred dollars. With the inflation in Germany, it was a small gold mine.

Born's friend had found a rarity: an American sympathetic to Germany. A few months later, Born traveled to the luxurious Adlon Hotel in Berlin, met Goldman, a heavyset gentleman of about sixty-three, and discovered what made this American different. Goldman, a first-generation American whose father had emigrated from Germany, did not believe that Germany was responsible for the war. In 1917, he had refused to contribute to an Allies' war loan, and the family firm, Goldman, Sachs & Co., forced his resignation as senior partner. By this time, Goldman had already financed the formation of such companies as Sears, Roebuck, & Co., F. W. Woolworth Co., the Studebaker Corp., and B. F. Goodrich. In retirement, he was an art collector and philanthropist—although always an anonymous one—doing what he could to support Germany. Born left the meeting with more research funds, enough to support Stern for a while longer, and with a new and lasting friend in Goldman.

About that time, Stern found a research partner. Walther Gerlach, a new lecturer in Wachsmuth's experimental physics department, wandered by Born's lab for a chat and discovered an atmosphere where "one sat the entire day, talked, and did physics." It took only a few visits for him to become an unofficial member. Born's first reaction was, "God be thanked. We have an experimental physicist who can help us here." Born and Stern wasted none of Gerlach's time—Stern working with him on space quantization and Born on lattice theory and electron affinity.

Born had a side interest as well—closely following the developing field of quantum theory. When writing to Lewis to thank him for finding a place for Paul Epstein at Caltech, he argued against Lewis's criticisms of Bohr's atomic model, telling him that reading Sommerfeld's *Atomic Structure and Spectral Lines* would dispel all doubts. "One thing shines out of the chaos: the classical relationship between the motion of charged particles (nuclei and electrons) and radiation is wrong and must be described by" a quantum formula. Born was a believer in the new theoretical direction if not actively pursuing the answers to questions posed by the new theory. "The true laws are quantum laws that unfortunately we know only little about."

An expansiveness was growing in Born. Hedi even mentioned in a letter to the Einsteins that Max was interested in lecturing in America to earn some extra income, in case they knew of any opportunities. A deep anxiety, however, about the growing hostility within Germany still preoccupied him. To Einstein, he predicted dire consequences from reparation payments.

> We are not going to pay as much as is asked for. But I can see the effect of power politics on the minds of the people; it is a wholly irreversible accumulation of ugly feelings of anger, revenge, and hatred. In small towns such as Göttingen, this is very noticeable. I can, of course, understand it. My reason tells me that it is stupid to react in this way; but my emotional reaction is still the same. It seems to me that new catastrophes will inevitably result from all this. The world is not ruled by reason; even less by love.

Serge Boguslavski, one of Born's favorite Göttingen students from before the war, wrote Born "a shattering view . . . from the land of the Bolsheviks" that reinforced this verdict. Boguslavski had finally gotten a letter out of Russia to Born with an urgent request for an official invitation for him to come to Germany. The conditions in Russia were more deplorable than in Germany, and Boguslavski was not only starving but had tuberculosis. Born wrote to everyone—Planck, Hilbert, Ehrenfest, Einstein—for help. When they all said there was nothing that they could do, Born knew there was no way of getting Boguslavski to Germany.

The last months in Frankfurt were hectic. Born was working with Fräulein Bormann to finish his experiment with silver atoms and trying to organize the institute for Erwin Madelung. Richard Wachsmuth had succeeded in placing Madelung first on the list to the Ministry for Born's successor with Stern second, and Madelung had accepted. Born reported to Einstein that Stern was so unhappy about "his prospects . . . under the current anti-Semitic conditions" that "he is thinking of going into industry, which I consider a crazy idea." Stern stayed in Frankfurt for another year, however, continuing his research with Gerlach—and Wachsmuth put merit above ethnicity when he nominated Stern for a Nobel Prize a few years later.

A piece of news from Copenhagen made Born consider dropping everything. On February 21, Landé received a letter from Niels Bohr reporting that he had just derived the entire periodic system from quantum theory. The next day, Born wrote excitedly to Franck, who happened to be in Copenhagen.

> Wouldn't that be grand! But the tragedy is that we understand not a word of how he does it. He only says that it is based on the correspondence principle. . . . Dear, good Franck, be a nice chap and write to us about it as well as

> you understand it, or ask Bohr or Kramers to write something clear. Otherwise, we'll explode from curiosity. Or let me come to Copenhagen in eight days. At such an exciting event I would like being close by. Bohr is an astonishing man.

Stern, equally enthusiastic, added, "Dear Franck, You must snatch the great secret away from Bohr and tell us immediately."

Born did not go north. He was too busy completing plans for the move to Göttingen and dealing with an unexpected complication—a robbery. They lost their silver, linen, bicycles, even Max's suit and shoes. He felt insecure in the house and had trouble sleeping. By April 1921, when they left their home and lovely garden, Frankfurt had begun to resemble chaotic Berlin. For Hedi, the move was particularly difficult. She was five months pregnant. She took the girls to stay in Leipzig while their apartment was renovated, and Max went to Göttingen alone.

SIX

THINKING HOPELESSLY ABOUT QUANTA

BEFORE LEAVING FRANKFURT, BORN HAD WRITTEN TO A FRIEND THAT HIS AMBITION WAS "to bring Göttingen physics to further heights." Once in Göttingen, excited to be with Franck, pleased with the eighty students in his seminar, renewed with energy and drive, he was ready to start the climb. A sense of change already pervaded the university. The old mandarins Hilbert and Klein were in decline along with their traditions. Instead of walks up into the hills at precisely 3 P.M. on Thursdays, new institute directors such as Born and Franck rode their bikes together across town to the physics building on Bunsenstrasse.

Hedi and the girls came to Göttingen in early summer. They settled into a large renovated apartment with a big backyard for the girls. Life was easier. Göttingen was in the midst of farmland, and although prices were high, food was more available than in Frankfurt.

Soon after, in a postcard to Einstein, Born announced the arrival of their home's most important new inhabitant—"A small boy, Gustav Born, came into the world on July 29th."

In mid August, while Hedi and little Gustav rested in Göttingen under the experienced watch of Minna Oberdiek, Hedi's own childhood nanny, Max traveled south to the Austrian village of Ehrwald. The first semester had exhausted him, and he was suffering from a sinus infection and asthma. He wanted to rest in the majesty of his beloved Tirolean Alps. He also set up an opportunity to meet his new assistant, Wolfgang Pauli.

Almost two years earlier, Born had been so impressed by Pauli's article on gravitation theory that he invited him to Frankfurt. Pauli declined, as he was working on his doctorate under Sommerfeld in Munich. When Born started looking for an assistant in Göttingen, Born again asked Pauli. In his recommendation to the university's Kurator, Born deemed Pauli "the greatest talent in the physics area that has emerged in the last years."

What made him such a brilliant physicist was his combination of strong scientific intuition with acute mathematical insight. At twenty-one, Pauli had awed the entire physics community with his depth and mastery in a review article on relativity for Klein's *Mathematical Encyclopedia*. Born called him the "infant prodigy."

Inviting Pauli to Ehrwald was a good way to prepare for Göttingen but a bad way to rest. Someone once rightly said of Pauli's bulging eyes that they missed nothing. He and Born got to know each other on daily walks along grassy slopes, breathing the mountain air and discussing physics—"very heavy physics," as Born later described it. Alpine peacefulness could not compete with Pauli's intensity.

Born began a new research direction within days after he and Pauli reconvened in Göttingen to start the 1921 fall term. Even though he thought "the quanta really are a hopeless mess"—Bohr's theory was breaking down with nothing to replace it—he and Pauli decided to tackle some quantum calculations. For his first step, Born borrowed techniques used by astronomers to estimate the orbits of planets when they encounter disturbances or perturbations. He wanted to find out if the same approach could be applied to the helium atom, a three-body particle with two electrons orbiting around the nucleus.

Born had kept up with developments in quantum theory, giving a joint seminar on it with Landé and Stern while in Frankfurt and studying Bohr's articles and Sommerfeld's book *Atomic Structure and Spectral Lines*. Except where his research on lattice theory with Landé (and earlier with von Kármán) crossed over into the area of quantum theory, Born had made no direct investigations. Pauli, however, was no neophyte quantum researcher. His dissertation under Sommerfeld applied Bohr's quantum conditions to the hydrogen molecule ion (H_2^+)—another three-body problem, as this ion has one electron and two nuclei. Pauli's predictions did not match experimental results.

These experimental results were in the patterns of spectral lines emitted by the elements. For some time, experimentalists had been able to heat individual elements then dispersed the emitting light, using a prism to break it into its component wavelengths. For each element the spectrum was a measurable and unique pattern of lines. Photographs captured arrays of single lines, doublets, triplets, and higher multiplets. These distinct line patterns were the only clues to the atom's inner life, and so to the understanding of nature's building blocks. Predictions from theoretical models had to match these records.

With a model of the nucleus and electrons of an atom visually resembling the solar system, Bohr had interpreted spectral lines as the result of an electron jumping up or down from one stationary orbit to another. These transitions involved either emission or absorption of energy. Bohr's model had led to the correct values for the spectrum of the simple hydrogen atom with its nucleus and single electron, but failed to explain more complex spectra, such as the three-body ion that Pauli had examined in his dissertation.

In the preface to *Atomic Structure and Spectral Lines* Sommerfeld wrote, "Since the discovery of spectral lines, no expert could doubt that the problem of the atom would be solved if one would learn to understand the language of the spectral lines."

Sommerfeld wanted to understand this language first by examining the lines' patterns then by attempting to discern a general formula to explain the observed values. In 1916, he had refined Bohr's circular orbit for the electron to the more sophisticated ellipse, introduced a fine structure constant, and found good agreement with the doublets observed in the lines of the hydrogen spectrum.

Born and many other physicists considered Sommerfeld's book "the Bible of the modern physicist." But Born's appreciation for Sommerfeld's accomplishment did not extend to the type of research it required. He judged Sommerfeld's efforts an accumulation of guesses. Instead, Born wanted to develop, in Hilbertian fashion, a general theoretical framework, asking questions that led to mathematical solutions as he had done in crystal dynamics. Born's letter to Pauli in 1919 inviting him to Frankfurt had stressed this idea.

> You regard the application of the continuum theory to the interior of the electron as meaningless because it is principally not a question of observable things. I have pursued just these thoughts for a long time, certainly until now without positive success, namely that the path out of the quantum difficulty has to be sought from quite principled points: one is not allowed to transfer the concepts of space and time as a 4-dimensional continuum from the macroscopic experience to the atomic world which demands obviously another type of number manifold for an adequate picture.

Born harkened back to the inconsistencies of Bohr's model and reiterated an idea he always emphasized, to throw out "superfluous elements and describe as simply as possible." He knew that the macroscopic continuum of the classical world did not describe the microscopic world of the electron and its discrete jumps. Pauli's presence in Göttingen was the ideal catalyst for Born to begin exploring basic quantum principles that Ehrenfest, Einstein, Bohr, and Ladenburg had individually introduced.

Although he was anxious to move ahead, Born was obliged to enter into the quantum maze slowly. He had other commitments. After the publication of

Born's first book on crystal dynamics in 1915, Sommerfeld had asked him to write a comprehensive article on the subject for the fifth volume of the *Mathematical Encyclopedia*. The upheaval of war delayed his start until 1919, and after that, the article became for Born "the bane of my existence." He once told Sommerfeld, "I hardly allow myself to go out. . . . The ghost of the article always stands on my back." At one point, Born's guilt about so many missed deadlines prompted him to suggest that Sommerfeld verify with Franck that he had really worked on nothing else.

Of course, as Born stewed, the research in the field swelled, setting him back still further. With funds from Artur von Weinberg, his Frankfurt benefactor, Born hired an assistant, Emerich Brody, a young Hungarian physicist, to help with the article, and gradually it began to take shape. Born's young doctorate students found themselves "sighing" under the burden of calculating and correcting drafts.

That fall, poor health also slowed Born's research. Having failed to recover from his asthma during the summer, he spent the fall sitting up whole nights breathless, coughing, choking, and finding a little sleep only in the morning.

By mid December, he was well enough to accept an invitation, together with Courant, Franck, Hilbert, and Runge, to Rogätz, the country estate of the industrialist Carl Still. Franck later described Courant as having "an affinity for millionaires," and Still was one of Courant's early successes. A wealthy manufacturer of coke ovens interested in mathematics and physics, Still had met Courant in 1920 while taking one of his courses. That year, he invited Courant, along with Born and Franck, to his home to meet his friends and raise money to equip Franck's planned lab in Göttingen. Still personally contributed much of the 68,000 marks that Born eventually raised. Now, a year later, the all-male party was off for three days to a hare hunt—"Poor hare!" Born exclaimed to Gerlach. These outings became a yearly tradition. The professors, who offered camaraderie but never took part in the shooting, each carried home either a hare or a fat goose, courtesy of the Stills. On this particular occasion, Courant, true to his reputation, also went home with 100,000 marks for the Göttingen Mathematical Society.

Hedi stayed home with the children, baking honey cake and batches of chocolate, anise, and butter Christmas cookies, the aroma of which filled the house and wafted over the family's elaborate Christmas tree and crèche. In a postcard to "Mama" Born (Bertha), she proudly noted that she had done all the baking herself. The front of the postcard featured Gustav, about whom the adoring mother wrote, "Here you finally meet Gustav. Is he not sweet? And doesn't he have a powerful brow and intelligent look?"

On Born's return from Rogätz, Pauli broke up the festive Christmas spirit with distressing news. Only three months into his two-year stay, he wanted to leave Göttingen in April and move to Hamburg. The Vienna-born Pauli was city bred, and he told Born that he could not stand life in the quiet, rural town. Born and Pauli had only just begun their research, but Born thought enough of his brilliance to lament to Einstein that he would never have another assistant equal to Pauli.

Pauli and Born were different spirits. Born was neat and prompt; Pauli was neither. Born was an early riser; Pauli had a penchant for late nights. With no night life in Göttingen, Pauli would work at his desk until the wee hours, "rocking slowly like a praying Buddha," then sleep late. Born had to send the maid to Pauli's lodging by 10:30 A.M. to wake him on the mornings that Born was sick, otherwise he would miss lecturing to Born's 11 A.M. class on electron theory. (That never changed. Years later, a scientist trying to contact Pauli reported, "Pauli was reached by telephone, but it being only 1:30 in the afternoon, he was not yet out of bed.")

A major difference between them, and probably a factor in Pauli's decision to leave, was their creative route to physical insights. Each man started from experimental evidence but one then turned to mathematics, the other to his intuition. Their scientific interests meshed; their approaches diverged. Six months later, Pauli did not hesitate to take a year's leave from Hamburg and accept an offer to work in quiet, wintry Copenhagen with Niels Bohr, whose physical intuitiveness was more in line with his own. Born's mathematical "formalisms" soon became the butt of Pauli's notorious sarcasm.

Born was beset not only by the prospect of Pauli's departure but by an increasingly grim economic situation for his students and assistants. The Borns—like other professors—had adequate income to cover food and heat for the winter (they had brought a wagon load of coal from Frankfurt), but worsening inflation left his students poverty-stricken, too cold and hungry to study effectively. Born told Paul Epstein at Caltech, "You probably have no idea how many worthwhile men are immersed in need here and what a worry we old ones (in physics, Franck and I) have with such people."

Emerich Brody, Born's assistant on the lattice dynamics article, was a case in point. The grant from von Weinberg was exhausted and, unlike Pauli, the ministry gave him no funds. Born, Franck, and Courant collected private donations and could only pay him a wage that barely kept him and his family from starving. Born could arrange a lectureship for him in Göttingen, but in the short run, he would earn no salary and in the long run, as a Hungarian Jew with "decidedly Eastern ways," Born knew that he would not be offered a chair. Born considered Brody's plight "humiliating" and wrote dozens of letters asking for help. He managed to keep Brody and his other assistants alive.

This New Year signified a special time of passage for Born. His beloved friend and second father, Alfred Blaschko, was dying from cancer. Born went to Berlin and spent two days at his bedside, talking about life, death, and expectations. When Blaschko died a couple of months later, Born felt that "in death as in life he was the example which I wish to follow."

Professor David Hilbert's sixtieth birthday at the end of January highlighted other changes. This banquet and celebration was nothing like the antics of his fiftieth birthday, when Born, von Kármán, and the rest of the *El Bokarebo* crew had written and performed satirical songs. (Born and Franck did invite Einstein to play the violin, but he had to cancel this "escapade into the Eldorado of erudition," as he put it.) This time, Born's contribution was the article "Hilbert and Physics" for a section of *Naturwissenschaften* dedicated to Hilbert. As his former teacher had grown ten years older, Born had gone from composing new lyrics for "Ching, Ching Chinaman" to describing Hilbert's insights into general relativity.

In spring 1922, Born's attention, at least as much as the still incomplete encyclopedia article would allow, was focused on quantum theory. From Hamburg, Pauli had continued their collaboration on perturbation theory applied to the helium atom. Born enthused to Einstein that they had confirmed an old claim of Bohr's about the orientation of the axes of its two electrons. But a month later, they discovered, just as Pauli had with his dissertation results, that theory did not coincide with experimental data.

Born would soon have the opportunity to discuss these findings directly with Bohr. The Danish physicist was coming to Göttingen in June as the first guest lecturer since the end of the war, funded by the prestigious Wolfskehl Foundation (a foundation set up to reward whoever might solve Fermat's Last Theorem, the interest from which now supported guest lecturers). Bohr was to give six lectures in two weeks on quantum physics, the most controversial and important research in physics.

The impact of these lectures was not lost on Hilbert. He saw them as a message to scientists, especially in France and Belgium, who continued to boycott German conferences and exclude German scientists. The message: Germany's science and scientists would not and could not be marginalized.

On Sunday, June 11, Niels Bohr arrived at the Göttingen train station. As a foretaste of the extraordinary energy of the next two weeks, three of Bohr's young colleagues proposed bicycling from Copenhagen to Göttingen for his lectures. (One of them, George de Hevesy, planned to ride on to his native Hungary afterward.) Late spring was the perfect season for such sporting spirit. Reflecting this enthusiasm, Bohr's lectures, coinciding with Göttingen's yearly *Handel Festspiele* (festival), soon became known as the *Bohr Festspiele.*

For the first session, close to a hundred scientists from throughout Germany filled Lecture Room 15 to capacity. The first rows were reserved for VIPs, such as members of the German Physical Society and eminent professors, including

Göttingen's own, Born among them. Sitting farther back were rising stars: Born's former assistants Wolfgang Pauli, Otto Stern, Alfred Landé, and Walther Gerlach; his future assistants Friedrich Hund and Pascual Jordan; Göttingen students Erich Hückel and Rudi Minkowski (the scribes who wrote out the lectures for the math reading room); and a fair-haired student that Arnold Sommerfeld brought in tow named Werner Heisenberg.

Bohr stood on the platform at the front, his head slightly tilted and his smile modest and warm. During the next two weeks, speaking softly in German flecked with a Danish accent and choosing his words carefully, he explained his theory. Friedrich Hund would report later that although his seat in the back and Bohr's low voice caused him to strain to catch the fundamental ideas, he came away with "an excellent overview of the connection between the spectra radiated by the atoms and the physical and chemical properties of the elements expressed in the periodic table." By the end of the lecture, which ran well past the dinner hour, all were scientifically nourished and physically famished, especially (as Hund recalled) owing to the harsh economic conditions of 1922.

Bohr made it clear that he did not have a finished theory. He was still struggling to understand the atomic structure through his "Correspondence Principle," which assumes that as an electron's transitional frequency decreased, for example, quantum conditions correspond to classical conditions. Hence, he could erect a bridge between the classical and quantum worlds.

At the end of another lecture, Heisenberg stood up and questioned some of Bohr's comments. For a young student to challenge a professor was rather presumptuous, but Bohr reacted by inviting Heisenberg for a long walk. They had not met before the lectures, but Bohr knew—and disapproved—of an atomic model that Heisenberg had developed to account for certain spectroscopic data.

When the two weeks were over, Bohr invited Heisenberg to Copenhagen. Born, however, recognizing the same potential, had already worked out an arrangement with Sommerfeld. Heisenberg would come to Göttingen in the fall to be Born's assistant while Sommerfeld was in America lecturing at the University of Wisconsin. Heisenberg would return to Munich and Sommerfeld the following summer to finish his Ph.D.

Just five days after Bohr's lectures, Born wrote a two-page article, "On the Model of the Hydrogen Molecule," for *Naturwissenschaften*. It was a bold announcement of his future intentions. "The time is perhaps past when the imagination of the investigator was given free rein to devise atomic and molecular models at will. Rather, we are now in a position to construct models with some certainty, although still by no means complete certainty, based on the quantum rules."

He was confident in his dismissal of the old models. As Born explained, there were new principles upon which to build as well as new analytic techniques such as the celestial mechanics that he and Pauli had employed. Clarity, energy, and a view to the future stamped his remarks. Born felt sure of himself. With his encyclopedia article nearly finished and a promising new assistant, he and his Göttingen Institute were poised to join Munich and Copenhagen as a major center of quantum research.

Only one black cloud sullied the horizon—the University of Berlin wanted Franck. Born wrote to Einstein that he lived in dread of the possibility of Franck moving. The thought of bringing the physics faculty to new heights without his close friend, whom he considered the premier experimenter in Germany, was an anxiety that stayed with Born for the next year and a half—until Franck decided not to go.

Tired from the hectic summer, Max and Hedi visited the Stills for five days in mid September. From there, they went to Italy for their annual vacation, paid with 22 pounds sterling Max received for the English translation of his book on relativity. Fearing this money would not take them far, they also took out a short-term bank loan in the beginning of August, when the exchange rate of marks to lira was 29 to 1. After enjoying Florence and Rome, they returned to Göttingen with about 800 lira. When they went to the bank to exchange the lira and repay the loan, they received a surprise: The new exchange rate of marks to lira was 208 to 1; inflation had paid for most of their trip. It was just the beginning of a wild ride.

Born and his new assistant, Werner Heisenberg, both arrived back in Göttingen in mid October, and Born launched a systematic plan to determine where Bohr's quantum theory could predict experimental results. He wanted to start with a more sophisticated version of the three-body problem by using the excited helium atom to explore the behavior of electrons in aperiodic motion. (Earlier research had assumed a system at rest or one with periodic motion.) The introduction of aperiodic motion, however, was a mathematical complication of a high order, so first Born had to expand on the perturbation theory he and Pauli had worked with earlier. Not having the background to undertake this expansion, he began a seminar for both himself and his students to learn this theory.

Every Monday evening, about eight assistants and advanced students met at Born's home to study Poincaré's celestial mechanics, a complex perturbation theory. Heisenberg at first did not see the value of this exercise. As with Pauli, his quick, intuitive nature contrasted with Born's methodical, careful approach necessary to master the mathematics. Just a few weeks after arriving in Göttingen, Heisenberg complained to his father.

> Even the physicists are actually interested much more in mathematics than in physics. The result is that one has a somewhat bored impression of all the physics here; no one has the initiative to try something new; they pick out mathematically interesting topics that are in most cases exhausted as physics.

As the seminar progressed, Heisenberg had a change of heart. Soon, he was pleased to tell his father, "For once I will learn correct mathematics and astronomy."

The seminar was not all work. Hedi served refreshments, and Heisenberg played four-handed duets with Born on the Steinway in the drawing room. (In later years, Gustav often would lie under the piano while they played.) Born discovered that Heisenberg was not only as clever and insightful as Pauli but, unlike Pauli, was an early riser and had considerable musical talent. Heisenberg was a frequent guest in the Borns' home. That December, when he turned twenty-one, he celebrated by playing Mozart and Beethoven concertos with Born.

Another milestone as 1922 ended was that, as Born wrote to his friend von Kármán, "the Encyclopedia article is done and printed." In three and a half years of intermittent work, the article on lattice dynamics had become a tome of some 250 pages, none of which, after so much work, could Born bear to cut. He published it separately as a revised edition of his 1915 *Dynamics of Crystal Lattices* under the title *Atomic Theory of Solid States.*

Born's letter also included an apology to von Kármán. Ludwig Prandtl, director of the Institute for Applied Mechanics in Göttingen, had been offered a chair at the Munich Polytechnic. Prandtl had made the institute the most prestigious of its kind in Germany, "a Mecca of applied mechanics," especially aerodynamics. In the opinion of Born and several others, it was obvious that von Kármán deserved Prandtl's chair. The university faculty, however, came to a different conclusion, based on the fact that von Kármán was a Hungarian Jew. (Prandtl eventually turned down the Munich offer.)

In the four-page letter, Born "gathered himself together" and frankly confessed his embarrassment and discomfort with the conflict—revealing in writing the very private side he rarely showed in person. He began with a brief history about his ordeal in getting himself and Franck—two Jews—to Göttingen, especially with three other Jews—Courant, Landau, and Bernstein—already in the mathematics institute. He then tackled the question of taking up the fight alone on von Kármán's behalf, since others on the faculty either did not know von Kármán or thought the effort hopeless.

> At the time, it had to be handled quickly and definitively. I had to decide if I wanted to carry the fight for you against the enemies of Israel. I felt sick about it. I simply did not have the strength in body and soul to take on this

> goal myself unequivocally. I had had enough of the molestations of the Lenard people . . . I wanted my peace. That was not nice of me. I do not want to make myself better than I am. I know exactly what you are and what I have to thank you for. Ten years ago, you boosted me up and brought in proper physics as I threatened to sink in the formalisms of Relativity Theory. From you I have also learned how one has to work. . . . [B]efore you, I remain, nevertheless, an unfaithful friend and I must come to terms consciously with my weakness. . . . I would be pleased to know what you think. Write me if you consider me still worthy of it. . . . Your unfaithful friend, Max Born

Von Kármán's response to Born, if there was one, did not survive. Courant also wrote to him about the ordeal, and to this von Kármán clearly did not reply. He was not one to share his feelings, in any case. Born and von Kármán's friendship, however, continued.

Not all German Jews, or even all members of Born's circle, shared his profound frustration at anti-Semitism. Einstein had written to Born, "[Discrimination against Jews] must be seen as a real thing, based on true hereditary qualities, even if for us Jews it is often unpleasant." Born contrasted their positions.

> [Einstein] often argued the case in favour of the suggestion that Jews ought not to press their claims in an attempt to obtain the more desirable positions, particularly academic ones, but should create jobs for themselves, to be filled from their own ranks. I was, as far as I can remember, not altogether of the same opinion; my family was amongst those who strove for complete assimilation and who regarded anti-Semitic expressions and measures as unjustified humiliation.

In the last couple of years, episodes of "unjustified humiliation"—such as those with Stern, von Kármán, Franck, and himself—had confronted Born and would continue to do so.

By the beginning of 1923, Born was politicking with Sommerfeld about Heisenberg, an important asset in Born's future research plans. When Heisenberg moved to Göttingen, it was with the understanding that he would return to Munich when Sommerfeld came back from America. Now Born had a different proposal. In a long letter to Sommerfeld, he described Heisenberg's virtues—exceptional talent, modest ways, zeal, enthusiasm, and good humor—and his popularity and value to the faculty. It was the beginning of his pitch that he needed Heisenberg more than Sommerfeld did. Born argued that after taking his Ph.D. exam in Munich, Heisenberg could habilitate with him in Göttingen and become his lecturer and collaborator. Sommerfeld was generous; he agreed to Born's plan.

In the meantime, Born and Heisenberg started to capitalize on insights gained from the seminar on perturbation theory. First, they had already applied Poincaré's theory to confirm Bohr's buildup of the periodic table. Then they systematically explored the excited helium atom, as Born had originally planned. After much calculating, they found no agreement with the experimental spectroscopic lines. Born clearly saw the need for a new quantum concept.

In March, Born shared these negative results—which he termed a catastrophe—with Niels Bohr. As a relative newcomer to the field who respected Bohr's position, Born sought his approval to publish. Bohr, of course, encouraged Born to go ahead and also acknowledged the "deep-going difficulties, which we meet everywhere, as soon as we have anything to do with systems with several electrons." Born wrote to Einstein that he found himself no "closer to the great mystery of the quanta" in spite of all his efforts, but he was "fairly sure though that in reality it must all be very different from what we think now." He published similar thoughts in a review article in July, writing that the "entire system of concepts of physics must be rebuilt from the ground up."

As Born and Heisenberg examined the orbits of the helium atom's electrons, French troops occupied the Ruhr valley because Germany had defaulted on its reparation payments. German workers in this economically important coal region protested France's action by refusing to work. The German government then printed money to continue paying the workers, further inflating an already weak mark. At the beginning of the war, its exchange rate with the dollar had been 4 to 1, and in July 1922, when the Borns went to Italy, 493 to 1. By January 1923, it had escalated to 17,792. Ten months later, in November 1923, the rate would be 4.2 trillion marks to 1 dollar.

The impact of inflation was devastating to a fragile and increasingly polarized society. As Born later observed, "The great majority, the backbone of the German bourgeoisie . . . became poor and therefore an easy prey to political agitators."

For Hedi Born and other housewives, inflation meant lining up once or twice a week to get her husband's salary in slips of paper representing millions or billions of marks and then rushing to long queues at the butcher shop, fish market, produce stall, or bakery to buy whatever was available before the purchasing power was halved or worse. For students, who were expending most of their energy merely to exist, it meant quickly buying food whenever money was available. For faculty, it meant increasing holidays, decreasing the salaries of assistants, and shrinking the institutes' budgets to make ends meet. A serious topic at the mathematics-natural science faculty meetings was how best to spend tuition paid by six foreign students. These funds—19 pounds, 10 shillings—increased in value as the mark inflated. After some debate, the faculty finally decided to spend 10 shillings for paper, put 4 pounds in a safety deposit box, and to deposit the rest in a bank account.

Richard Courant's ingenuity provided the only slight glimmer. He lent the university an electrically driven adding machine that could calculate up to nineteen decimal digits, just enough for the administration to multiply the weekly increase in salaries and budgets by a special inflation factor supplied by the government. In return, he received the confidential factor early and used the knowledge to buy books and supplies before the change was announced.

Rocketing inflation destroyed most personal wealth overnight. The Born-Kauffmann families already had lost most of their fortunes by investing in war bonds, but the Kauffmann textile works, in which Max still owned stock, continued to hold on. Thirty-four hundred workers were spinning, bleaching, and dyeing yarn and fabric during what the company labeled a "bitter and disastrous" year. After observing one day that the stock price was rising, Hedi took the train to Berlin, sold the stock at a profit, and returned victorious. They invested the funds in a three-story apartment house on Wilhelm Weberstrasse. Hedi's father, who had returned to Göttingen to be near his children, moved into one of the three apartments. Max and Hedi had to wait a few years until another renter moved out.

Their celebration was short-lived. With their only remaining asset—a mortgage held on a house they had sold in Breslau—the Borns met a fate common to millions of middle-class Germans. The final mortgage payment of 50,000 marks was due to them at the peak of inflation, yet because the German Supreme Court had ruled that "a mark is a mark," Born one day opened an envelope containing a few postage stamps as payment in full.

The difficult times did not slow the social pace at the Borns' apartment on Planckstrasse. Assistants, students, and friends such as Franck and Courant, who lived nearby, came over for dinners and parties. Hedi enjoyed mothering young assistants and, at one point, organized dancing lessons for them.

The pretty little Born girls were growing up too. With ribbons in their hair, Irene and Gritli rode with the family in a horse-drawn carriage—cars were scarce in Göttingen—to Grandfather Ehrenberg's on Sundays. They curtsied on arrival and ate silently during dinner, but their formal good manners masked a devilish side. By the time Gustav (known as Buzi) was three, his sisters were whisking him into the woods to dip him in the cold waters of a nearby little stream. Their mother considered him very delicate, but his sisters decided that she coddled him. He needed to toughen up. They threatened Buzi with reprisals if he tattled, and he never did. As they grew old enough, they all took music lessons together from Nina Courant. Hearing Bach's double concerto for cello and violin had decided their father on the girls' musical direction. Irene was to play the cello and Gritli the violin. Gustav later joined them, first on the piano, then on flute.

Hedi spent long hours alone lying on her bed writing plays, poems, and voluminous letters. She insisted on quiet from the children and Max's piano playing. A bright and gifted woman—Einstein compared her poems to Goethe's in his later period—in Göttingen, she began the search for self-expression. In Berlin, she had used her wits for the family's survival; in Frankfurt, she had mainly taken care of the household and enjoyed the social whirl. Now she focused on her creativity. Episodically, she sent long letters filled with discussions of philosophy or social issues to famous writers, especially those with Nobel Prizes such as Selma Lagerlöf and Romain Rolland. (She also wrote to André Gide, but before he won the prize.) All responded briefly, each pleading fatigue and overwork and apologizing for the delay.

Max spent the summer of 1923 buried in his research, often working late into the night. When he was not battling asthma and bronchitis, this was his best way to escape from the depressing reality of daily life. Between the distractions of visitors and frustrations of young colleagues such as Gerlach seeking job advice, solitude often was hard to find. At one point, he complained to Einstein,

> From tomorrow I am going to feign death and refuse to see anyone. It is not really that I have anything special on. As always, I am thinking hopelessly about the quantum theory, trying to find a recipe for calculating helium and the other atoms; but I am not succeeding in this either.

To Born, music was a solace nearly as important as work. He practiced "systematically" and was making some progress, but he had trouble finding regular partners with whom to play.

In late fall, the government altered its spending habits, limited its printing of money, and in the change most obvious to the public, introduced a new currency, the rentenmark, which was backed by bonds on German industry. One new rentenmark equaled millions of the old deutsche marks. David Hilbert remarked skeptically, "One cannot solve a problem by changing the name of the independent variable." But he was wrong. Prices returned to expected levels, economic stability returned, and opportunities opened up quickly for the entire country, including Born. Springer Publishers, with whom Born had published a few books, signed him and Franck to edit a series written by German experts on the structure of matter. The Kaiser Wilhelm Institute began granting Born regular support for his students' and assistants' research. An emergency committee for scientific research provided funds for Heisenberg's upkeep. The Rockefeller Foundation, which made its first grant of $7,500 to Göttingen's physics institutes through its new International Education Board, provided a new lab assistant, more supplies, and an extension to the building.

A problem that the currency stabilization could not solve, however, was that of the angry citizenry and growing instability of the German state. Born saw this, and wrote to Einstein.

> The follies of the French sadden me, because they strengthen our nationalism, and weaken the Republic. I give a lot of thought to what I could do to spare my own son the fate of participating in a war of revenge. But I am too old for America and, moreover, the war hysteria seems to have been even worse there than here.

In September, Heisenberg returned to Göttingen as planned to do his Habilitation under Born. While Heisenberg continued to work on his own quantum model, Born offered a new course for the 1923–1924 fall term called "Perturbation Theory Applied to Atom Mechanics." His lecture notes later became his first book on quantum theory—a book he called volume one, with the "bold" promise of a volume two when the final formulation had been found. True to his word about the demise of old quantum theory earlier in the year, he began the lectures by proposing a major new approach: to incorporate the electron's transitions from one stationary state to the next by mathematically acknowledging the discrete nature of the jump. Rather than use differential equations, which assume continuity, he replaced them with difference formulas for discontinuities, thus making the jumps an integral part of the theory.

The moment was ripe in Göttingen for this idea. Not only was Courant holding a seminar on difference formulas that particular semester, but the idea of continuity and discontinuity had regained focus in explorations of the wave-particle duality of light. American experimentalist Arthur H. Compton had recently shown that light, which everyone thought to be a wave, behaved like a particle. Years earlier, Einstein had theorized about this, but here was proof—more than enough to stir Born's creative imagination.

Thus began the transition from old to new quantum theory. Heisenberg reported it to Pauli in an October 9 letter as he outlined how he was incorporating these ideas into his own quantum model. He also wrote to his father, who—fearing the new model was deficient—pressed him not to publish right away. Heisenberg promised that he would listen to Born's advice as well as Bohr's, of whom he wrote, "I realize ever more that Bohr is the only person who, in the philosophical sense, understands something of physics."

Bohr had captivated Heisenberg at the *Festspiele* in Göttingen, but apart from a brief correspondence, they had had little contact. More important for Heisenberg in deepening his sense of Bohr as a physicist, perhaps, was Pauli, who had just finished up a year in Copenhagen. Pauli had been one year ahead of Heisenberg in Sommerfeld's program, and Heisenberg regularly sought his scientific judgment. Bohr's insights into physics now compelled Pauli as Born's had not.

Bohr's strength was philosophical insight arising from deep intuition; Born's, a philosophical outlook that searched for more general relationships and fundamental laws through mathematics. They were pursuing a similar goal—understanding the behavior of electrons and atoms—but they asked different questions in search of the answer. Bohr drew on his intuition to track down the *why* of physical mechanisms. Born felt that "mathematics was cleverer than we are." One of his students called math his royal road, "'a Via regia,' which led him to the uncovering of nature's secrets." Pauli and Heisenberg gravitated to the Bohr approach, but they integrated ideas from both men, their communication providing a tie that allowed for an important synergy between the four physicists.

Parallel recollections from Born and Bohr show that their contrasting styles had a deep root. On essay day at school, young Max Born stared blankly at the sheet of paper on his desk, too paralyzed to write down the words he had memorized the previous evening. Rote learning, time pressure, and complex German grammar immobilized him. As his schoolmates scribbled away, his mind—and paper—remained empty. Later, he relived these torturous moments in a recurring anxiety dream. Schoolboy Niels Bohr, however, had too much to write. Trying to describe ships arriving in Copenhagen harbor, he had so much to say, wanting to make every thought exact, that he did not know where to start. On essay day, he took ill and stayed home, too overwhelmed by his rush of ideas to write. Weak in memory skills, Born used spatial symbols as a route to abstraction—mathematics, instead of words. Bohr talked ideas out in his head; he strained over verbal nuance and subtlety. He used words to form abstract ideas.

In March 1924, Born spent much of the month in bed with his regular spring asthma attack. For two of those weeks, Heisenberg was in Copenhagen for his first visit with Bohr. He returned to Göttingen with news of a new Copenhagen theory, the Bohr-Kramers-Slater (BKS) theory of radiation. Born was excited. He wrote to Bohr, "Although I have only Heisenberg's brief oral report, I am quite convinced that your new theory hits the truth." In this letter, he also addressed Bohr's request for Heisenberg to work in Copenhagen for the winter. Born did not want to lose Heisenberg, whom he described as a "dear, valuable, very clever person who has grown in my heart." Since he himself had received an invitation to lecture in America in the winter, and considering Bohr's wish "decisive," he graciously agreed to the proposal.

The BKS theory originated as a response to the discovery of the Compton effect. In 1923, John C. Slater, a young Harvard postdoctoral fellow, arrived in

Copenhagen at Christmastime with the idea that an atom in an excited state consists of an orchestra of virtual oscillators that produces virtual radiation of all possible frequencies that it can emit when in that state. Then, when it drops to a state of lower energy, it emits a light quantum that is absorbed by another atom consisting of a similar orchestra of virtual oscillators, conserving energy and momentum in the process. Bohr, hearing about Slater's virtual radiation idea, was enthusiastic, but because he had never accepted Einstein's light-quantum hypothesis—indeed, he was adamantly opposed to it—he rejected Slater's contention that energy and momentum are conserved in each emission and absorption process. He insisted instead that energy and momentum are conserved only statistically, over a large number of absorption and emission processes. He transformed Slater's idea, convinced his Dutch assistant Hendrik Kramers of his point of view, and wrote up the resulting BKS paper without allowing Slater to contribute a single word—all of which Slater resented bitterly until the end of his life. Not surprisingly, when Einstein heard about the BKS paper and Bohr's insistence on statistical conservation of energy and momentum in absorption and emission processes, he declared to Born that were this the case, he would rather be a "cobbler or even an employee in a gaming-house."

Bohr's assumption about the conservation of energy and momentum proved to be wrong, although Bohr tenaciously held on to it for another year. Slater's idea, however, about a virtual radiation field held great promise, and Kramers recognized its potential value. He used it to replace a formula that Rudi Ladenburg had published in 1921 in order to create a theory of dispersion for the interaction between radiation and atoms. Kramers's new theory was an important step in reconciling the tug-of-war between the classical, continuous theory of dispersion and the quantum, discontinuous theory. His formulation relied only on atomic transitions—which are observable as spectral lines—rather than the nonobservable orbits of electrons in atoms.

New insights were quickly building one atop the other. In July, following up on Kramers's work, Born published "On Quantum Mechanics," thereby coining a defining phrase in the field. Also, avoiding the traps of the BKS theory, which Einstein had convinced him to steer clear of, Born applied Kramers's dispersion approach to the interaction of radiation with atoms. Like BKS, he used virtual oscillators to replace atoms; and, since an oscillator was transiting between two energy states, he represented the transition frequency as the difference between that in initial and final states, applying the mathematical technique of Fourier analysis to the transition frequencies. His results, which constituted a major step forward, laid out a quantum discretizing rule that relied on averaging over transition frequencies. It replaced differentials (for continuous functions) with differences (for discrete ones).

Prodded by Born's intense focus, Göttingen was flourishing; Copenhagen was doing the same under Bohr.

Born knew that the advances were still short of providing a complete answer to the question of the spectral lines, but he sensed the end in sight. Writing to Landé about the specifics of his theory, he said, "I have a gentle hope that we soon will get to the bottom of it." Born had once written to Einstein, "It is odd that so many people have no feeling for the intrinsic probability of a theory." Born, it turned out, had a good nose.

On July 28, 1924, Heisenberg successfully gave his habilitation lecture, "The Historical and Logical Development of Quantum Theory," having used his most recent work (which incorporated Born's and Kramer's new ideas as well as some of his own) as his Habilitation thesis. He left to spend a month in Munich and was in Copenhagen for his seven-month stay by September. His arrival coincided with the absence of Bohr's assistant, Hendrik Kramers, allowing Heisenberg to be Bohr's assistant for two weeks. When Kramers returned, he brought with him a new formulation of his earlier dispersion theory, which incorporated both Slater's virtual radiation field and Born's discretizing rule. He set to work refining the physical theory and writing a paper about it.

Kramers was imposing—older than Heisenberg by five years, scientifically accomplished, multilingual, musical, aloof, and most important, Bohr's confidante. He also did not think much of Heisenberg's new atomic theory. Their relationship was not a comfortable one for Heisenberg. Kramers's personality and position brought out a different side to Born's "dear, valuable man"—a fiercely competitive one.

As Kramers finished up his article on dispersion, Heisenberg decided to explore Kramers's theory with a more mathematical approach. His results soon convinced him of an alternative interpretation concerning the virtual oscillators, and he brought his ideas to Kramers. Kramers disagreed. Undeterred, and seemingly indifferent that it was Kramers's theory, Heisenberg continued to press his point. First, he insisted that Kramers should incorporate his new ideas into the article. Then he added a demand to be coauthor.

Unable to resolve the dispute, the two took their differences to Bohr. As dusk descended on a cold December day, Bohr listened to a long debate between Kramers and Heisenberg—his reserved assistant versus his articulate guest. It was no contest. In Bohr's judgment, Heisenberg was right. Kramers had to rewrite the paper to include Heisenberg's interpretations, which Bohr thought were the correct ones, and add Heisenberg's name to the paper as coauthor. Bohr later told a colleague that Kramers should have published the paper by himself, and Kramers later told a student of his that he put Heisenberg's name on it purely as a courtesy. In any event, the methods of the final paper, which used Fourier analysis and the discretizing rule, owed much to Born's previous work.

In Göttingen, Born, having postponed his trip to America, had begun a new project with his student, Pascual Jordan. Mathematically sophisticated, Jordan had earlier helped Born with the encyclopedia article. Their new objective was to introduce into Planck's original 1900 quantum theory the concepts of transition quantities. From there, they planned to investigate atomic collisions.

At the end of February, Born traveled to Copenhagen with Hedi for a ten-day stay to lecture on the physical and chemical properties of crystals. Soon after they returned to Göttingen, Max took off for the Swiss Alps to recover from asthma. In between skiing, skating, relaxing, and thinking in the plentiful sunshine, he wrote to Hilbert that he was better. It was a respite that he would value in coming months.

When Born returned from Switzerland at the beginning of April, he and Jordan continued with their work. They discovered that the number of quantum jumps (the transition probabilities) related to the observed spectral line intensities, the square of their amplitudes. With Heisenberg, who had just returned from Copenhagen, Born and Jordan hypothesized that perhaps one could formulate transition amplitudes (later interpreted as the matrix elements of the position and momentum operators), which might be central to a new theory.

One striking feature of their paper on the Quantum Theory of Aperiodic Processes was its emphasis on observability. "A postulate of great reach and fruitfulness states that only such quantities should enter into the true laws of nature which are in principle observable and measurable," it stated. For his inspiration, Born reached into the past, noting Einstein's realization that it was impossible to measure the absolute simultaneity of two events at two different places. Observability was not a new idea. Pauli had talked about it for years, as had Born, and in his dispersion paper with Heisenberg, Kramers had included only observable quantities—but Born and Jordan gave it new emphasis.

Heisenberg did not take part in their actual research, nor did he share the content of his work with Born and Jordan; but he wrote to Pauli that he was examining the hydrogen atom using Born's discretizing rule and virtual oscillators to explain the observed intensities of spectral lines. When Heisenberg could not get his formula to give him the known intensities, he tried a simpler project: working with anharmonic, or aperiodic, oscillators. (Although such oscillators can be represented as a sum of periodic functions, they are themselves not periodic.)

The ideas and tools he used were all around him. Some came from Born's paper in 1924—applying Fourier analysis to the oscillator's transition frequencies, a technique that Kramers and Heisenberg had used in their dispersion paper in

December. Some emerged from his discussions with Born and Jordan on their current work—transition amplitudes, intensities of the spectral lines, and observability. In between his paper with Kramers and his work in Göttingen, Heisenberg had wavered about observability, but either the talks with Born and Jordan, debates with Pauli, his own pragmatism, or all of these together renewed his interest.

Heisenberg took these ideas and combined them with physical insights directed by Bohr's correspondence principle (relating quantum equations to classical ones) and by including transition amplitudes into the equation of the motion of an electron. From all of this, he worked out a strange multiplication rule. He knew that this was not a quantum-mechanical formula—it did not even include Planck's constant—but he was not sure exactly what he had developed.

Heisenberg stopped his work about the same time as Born and Jordan finished their paper on aperiodic processes. They sent it off for publication in the second week of June, just a few days after Heisenberg took off for ten days to the island of Helgoland in the North Sea to recuperate from a bad bout of hay fever.

With long walks in the clear air and invigorating daily swims, Heisenberg's mental clarity and his hay fever improved. Always thinking about physics, he suddenly saw that holding energy constant, in accord with the principle of the conservation of energy, was the key. Then, using his multiplication rule, he derived values that agreed with the observed spectral line intensities. On his way back to Göttingen, he stopped off to see Pauli in Hamburg to talk about his new results. Pauli was skeptical. Once back in Göttingen, Heisenberg continued to refine his ideas.

On or about July 10, Heisenberg handed Born a paper containing these fresh insights. He asked Born to review it and decide if it was worth publishing; he really did not know whether or not he was on to something. He sent a copy of the paper to Pauli and wrote his father that his work was not going well. Then, with Born's permission, he took off to lecture at the Cavendish Laboratory in Cambridge.

Born was extremely tired and set Heisenberg's paper aside. When he looked it over a few days later, he realized that Heisenberg had created multiplication rules for the transition amplitudes. Not recognizing the mathematics, Born pondered over it so intensely that he barely slept. Suddenly, he realized what Heisenberg had done—unknowingly but brilliantly. His notation represented matrix multiplication.

On July 19, Born took the train to Hanover for a meeting of the German Physical Society. He saw Pauli and asked him to collaborate on Heisenberg's formulation. Pauli refused, saying sarcastically, "I know you are fond of tedious and complicated formalism. You are only going to spoil Heisenberg's physical ideas by your futile mathematics." Born's response, if he made one, is lost.

After returning to Göttingen, Born noted the assignments for Jordan and his other assistants in the July 23 entry of his Daybook, specifying for himself "Heisenberg's Quantum Mechanics." Under "new problems" he wrote, "Connection between de Broglie's theory and the Duane-Compton derivation of [X ray] interference." With that entry, he stopped writing in his Daybook—and he stopped thinking about de Broglie's wave theory of the electron for some time to come. For about the next four days, Born studied Heisenberg's article sufficiently to send it off to the *Zeitschrift für Physik*. From there, he did not look back.

What he realized from tracing Heisenberg's logic, simplifying the formulae into matrix notation, and rewriting Heisenberg's version of Bohr's quantum conditions for p (momentum) and q (position) was that the product of pq did not equal qp—they did not commute. Born found that Heisenberg's formula determined the value of the diagonal elements $pq - qp$, which were equal to $h/2\pi i$. Wondering about the value of the off-diagonal elements in the matrix, Born began to expand beyond Heisenberg's original concepts. He arrived at what he considered the only reasonable conclusion, that the other elements equaled zero, giving him the equation $\mathbf{pq} - \mathbf{qp} = h/2\pi i\mathbf{I}$. It was the fundamental commutation law of quantum mechanics, from which all of quantum mechanics can be constructed.

Born was proud that he was "the first person to write a physical law in terms of non-commuting symbols." Almost fifteen years later, he cited his discovery of this formula as "the climax of my research."

At the time, he was also quite pleased to tell Franck of the discovery, in part because whenever he and Franck collaborated, Franck always checked first with Bohr in Copenhagen before proceeding. This time, however, the solution to one of the greatest mysteries in science did not come from Copenhagen but from next door. Heisenberg, it appeared, had provided the last basic piece in the quantum puzzle.

Unable to prove the commutation equation he had worked out, Born invited Jordan to help him. Jordan, whom Born later described as "of the Heisenberg type—head and shoulders above the others" quickly came up with the proof that the off-diagonal elements are zero. In the next two weeks, Jordan and Born established the soundness of Heisenberg's formulation, and they continued the collaboration by mail when Born went on vacation in August. Born and Jordan's expanded formulation proved that conservation of energy meant that the off-diagonal elements in the "energy matrix" are zero, and that the Bohr-Sommerfeld conditions in the old quantum theory corresponded to Born's new commutation rule.

The *Zeitschrift für Physik* received Born and Jordan's paper on September 27. They sent a copy to Heisenberg, who was in Copenhagen. Heisenberg, showing it to Bohr, said, "Here, I got a paper from Born, which I cannot understand at all. It is full of matrices, and I hardly know what they are."

Born, Jordan, and Heisenberg then started on what became the famous "three man" paper. They collaborated long distance until Heisenberg came to Göttingen in mid October to finish the article before Born left for a lecture trip to the United States. By then, Heisenberg had mastered matrices.

The final version of the paper gave a logically consistent formulation of quantum mechanics. The abstractness of the matrix algebra, however, involving infinite matrices, made it difficult for many to grasp the new ideas. Pauli wrote to a friend, "First of all one must try to set Heisenberg's mechanics free from Göttingen's flood of formalistic scholarship and expose its physical core better." Yet Born had seen what Pauli had originally not seen. In so doing, Born had accomplished what he had set out to do—namely, to "bring Göttingen physics to further heights."

As the three physicists completed the article, Augustus Trowbridge, a physicist from Princeton and the European representative for the Rockefeller Foundation's new International Education Board (IEB), arrived to evaluate physics faculties throughout Europe. Through the IEB, the Rockefeller Foundation had begun offering fellowships for young scientists to study in foreign countries.

Trowbridge caught the new excitement in Göttingen, that, to him, had replaced Berlin's old "eminence" in the pure sciences.

> The center for physics and mathematics seems to have been shifted to Göttingen. This is a return to importance in mathematics, but a distinct new shift in the case of physics. . . . [In] Germany it seems to be Born first in Mathematical Physics with Sommerfeld and Planck among the older men in the running. (Einstein is very busy with visitors.) Franck seems to be distinctly first on the experimental side of Physics in the eyes of his colleagues in and outside of Germany.

Göttingen was the only European university that Trowbridge credited with two first-ranked professors—Born and Franck—together with two second-ranked ones—Robert Pohl and Ludwig Prandtl. Trowbridge boosted his assessment of mathematics in Göttingen because of its close association with the physics department.

Just before leaving on his trip, Born wrote to Bohr about the new theory, explaining his mathematical beliefs as well as expressing his admiration for the skills of Bohr and Heisenberg.

> I am so glad that Heisenberg's idea pleases you. I believe with complete certainty that it signifies great progress and that the form, which Jordan and I have given it, is in a certain sense somewhat final, so far as one can say that at all in physics. For me, the possibility of this formulation has an entirely personal charm. Since my student times, I suffer with an idée fixe, that is to say, that all significant laws of physics must find their adequate formulation as invariants of linear substitutes. . . . Everything I did myself, e.g., the contribution on lattice theory of the crystal, always ran out of this. . . . Certainly such a one-sided type of consideration seems really ridiculous to you if the formalism is quite narrow for the actual physical connection. I am conscious that I lack that physical intuition which you or Heisenberg have, and that all that I can contribute to the promotion of the matter is the adoption of the physical measurements into a mathematical scheme. That is why I am so pleased that Heisenberg's theory fits so beautifully with my fixed idea and through it, one can get an overall view, as it appears to me, of all the theoretical possibilities quite well. I wonder how far the formulated principles reach; it is quite possible that new principles must come in addition. But that the skeleton of the formulas is reasonable, I am almost convinced of it.

Born maintained this modest estimation of his contribution for years to come. Bohr saw it in more balance, writing to Carl Oseen in Sweden that "brighter days" were ahead "because of the development of the new quantum mechanics which, resting upon ideas of Kramers and especially of Heisenberg, has been shaped into such a wonderful theory by Born."

In the wake of this great discovery, Max left Göttingen with Hedi on October 26, the children staying behind in the care of Grandfather Ehrenberg and his housekeeper, Fräulein Redlich. As the *Westphalia* sailed from Hamburg on the 29th, Born had reservations about departing. He knew that the research would continue unabated, and he worried about missing the action. As he would later tell his student, Maria Göppert, "it was just the wrong time to go."

SEVEN

BUT GOD DOES PLAY DICE

ONE MORNING IN DECEMBER 1925, ABOUT A MONTH AFTER HE HAD ARRIVED AT THE Massachusetts Institute of Technology (MIT), Max Born opened an envelope containing a scientific reprint. The title, "The Fundamental Equations of Quantum Mechanics," amazed him. The author's citation read, "P. A. M. Dirac, 1851 Exhibition Senior Research Student, St. John's College, Cambridge," an obviously young British physicist and one Born had never heard of. Reading through it, Born saw that the unknown Dirac had formulated a theory of quantum mechanics similar to the one Born had just completed with Jordan and Heisenberg. Born immediately recognized that although the author was a "youngster . . . everything was perfect in its way and admirable." Rather incredibly—and working all alone—Dirac had developed, written up, and sent to the *Proceedings of the Royal Society* the essentials of quantum mechanics nine days before Born, Jordan, and Heisenberg had sent in their own version to the *Zeitschrift für Physik*. Born could not imagine how he had accomplished this feat. Sitting in his office at MIT, all alone, thousands of miles—and an ocean—away from Göttingen and colleagues, Born must have felt at a loss. Later, he called Dirac's article one of the greatest surprises of his academic career.

In time, the mystery of who Dirac was and how he came to this precocious and surprising work was cleared up. Cambridge mathematical physicist Ralph Fowler had received the galley proof of Heisenberg's breakthrough article on quantum theory in mid August and sent it on to his lanky, taciturn student, Dirac, with the note, "What do you think of this? I shall be glad to hear." Three months later, Dirac, who had studied engineering and mathematics before coming to physics, answered the query with his own formulation of quantum mechanics. He had not recognized Heisenberg's strange multiplication as matrix multiplication, but he had realized its noncommutative nature. To address it, he had developed his own set of mathematical rules to develop his theory.

Dirac and Born were to become close friends in the small network of quantum researchers; but before that opportunity presented itself, Born had many miles to cover preaching the new quantum gospel to the neophytes of the New World.

One month earlier, on November 11, after fourteen days at sea, the Borns had begun their first day in New York when the ship's horn blew the 5 A.M. wake-up call. The *Westphalia* lay at anchor outside the harbor while immigration officers counted passengers and made lists. As the Borns looked out at the city, they saw a thick fog punctuated by skyscrapers arising as "gray ghosts out of haze and smoke." At 10:30, the steamer docked at the pier. The Borns descended the gangway, entered a long gray hall built of iron, then waited by the letter *B* for their luggage before going through customs. The secretary of Born's benefactor, Henry Goldman, met them promptly to help with details, then drove them to the Astor Hotel.

Hedi found herself in a "cauldron," a boiling pot of a city filled with noise, speed, and confusion. The New York City of 1925 was one of tearing down and building up, traffic congestion and impatient drivers, luxury and poverty. Beneath the Borns' hotel window, Hedi watched an old house being wrecked to make room for a skyscraper. She counted off the stoplights at the end of every block and noted the strategies of pedestrians for crossing busy streets; and she marveled at the extravagance of the Astor Hotel: clean sheets daily, several changes of hand towels a day, flagons of ice water at meals, 100 bellhops ready for service, shopping boutiques inside the hotel, and telephones rather than bells to call for service. The Borns paid 5 dollars a day—half the standard rate because the manager was a friend of the Goldmans.

The Borns spent three days strolling through the streets of New York, interspersed with a number of lunches, teas, and dinners with the Goldmans in their sumptuous apartment on Fifth Avenue. Their collection of paintings and sculptures by Rembrandt, Titian, Holbein, Rubens, Hals, Van Dyck, Cellini, Della Robbia, and Donatello—all lovingly described by Goldman even though he was almost blind—rivaled the neighboring Metropolitan Museum of Art. On the first day, Hedi had felt she was in a lunatic asylum; by the third, she was sorry to leave for Boston.

Boston was different. The Borns settled into a two-room apartment located next to a noisy railway track. While Hedi explored the city, Max spent the day at MIT, arriving home at about six. He had to give a total of thirty hours of lectures—ten on lattice theory and twenty on atomic structure that included the new quantum theory—to the institute's faculty and students. To make sure that he could deliver them without halting his speech, he had two physics assistants translate into English the thirty lectures he had written out beforehand in German. On the invitation of the director of Harvard's Jefferson Laboratory, Theodore Lyman, he gave five lectures there as well.

The lectures on lattice theory were straightforward, but the quantum ideas were virgin, never before presented or discussed. When he began the lectures, only Heisenberg's original article was in print. Born carefully took his audience from the early work of Bohr and Sommerfeld through Heisenberg's ideas on quantum multiplication and on to the work that first he and Jordan and then the three together had done to formulate quantum mechanics. He had no way of knowing how these ideas would be received, but contemplating the work away from the frenzy of Göttingen, he believed that "the new theory is really quite reliably based," and that he could lecture on it "without qualms."

Born's confidence came mostly from educated intuition. Even though quantum mechanics explained some spectroscopic data, such as the Balmer Series (the name for the hydrogen spectrum's four visible spectral lines), more calculations clearly had to be undertaken. Consequently, he summed up his MIT lectures modestly by identifying the results as "a first step," saying that "only a further extension of the theory, which in all likelihood will be laborious, will show whether the principles given above are really sufficient to explain atomic structure."

Born had spent three relentless months uncovering quantum secrets in Göttingen, and leaving his research and collaborators behind had been frustrating. Yet within ten days of arriving in Boston, he found a new collaborator. Norbert Wiener, a young MIT mathematician whom Hilbert and Courant had already invited to Göttingen for the summer, had developed a more precise method of Fourier analysis. With it, Born thought they could apply quantum theory to the problems of the continuous spectrum, such as collision theory—as when atoms collide—problems that seemed impossible to handle with matrices. Watching Franck's collision experiments in the lab back in Göttingen had triggered Born's interest not only in the interaction of matter with matter but also of matter with radiation. In the three-man paper, Born had outlined the problem as well as an approach for replacing an element of a matrix referring to two discrete states by a function of two continuous variables. With the rules for the transition unknown, he could go no further. Using Wiener's new method, they began immediately to generalize matrices as linear operators. Their article—ready before the Christmas holidays—was the first article on the new quantum mechanics written in the United States.

Max and Hedi took advantage of their time together to absorb American culture, but they were not ready for certain aspects of brash, bustling America. The Boston Orchestra's modern pieces were "grotesque, silly stuff"; they found the overture from *The Magic Flute*, a Beethoven symphony, and "Liebestod" from *Tristan und Isolde* more to their liking. The movie palace they attended was overwhelming, like a coronation room from the Old World. Then there were football games—40,000 people in a huge open arena to see twenty-two men in a sport that made rugby look like a graceful round dance. Max stopped short of calling it "madness" because "one should not judge from one's own standpoint," but it seems that is what he actually thought.

The Borns had planned a five-month journey, three at MIT with two more available for lectures elsewhere. Once in the United States, their schedule filled up quickly. Born and quantum mechanics were in demand. From Boston to California and back, major centers of physics wanted a presentation. Being a newcomer to America, Hedi was excited about the trip west, but at the end of December, she got a bad case of food poisoning and ended up in the hospital. When she could not fully recover, she took a ship back to Germany in mid January. As Max watched her train pull out of the station, he wistfully thought of the lovely travel adventures they had had and the important place she held in his life. Then he gathered himself together and decided to enjoy the rest of the trip as best he could on his own.

On January 22, 1926, Born gave his last lecture at MIT before an audience of about 1,000 people. His stay in Cambridge had been a success, a fact underscored a few months later when the printed version of his lectures appeared: *Problems of Atomic Dynamics*, the first book on the new quantum mechanics. Edwin Kemble, a young Harvard professor, reviewed it.

> It is a happy experience for American physicists that Professor Born was engaged to deliver these lectures on atomic dynamics just as the first accounts of the new matrix mechanics were appearing in Germany. The prompt publication of the text of the lectures with their summary of the first results obtained by this method should be of great service in helping us to keep up with the stream of thought in a field in which we have been prone to lag behind.

The promise of quantum mechanics excited scientists. With the physical meaning of the theory still unclear, however, and matrix algebra—the mathematical tool—strange and wholly unfamiliar, many did not see it as the final answer. Kemble added, "it would perhaps be rash to say that the year 1925 marks the beginning of a new era in physics." In only a few months, the shortsightedness of Kemble's conclusion became obvious.

On the evening of January 22, Born boarded a sleeper train for the beginning of an immersion into American sights and sounds. He arrived at his first stop, Schenectady and the General Electric Company, at 6:25 A.M. and, after resting for a few hours, got his bearing with a drive through the town. He was pleased to find an American version of home: old houses clustered around a river surrounded by hills and valleys. The atmosphere at the GE labs was equally comfortable, particularly the dynamism and collegiality of company scientists—like his old friend Irving Langmuir. By the time Born left, he regarded General Electric "the center of American physics; more even than Harvard."

Born's three-day visit set the norm for future stops: a couple of intense days spent lecturing, dining, and visiting laboratories, always discussing physics—and mostly quantum physics. General Electric, though, held a special treat: gadgetry, such as state-of-the-art radio speakers that dazzled him. When Born heard the music of Chopin and Debussy fill Langmuir's living room, he thought that it sounded more like a live orchestra in the next room than a broadcast from New York. Ever-present movie cameras captured the youthfulness of the forty-three year old. At nearby Lake George, Langmuir filmed Born as he prepared to learn a special kind of skate sailing, where one donned ice skates and held on to a sail that gathered the wind, pulling the skater across the frozen lake. Born had some success mastering the sport, but for one wobbly moment that knocked him off his skates, the 10 centimeters of snow on the ice cushioning his fall. After half an hour, his arms were too tired to continue, but he was, nonetheless, sufficiently excited to want to introduce the sport to Göttingen.

Born was delighted that Langmuir had completely understood quantum mechanics and deemed him "a remarkable mix of the deepest thoughtfulness and the wildest energy." Langmuir's fast pace—both in walking and thinking—wore Born out when they toured a GE plant that produced huge turbines, but Born still had time to feel "the pulse of our times." Industrialization, as represented by the size of GE's 75,000-kilowatt turbines, appeared to him as one of the forces that simultaneously "feeds off the lives of the majority of our contemporaries and in reverse rules their lives." This was true, he thought, whether one "condemns it or is inspired by it." Wanting Hedi to see the future as well, he promised to take her to the Krupp works one day.

After Schenectady came Ithaca, Buffalo, Chicago, Pasadena, Berkeley, Madison, New York, Princeton, and Washington, D.C.—in total, twelve major universities and research centers, where Born introduced thousands of professors, students, and the lay public to the new quantum theory. The influence of the new ideas was not limited to physicists. In Berkeley to deliver the Hitchcock Lectures, he had the chance to introduce them to his old friend, the chemist Gilbert Lewis.

Born grew more confident in the limelight. At a luncheon in Berkeley, university president William Wallace Campbell praised Born and asked him to say a few words. "What could I do?" he wrote to Hedi. "I spoke as well and as humorously as allowed by the dry fellow that I am. Since I succeeded in raising a broad smile on the stoic face of Gilbert Lewis, perhaps it wasn't so bad."

Born's lectures on quantum mechanics entranced students along the route like the Pied Piper's flute (and aptly so, as Göttingen was not far from the legendary

Hamelin). Even those such as the young chemist Linus Pauling, who at the time saw no way of applying quantum mechanics to his research, eventually made the trip to study at "the fount of quantum wisdom," as Karl Compton called Göttingen. Born's U.S. lectures, the continuing developments in quantum physics, and the Rockefeller IEB fellowships pulled them there.

At Princeton, almost at the end of the tour, the faculty and students were so eager to learn the new quantum theory that the departments of physics, astronomy, and mathematics prepared for Born's visit by studying the theory jointly. The March 19 *New York Times* lauded the lectures in an article entitled "His Dynamic Theory Rivals Einstein's." It quoted Karl Compton, the chair of the Princeton physics department, on Born's success "in stimulating and developing his younger colleagues and students who have made Göttingen a great center of activity in modern physics."

Born caught the flu in Princeton, traveled on to Washington, D.C., and there had to stay in bed. He never gave his last lecture, which was scheduled for the National Bureau of Standards. The flu plagued him so much that he feared missing his ship, but he made it to New York on March 23 with a day to spare.

In the two-month tour, Born had traveled more than 6,000 miles, covering the best and worst of the country. The best were clearly the thunderous power of Niagara Falls—"magnificent at night"—and the grandeur of the Grand Canyon—"suddenly a schism, this breathtaking precipice, rugged and torn, as no fantasy can imagine it." These impressions he balanced with the disparity in living conditions in urban America, such as Chicago, where "huge steel mills surrounded by horrible worker housing, without tree or bush, were followed by the real suburbs, even more desolate and dirty." Born found that the train ride back and forth across the country afforded him a perspective on the American soul—his traveling companions being lively, uncritical of the American way of life, and frequently interested in talking about money. On the train out of Chicago, while answering questions about being German, he learned that nearly half the people on the train were also of German descent.

Now, his thoughts, which were "flying ahead of his body," were only of the homebound trip and, once there, staying put for a while. He wrote to Hedi,

> Woe to anyone in Göttingen who speaks to me in the next three months of a repeat of such an American trip! I will beat him down. It was very lovely—but I have had enough! Enough of traveling around, enough of holding lectures, enough of the Americans. I will be happy to be with you all!

In Pasadena, as he chased the four-year-old daughter of one of his colleagues around the garden, Born had realized how much he missed his own "sprouts." Now he wanted to make sure that Hedi waited for him to have the Easter egg hunt.

Born's welcome home party gathered all the physicists and mathematicians to Planckstrasse. Souvenirs from the trip literally paraded around the house—twelve-year-old Irene and eleven-year-old Gritli in their Indian headdresses, five-year-old Gustav in his cowboy suit. The star attraction, though, was the electric train. As a treat for the guests, father and son proudly laid the tracks in the sitting room and set up the engine and cars. As everyone watched, Gustav did the honors and flipped the switch. The response, to Gustav's disappointment, was silence and stillness. His sobs filled the room as the physicists got to work, dithering over the motor and wires, trying to figure out why, comparing theories. Hedi, having no patience with physical theories, sat on the floor and systematically examined each piece of track. When she located the bad one, she took it out and started the train.

Born had one disconcerting moment after returning—his discovery of a paper by Pascual Jordan at the bottom of his suitcase. Just before Born's departure, Jordan had given it to him for possible publication in the journal *Zeitschrift für Physik*, of which Born was an editor. Born had packed it, intending to read it on the trip. When he pulled it out and finally read it, he saw that Jordan had discovered the important statistical laws that Enrico Fermi had just published in the *Zeitschrift für Physik*. Shortly, Paul Dirac made the same discovery of what became known as the *Fermi-Dirac statistics*. These laws describe the statistical distribution of identical particles of spin $^1/_2$ (now called *fermions*). They follow Pauli's exclusion principle: that only one such particle can occupy an energy state at a time. These laws, as applied to electrons, aided in the development of the field of electronics. Amid the serious problems that rocked his future relationship with Jordan, Born always felt guilty, even "ashamed," that he had robbed Jordan of his due.

The Fermi-Dirac statistics was one of several new developments in quantum mechanics that had occurred while Born was away. As eager as he was to get back into research, however, he had to balance it with a weighty decision. Would he take one of the four job offers in the United States? MIT, Cornell, and the University of Wisconsin were all trying to outbid one another. In addition, during an informal chat in Born's hotel room in New York, General Electric president Gerald Swope had pledged financial support for Born to pursue research at GE or another place of his choosing. Born had assured Swope he would seriously consider the offer.

Born did not want to leave Göttingen, but in a letter to Theodor Valentiner, the university's Kurator, he cited strong reasons for going—satisfaction from organizing theoretical physics in America, greater financial benefits, and the

American way of life as compared to the "always depressed conditions in Europe." This was not an idle remark. The previous October, when IEB representative Augustus Trowbridge had visited Göttingen, he described professorial life as "living hand-to-mouth"—frugality, shabby dress, no financial savings, no fine meals, and third- and fourth-class train tickets. He had "horrified" his hosts by suggesting they take a cheap horse taxi rather than walk half a mile. When the Trowbridges left Göttingen, they were too embarrassed to use their first-class train tickets and instead "slunk into a 2nd class compartment as far as the first junction."

Max had another good reason for leaving. Eight years after the end of the war, the League of Nations in Geneva had yet to admit Germany. He told Hedi, "When the gentlemen in Geneva act so foolishly, I have to consider a retreat from this crazy Europe." The depth of Europe's political instability weighing on him, he relied on advice from Carl Runge. Runge must have reassured him, because afterward, Born began to use the American offers as leverage to wring concessions from the Ministry of Education.

Born's main demand was to reverse recent serious damage to his institute—the loss of Heisenberg to Copenhagen. While Born was in the States, Bohr had written to him that Kramers had accepted a chair in Utrecht and that Bohr wanted Heisenberg as his new assistant. Born wrote back that he understood Bohr's wish only too well. "To lose him is for me a very great loss—we have worked together so beautifully and I like him personally so much—but it is also a difficult loss for our institute and university." Born wished he were in Göttingen so that he could secure a similar position for Heisenberg, but he graciously conceded defeat. "In any case, I must hand off Heisenberg to you, since for him it is still splendid to be recognized so young and to be allowed to work with you." Now Born saw a possible way to remedy the situation.

Courant, the dean of the mathematics-physics faculty that year, lobbied Kurator Valentiner, emphasizing that Born's departure "would be a great loss and would especially endanger the development of all our big plans. Born is one of our colleagues upon whom the international reputation of our university is based." Some Göttingen professors expressed concern that the loss of Born could jeopardize funding from the Rockefeller Foundation.

For two and a half months, the politicking went back and forth while the Americans continually improved their offers. Ultimately, Göttingen—science and homeland—won. On July 4, Born wrote to Valentiner that he was rejecting the American offers; but he also laid out his disappointment over the Ministry's failure to address his request for Heisenberg. Comparing his own gains—mostly financial—to those reaped by colleagues in similar situations—that often included adding assistants or a chair—Born felt underappreciated. A ministerial official assuaged Born by pleading poverty and promising to recall Heisenberg soon. Born backed down; Heisenberg, after all, was only on temporary leave from Göttingen, so there was reason to believe he would return.

Born's decision to stay genuinely pleased Valentiner. His draft letters, handwritten in the old German script of "Sütterlin" with characteristic purple pencil, had argued in Born's behalf. Unbeknownst to Born and perhaps to Valentiner, however, someone made a comment in the margin of Courant's supporting letter next to the sentence, "Born is one of our colleagues upon whom the international reputation of our university is based." In the Kurator's familiar purple pencil—but not his handwriting—someone had written with special emphasis "*Jude!*" Midway through the decade, a noticeable current of anti-Semitism in Göttingen was becoming bolder.

Indeed, Göttingen had a longstanding history of anti-Semitism. Retired civil servants, faculties such as agriculture, and certain students made the university town fertile recruiting ground for the National Socialist German Worker's Party. In May 1922, the Nazis had started a chapter, one of the first in the region. Its members were particularly resolute. In 1925, a twenty-two-year-old chemistry student named Achim Gercke began to compile a list of all German professors of Jewish origin. He called it "The Archives of Racial Statistics for Employment Classes," the purpose of which was to provide "a weapon in hand that should enable the German Reich to exclude the last Hebrew and all mixed race from the German population in the future and expel them from the country." For months, Gercke and a small band of collaborators culled names and religious affiliations from the university library's collection of Gymnasium and university documents as well as from newspaper articles and death notices. Their immediate goal was to discover all the professors in Göttingen with Jewish ancestry. (By 1934, Gercke, the new director of the Racial Service in the Ministry of Interior, had amassed a card index containing the names and family trees of about 3 million Germans with any hint of Jewish ancestry.)

Gercke kept the research secret, even from party members. Certainly, professors were unaware of such activities. Born and Franck, however, glimpsed the beginnings of extremism when, in 1927, Fritz Haber and Paul Ehrenfest were blackballed for membership in the Göttingen Academy of Science because "it could not be proved they were of pure Aryan descent."

Just at the end of Born's U.S. trip, the Austrian physicist Erwin Schrödinger had published a rival quantum waves mechanics. Based on the wave theory of matter introduced by the young French physicist Louis de Broglie, Schrödinger's theory seemed to conflict with Göttingen's particle theory and discontinuity. Schrödinger's waves, like other waves, were continuous. Electrons moved around a nucleus in specific orbits consisting of an integral number of electron wavelengths. With his wave equations Schrödinger could derive the correct formula for the hydrogen Balmer series of spectral lines· His analytic tool of differential calculus was well understood by physicists—especially students—as opposed to Göttingen's impenetrable matrices. Many physicists quickly dubbed Schrödinger's formulation the true one.

Born learned of Schrödinger's work only when he returned, but he was familiar with de Broglie's matter theory of waves. The previous summer, just after he and Jordan had finished their work on aperiodic fields preliminary to studying atomic collisions, Born had written to Einstein that he thought de Broglie's theory could be "of very great importance." What attracted him most was the "mysterious differential calculus on which the quantum theory of atomic structure seems to be based." Struggling with the article Heisenberg handed him, however, had drowned out these thoughts. His mind became immersed in matrix algebra. Collisions still enticed him, as his discussion in the three-man paper showed, but ideas about differential calculus receded and did not resurface, even when he teamed up with Norbert Weiner to take another look at collisions.

Now, after studying Schrödinger's article, Born realized that he and Wiener had stopped their research one step short of discovering wave mechanics. Using complicated and nontransparent operator theory, they had developed a formula equivalent to Schrödinger's commutation law but applied it only to the variable time instead of to both time and momentum. Born later wrote that wave mechanics "was the most outstanding example of my being quite close to an important discovery and letting it slip by."

Whatever his frustration over this lapse, Born quickly saw the usefulness of Schrödinger's wave mechanics for examining collisions. Fortunately, there was no theoretical reason against using this form of quantum theory. Independently, Pauli and Schrödinger had recently showed that the two very different formalisms—that which described waves and that which described particles—were mathematically equivalent. This amazing development settled any debate about which form of quantum theory was correct.

Born worked on his new collision theory for the next two months and submitted a four-page paper for publication at the end of June. The abstract at the beginning stated, "Through the investigation of collisions it is argued that quantum mechanics in the Schrödinger form allows one to describe not only stationary states but also quantum jumps." Born wanted to see if Schrödinger's wave equation could describe the collision of a free particle, such as an electron, with an atom. He set it up to describe the interaction of the joint system and used perturbation theory to find an expression for the wave function of the scattered particle. He found that indeed this system gave him a solution—to his mind, the only possible one.

> One gets no answer to the question, "what is the state after the collision," but only to the question, "how probable is a specific outcome of the collision" . . . Here the whole problem of determinism comes up. . . . I myself am inclined to renounce determinism in the world of atoms. But that is a philosophical question for which physical arguments alone are not decisive.

Born had been looking for electron jumps, and he had found them. In this preliminary paper, he had also found a probabilistic—not a deterministic—outcome of the collision. He was not particularly surprised. "Statistical considerations" were commonplace, he later said, something with which he was comfortable. Hence his contemplative statement—hardly one that seemed destined to herald a revolution in science (or to help complete the quantum revolution, as it also did). Rejecting determinism, to his mind, required a greater breadth of consideration.

In his next paper, however, he warmed to the importance of the idea, explaining more fully that electron waves were not continuous clouds of electricity, as Schrödinger interpreted them, but rather represented the probability of finding a particle in a certain place after a collision. (Specifically, the absolute square of the wave function of the scattered particle gave the probability of finding the scattered particle in a given direction with respect to that of the incident particle.) These probabilistic waves were not physical waves like Schrödinger's; they were "ghost fields" inspired by Einstein's earlier representation of waves as guides for light quanta. In stating his hypothesis in the introduction, Born noted somewhat of a paradox—"the motion of particles follows probability rules, but the probability itself propagates in conformity with the law of causality. . . . This means that a knowledge of the state at all points at one moment determines the distribution of the state at all later times."

Causality had been the basis for the laws of nature for centuries. As one scientist said, "The principle of causality is . . . the general expression of the fact that everything which happens in nature is subjected to laws which hold without exception." Born's statistical interpretation of Schrödinger's wave function—an essential element of quantum mechanics—was causality's death knell. Its counterpart, acausality, was not a new idea: the Bohr-Kramers-Slater theory had already tried to dispense with it two years earlier; but BKS was the outcome of a statistical assumption, an artifice to avoid quanta. Born's result was different—his discovery was derived from a logical basis.

Born anticipated that some physicists would "assume that there are other parameters, not given in the theory, that determine the individual event"—that is, that something else lay deeper that would reestablish cause and effect.

His anticipation was warranted, though initially he was attacked by one of his own. In a letter lost and forgotten, except in Born's memory, Heisenberg called him a "traitor" to quantum mechanics because Born had used Schrödinger's wave equation in his analysis. Heisenberg, seeing a preference for wave mechanics in the physics community, voiced his distaste to both Born and Pauli. "The more I think about the physical portion of the Schrödinger theory, the more disgusting I find it." Schrödinger's classical interpretation of the behavior of electrons did not sit well with Heisenberg.

Heisenberg's pique with Born was short-lived; but Schrödinger had no affection for Heisenberg's ideas or Born's. In a footnote to his paper on the equivalence of the two theories, Schrödinger wrote, "No genetic relationship whatsoever with Heisenberg is known to me. I knew of his theory, of course, but felt discouraged, not to say repelled, by the methods of transcendental algebra, which appeared difficult to me, and by the lack of visualization." Schrödinger found Born's interpretation of the wave function so unsettling that at one point he wished that he had never written his original article. What fostered dissent was not the mathematical formalism but the physical interpretation. Schrödinger's waves were real; in Born's new theory, they were probabilistic. The Göttingen group believed in quantum jumps; Schrödinger thought them nonsensical. In Copenhagen that fall, Bohr so relentlessly browbeat Schrödinger about the existence of quantum jumps that Schrödinger wound up sick in bed. Undeterred, Bohr sat at his bedside and continued to bombard him in a forced attempt to alter his view.

Schrödinger was far from alone in his opposition. Einstein, Planck, and von Laue agreed with him. To Einstein, Born's statistical interpretation was "certainly imposing. But an inner voice tells me that it is not yet the real thing. The theory says a lot, but does not really bring us any closer to the secret of the 'Old One.' I, at any rate, am convinced that *He* does not play dice."

Born and Einstein (as well as Bohr and Einstein) argued this point for the rest of their lives.

For the next several months, Born, Pauli, and Heisenberg wrangled with the physical meaning of the statistical interpretation. By October, Heisenberg realized the potential of Born's probabilistic interpretation when Pauli made it more profound. Using Born's work, Pauli wrote Heisenberg about a "dark point": When electrons collide, their momentum is assumed to be controlled and their position uncontrolled. One could not "inquire simultaneously" about position and momentum. Heisenberg responded, "I am *very* enthusiastic about your observations about the collision process because one understands the physical sense of the Born formalism much better than before."

Four months later, Heisenberg wrote a fourteen-page letter to Pauli that contained all the basic elements of what came to be the uncertainty principle, holding that the more precise the measurement of the momentum of a particle, the less precise the measurement of position and vice versa. He extended Pauli's idea to show that both measurements could not be simultaneously precise because of interference by the measurer.

In summer 1926, as Born worked on collision theory, a foreign invasion overtook Göttingen. The first wave had a large contingency of Russian physicists, then Paul Ehrenfest and his wife from Leiden, and lastly, Augustus Trowbridge for a follow-up visit. When Trowbridge arrived on July 2 at 6 P.M., he went immediately to a meeting of Göttingen's Academy of Sciences and heard Born give a "good" paper—actually, his first presentation on collisions—and from there on to Franck's for supper. With a front-row seat for the performance of Göttingen's medley of personalities, Trowbridge observed Hilbert's precarious health, Runge's charm, and Courant's tact. He reconfirmed his previous opinion of the high standing of Franck and Born and thought Born had "gained much in 'savoir-faire' during his visit last year in the USA." He kept the same impression of Göttingen's mathematics, mathematical physics, physics, and chemistry as the year before; "it is about the best [I have] seen on the continent."

Others shared this view. A second wave invaded in the fall. Before it began, Born took off with the children for a ten-day stay on an island in the North Sea. Hedi stayed home because she still suffered from stomach problems that had started eight months earlier in Boston.

Scores of students and distinguished researchers streamed into Göttingen and filled the town; at lunch, supper, and parties everything was physics. One of Born's students, Max Delbrück, described the scene as "people who were highly bizarre, genially mad, unworldly, and completely, decidedly difficult in their behavior towards their fellow man"—and thanks to American fellow H. P. Robertson, singing "Oh my darlin' Clementine." Visitors stepping off the train found a community of brilliant oddballs in a madhouse.

Ingenious ideas gushed from these Wunderkinder. The mathematician John von Neumann, who worked with Hilbert, became caught up in working out the formal quantum system of Born's statistical interpretation of the wave function. Pascual Jordan—and independently, Paul Dirac—integrated matrix mechanics and wave mechanics into a combined transformation theory. Eugene Wigner worked on group theory. The energy was contagious.

Even the social side had a distinctly Göttingen flair. Replacing the stiffness of the German "Herr Professor Doktor," the genial Born, Franck, and Courant often invited fellows and students home for supper. After seminars in his home, Born and participants hiked up into the hills or over to the little village of Nikolausberg to have supper at a country inn; and they held parties in the smoky, rustic restaurants around town. Fueled by an abundance of German beer, the professors and their fellows bobbed for apples or raced with a potato on a spoon. One student spied Dirac, who had lost the potato spoon race, practicing by himself.

In December 1926, hundreds of students, faculty, and townspeople marched through the meandering streets and alleys, flaming torches held high, the night sky lit up in grand Göttingen tradition. James Franck and his former research partner, Gustav Hertz, had won the Nobel Prize (reserved from 1925) for their research on electron collisions. The throng of well-wishers eventually collected in the market square in front of the medieval Rathaus for speeches, food, and drink. Born's excitement about Franck's recognition merged with his pride of having brought him to Göttingen and with his deep appreciation.

Just before Franck's Nobel Prize, Max Planck's chair at the University of Berlin became vacant. The doyen of Germany's theoretical physicists, who had given Born his first position twelve years earlier, was retiring. During the summer, the Berlin faculty committee had discussed options and received recommendations from other faculty who proposed Sommerfeld, Born, and Schrödinger, in that order. Sommerfeld, senior to Born by fourteen years, was the obvious choice, but it was not clear that he would leave Munich; second place therefore was very important.

At the November 2 meeting, the committee of the Berlin faculty gathered to make its final recommendation. With Born listed in second place, Fritz Haber moved for the meeting to adjourn, but Walther Nernst objected. Max Planck took this opportunity to request that the order of the names for the second and third positions be switched, making the final recommendation Sommerfeld, Schrödinger, and Born. In the letter to the Ministry, the faculty remarked on the difficulty of deciding on second and third places. They cited Schrödinger's "deeper originality and a stronger creative strength" but noted that "on no account should the merits of Herr Born be diminished." The letter did not mention that Planck and Schrödinger shared views on causality opposite to Born's, but this difference may well have played a role in Planck's decision.

Planck wrote a short, friendly letter to Born about the decision. Whether Born's great respect for Schrödinger reduced his disappointment at being placed third is unknown. His response to Planck acknowledged the honor of being included and the improbability of the call. He also noted a dilemma, if for some reason he were called: A factor in Franck's rejecting the call to Berlin two years earlier had been his relationship with Born. "Now," wrote Born, "I cannot break this loyalty to him." In the event he received the call, Born suggested that they call Franck as well; but no quandary arose. Sommerfeld stayed in Munich, Schrödinger succeeded Planck, and Born remained in Göttingen, lecturing and working with assistants and students to the extent that he complained he had no time for research.

The large number of students, their exceptionally quick minds, and the rapid developments in quantum theory put tremendous pressure on Born, whose forty-three-year-old brain was trying to keep up with those in their intellectual prime. Even before the quantum breakthrough, he had lamented, "I'm too old now, it will have to be one of the younger men [who solves the quantum puzzle]." Yet, owing to his success, he found himself, as Max Delbrück phrased it, "suddenly in the center of the madly funny whirlpool of the breakthrough of quantum mechanics where ideas tumbled in a mess in breathtaking tempo." As Born pushed himself to keep up, the stress began to exact a toll in the student-teacher relationship.

Born always worked closely with his students and assistants. He worried about their well being and supported their work, and they appreciated his inspiration and efforts on their behalf. To his assistants he gave a free hand to pursue their interests. He let his superstars stretch past him; to those less gifted, he patiently handed out respectable but doable assignments, often in the field of solid-state physics. The students felt Born's reserve and often went to Franck—the more amiable of the two—with problems, but they knew that Born tirelessly watched out for their interests.

In fall 1926, some of Born's new students as well as American fellows of the IEB perceived him as "unapproachable and unsympathetic." They wanted to work "independently." H. P. Robertson, the "Clementine" devotee, took offense because Born considered his ideas on the expanding universe "rubbish." Walter Elsasser thought Born "always seemed distracted and to be only half-listening" when asked for help. George Uhlenbeck simply thought that Born was easily intimidated by the intellects around him.

The pace and sheer number of research fellows overloaded Born, who was already exhausted from two strenuous years of research and travel. It was impossible for him to develop the same close working relationships that he had fostered earlier; the stress turned directness into bluntness and reserve into remoteness. Also, twenty-six-year-old Werner Heisenberg accepted an Ordinarius chair at Leipzig, so he would not be back to lighten Born's load.

Born did, however, work closely with one new student. In October 1926, an American Ph.D. student named Robert Oppenheimer appeared in the midst of Göttingen. A twenty-two-year-old New Yorker with a degree from Harvard and graduate work at Cambridge added his own special splash with "a very elegant kind of stutter, the 'njum-njum-njum' technique." He had met Born in Cambridge the previous July, when Born was there to lecture on collision theory at the gathering called the Kapitza Club. The topic, "On the quantum mechanics of collisions of atoms and electrons," coincided with Oppenheimer's interests. Since Oppenheimer did not like Cambridge, Born invited him to work in Göttingen.

For two semesters, Oppenheimer, like Born's former student collaborators, spent many hours at his professor's house. Almost from the beginning, Born experienced him as IEB fellow E. U. Condon described "extremely nervous, tense, and intense"—in short, a stressful person to work with. (Condon eventually transferred to Munich because he felt Born's work with Oppenheimer left no time for others).

Oppenheimer was impeccably mannered and cultured and, at the same time, arrogant and intellectually aggressive. Some of his fellow students, awestruck by his quickness, considered him the center of the group. One later wrote, "I felt as if he were an inhabitant of Olympus who had strayed among humans and was doing his best to appear human." Others, who suffered withering condescension if they made what he considered a silly remark, kept their distance. Born, who later described Oppenheimer as a person who oscillated between "arrogance and a somewhat unpleasant modesty," kept his feelings to himself at the time.

Oppenheimer started working with Born on defining the laws that govern the collision between a nucleus and α-particles (particles emitted in radioactive decay)—work that they eventually abandoned. Their major work together was the paper "On the Quantum Theory of the Molecule" from which came the Born-Oppenheimer approximation—a method using perturbation theory to justify the assumption to fix the position of the massive nucleus when calculating the motion of the much lighter electrons. Oppenheimer later said that he came up with the idea, and after writing up a simple four- or five-page article, showed it to Born. Born, who was aghast at Oppenheimer's style, rewrote the article.

In its final form with proper theorems, it comprised twenty-seven pages. Oppenheimer felt Born's lengthy treatment was unneeded and obscured the main ideas, but when Born mentioned Oppenheimer's writing style to Edwin Kemble, who was visiting from Harvard, Kemble reported back, "Unfortunately Born tells me that he [Oppenheimer] has the same difficulty about expressing himself clearly in writing which we observed at Harvard."

Oppenheimer also participated in Born's seminar. The open and free-flowing format was the perfect setting for Oppenheimer, who some saw as "a man conscious of his superiority." Striding to the blackboard and taking the chalk from the speaker, whether it was Born or a student, Oppenheimer would assert, "No, that is wrong," or "That is not how it is done." Usually being right did not ingratiate him.

One morning Born found on his desk a large sheet of parchment lettered in ornamental script and signed by the seminar's participants. It was a "request" for Born to stop Oppenheimer's disruptive behavior or the others would boycott the seminar. Born knew that he had to take the petition seriously but felt uncomfortable confronting Oppenheimer directly; that was not his style. So he worked out a different tact. When Oppenheimer came to work at Born's home a few days later, the paper lay conspicuously on his desk. During their discussion, Oppenheimer's eyes strayed to the strange-looking document, attempting to read it without appearing obvious. Born excused himself to respond to a prearranged call from Hedi, and when he returned, he found a different Oppenheimer. Neither man mentioned the petition, but the interruptions in the seminar ceased. Later, Born wondered if Oppenheimer had "harbored a secret resentment or embarrassment" over the incident.

Born had no question about Oppenheimer's brilliance, and in letters and reports he praised his abilities. On his dissertation Born bestowed the highest rating, "with distinction," and noted that Oppenheimer had chosen a very difficult topic. At his oral exam, on May 11, Born and Franck were both questioners. Franck took only twenty minutes, but it was long enough. On leaving the exam room, he exclaimed, "I'm glad that is over. He was on the point of questioning me." After his exam, Oppenheimer went to Leiden to work with Paul Ehrenfest. Before leaving, he presented Born with a gift of a first edition of Joseph Louis Lagrange's famous 1788 mathematical text *Mécanique Analytique*. Lagrange was the discoverer of one of Born's favorite mathematical tools, the calculus of variations.

Oppenheimer was going to an entirely different environment. The theoretical physics at the two universities reflected the personalities of their professors, and except for sharing a love of music, Born and Ehrenfest were opposites. Once when Ehrenfest was visiting Göttingen, a student presented a problem using wave mechanics. Born jumped up and demonstrated how matrices could also solve the problem. When Ehrenfest questioned this approach, Born replied that since both approaches were equivalent, it was a matter of habit, to which Ehrenfest answered, "a bad habit." There was a healthy professional tension between them. Students thought the nonmathematical Ehrenfest was afraid of the formulas and calculations that Born enjoyed. They saw Born as worried that if he made a calculation error, Ehrenfest would show him up with a simple model. For some students, Ehrenfest filled in the physical side of the theory that Born's mathematics sometimes glossed over.

Where Born was reserved and quiet, Ehrenfest was lively and excitable, always hamming it up for his audience. The students enjoyed his spirited teaching, but they also saw his intense side—the animation would swing to depression.

Born wrote to Ehrenfest soon after Oppenheimer's departure. The body of the letter, typed by his secretary, outlined plans for the rest of the summer and into fall 1927: two weeks in Switzerland to rest; the Volta Congress at Como in early September; work on an optics book; a visit to Ehrenfest; a couple of days in Bristol to receive an honorary degree; then to Brussels for the Solvay conference in October. The second part of the letter—a lengthy PS—Born wrote by hand because, as he explained to Ehrenfest, "the typist does not need to know." The confidential subject was Oppenheimer.

> Oppenheimer, who was with me for a long time, is now with you. I should like to know what you think of him. Your judgment will not be influenced by the fact I openly admit that I have never suffered as much with anybody as with him. He is doubtless very gifted but without mental discipline. He's outwardly very modest but inwardly very arrogant. Through his manner to know everything better and to continue any idea you give him, he has paralyzed all of us for three-quarters of a year. I can breathe again since he's gone and start to find the courage to work. My young people have the same experience. Do not let yourself keep him for any length of time. Stop! You are supposed to give me your opinion. Perhaps I just got very nervous.

Born soon learned from Ehrenfest that he did not mind Oppenheimer, but this opinion did not alleviate Born's concern.

> Your information about Oppenheimer was very valuable to me. I know that he is a very fine and decent man but one can't help it if someone gets on your nerves. My soul was nearly destroyed by this man even though I am generally not like a mimosa. I warn you against a lengthy get-together.

Born was becoming more like the fragile Mimosa than he realized, and would continue to do so.

Born had a rigorous fall schedule beginning with the Volta Conference. More than seventy physicists from thirteen countries, including Germany, England, and France, convened in Como, Italy, to commemorate the centennial of the death of the inventor of the eponymous voltaic cell, Alessandro Volta. It was one of the first such international gatherings since the end of the war, and the Germans were still cautious; the League of Nations had only admitted Germany one year earlier. Wanting to make sure that Germans in general were invited and not just a token few, Max Planck had hesitated before accepting the invitation. He had no need to worry. New politics were rapidly unfurling. Italy, now a fascist nation under Benito Mussolini, was praised by several speakers, while Russia's Red flag with the hammer and sickle hung high in the hall among the other banners.

"Dull" was Born's impression of the conference. He gave a lecture on the statistical interpretation of the wave function and listened to a banal discussion afterward. He was so bored during one session that he slipped out from the darkened hall during a slide show and bumped into Ernest Rutherford, discoverer of the nucleus, and Francis Aston, discoverer of many isotopes, doing the same. Having agreed the meeting was "unbearable," the three hired a taxi, drove around Lake Como to the Swiss border, and had a pleasant lunch. For Born, this side excursion stood out as the most significant event of the conference. He forged a friendship with Rutherford that would open up a fateful opportunity.

Niels Bohr delivered a speech at the conference entitled "The Quantum Postulate and the Recent Development of Atomic Theory." The quantum postulate, as Bohr explained it, "attributes an essential discontinuity" to any atomic process and implies that when one observes atomic phenomena one affects them. Bohr built up to this postulate by tracing the contributions of Schrödinger, de Broglie, Heisenberg, Born, and others, and threaded a description of Heisenberg's new principle of uncertainty through it. Upon all this, in a vague fashion, Bohr placed his own new overarching concept of complementarity: his explanation for the haunting wave-particle duality, where waves and particles were not contradictory but rather complementary, mutually exclusive but absolutely necessary to each other.

Born, Kramers, Heisenberg (twice), Fermi, and Pauli all commented. Born rallied to Bohr's interpretation. Heisenberg, who had not presented a paper at the conference, took the opportunity to discuss uncertainty, emphasizing the importance of the role of the observer and its intimate connection to Bohr's new synthesis of wave and particles. Bohr's ideas remained somewhat vague.

After the conference ended, Born had a brief interlude at the University of Bristol to receive his first honorary doctorate, bestowed at the opening of a new physics laboratory. (A sign of age, he remarked.) Then he went on to Brussels where a subset of those at the Volta Conference met for the Solvay Conference. It was the fifth in a series that had been held since 1911 to discuss basic concepts in physics and the first since the war to which German physicists—as opposed to only Einstein—were invited. The twenty-nine participants included all the European and American leaders in quantum theory.

During the six days of the conference, the participants sat in a meeting hall listening to lectures, Born sitting next to Madame Curie and whispering translations of those given in English or German. In their joint presentation, Born and Heisenberg reviewed the basic features of quantum mechanics. Their thoughts were unequivocal and definitive.

> We regard *quantum mechanics* as a complete theory for which the fundamental physical and mathematical hypotheses are no longer susceptible of modification. . . . Our fundamental hypothesis of essential indeterminism is in

> accord with experiment. The subsequent development of the theory of radiation will change nothing in this state of affairs.

The presentations and the discussions were intense. During breaks, the physicists relaxed and chatted before Langmuir's movie camera—Bohr animatedly conversing with Schrödinger, a youthful Heisenberg flashing a cocky grin, Ehrenfest making faces. By day's end, they exited through the ornately grilled door with pinched smiles and pensive frowns, ready to have dinner and resume their debates—Einstein, Planck, and Schrödinger versus Bohr, Born, Heisenberg, and Pauli.

The Solvay Conference became the springboard for the Copenhagen Interpretation, a philosophy based on complementarity, which Bohr had unveiled at the Volta Conference, and Heisenberg's uncertainty principle. The new philosophy cemented the importance of discontinuities embodied in matrix mechanics and the observer's role in quantum measurements. These ideas eventually became celebrated in thought experiments such as Schrödinger's, which takes the Copenhagen idea—that the state of an electron is unknown before it is measured and only becomes defined through measurement—and applies it to a poor cat. Through quantum logic, a cat in a box, whose fate is linked to a quantum measurement, can exist both dead and alive before the box is open to check on its true fate. Poems, such as "Idealism," one Born enjoyed, played with the idea too.

There once was a man who said: God
Must think it exceedingly odd
If he finds that this tree
Continues to be
When there's no one about in the Quad.
Ronald Knox

Dear Sir, your astonishment's odd,
For I'm always about in the Quad.
And that's why the tree
Will continue to be
Since observed by yours faithfully, God.

Believers in the Copenhagen Interpretation, especially Heisenberg and Pauli, sought converts to its fundamental principles, which they neatly wrapped up within this designation. Not being a member of the Copenhagen circle, Born was omitted. In his highly influential book *Physical Principles of Quantum*

Theory (1929), Heisenberg stressed Bohr's contribution and gave only one mention to Born. Pauli's 1933 *Handbuch der Physik* article gives Born's interpretation of the wave function a footnote. Textbooks on quantum mechanics written by former Copenhagen students make little or no mention of Born's contributions to the formulation of the basic theory. When they do cite Born, it is in regard to the Born approximation, a mathematical trick he developed in his second paper on collisions. His contribution of the statistical interpretation of the wave function, which established the basis for uncertainty, became wholly subsumed.

Heisenberg and Pauli had definitely developed an attitude toward Born's "formalistic approach" to physics. A few years later, their criticism discouraged the young physicist Felix Bloch—whom Born deemed "the best student of my friend Heisenberg"—from becoming Born's assistant. He went to Kramers in Utrecht at the last minute, even though he had already signed a contract to work with Born.

Pauli and Heisenberg had wrangled with the significance of the statistical interpretation of the wave function in the summer and fall of 1926, but Copenhagen adherents later responded that it was obvious. "We never dreamt that it could be otherwise," said Bohr. Born's contribution and place in the theory's hierarchy were lost within the "spirit" of Copenhagen.

EIGHT

DARK FUTURE

AUGUSTUS TROWBRIDGE VISITED GÖTTINGEN AT THE END OF MARCH 1928 AND INTERVIEWED Born about the progress of the IEB Fellows. His report to headquarters told as much about Born as it did about the students.

> Professor Born terribly overworked, largely because of tremendous influx of foreign students of all kinds. . . . Born evidently does not like to refuse any man and has some rather worthless men with him, wasting their own and his time. . . . Born teaching 15 hours a week, which is far more than average of the better men, even in the American universities, where conditions are notoriously bad.

Trowbridge advised him to be "hard-hearted" about whom he accepted.

Born had always impressed people with his efficiency and adroitness at juggling competing demands, but long hours of work were beginning to tell. In fact, he may have already dropped a ball. That spring Hedi went away to live quietly by herself for awhile. Tension was in the house with no outward indication of why.

In August, Born spent ten days at the Sixth Congress of the Association of Russian Physicists, part of which took place aboard the *Alexei Rykow* as it steamed down the Volga. Leaving the floating congress feeling tired and suffering from asthma, he went home instead of attending a mathematics conference in Italy. He needed rest, but it was not to be. Almost immediately upon his return, Ehrenfest came to visit from Leiden. After Ehrenfest's departure, Hedi went on a holiday with Margot Einstein, Albert Einstein's stepdaughter, while Max traveled to the South Tirol with Rudi Ladenburg. In a reversal, as Max came home, Hedi left for Bolzano.

Something unpleasant had happened during Ehrenfest's stay. In a later letter to him, Hedi alluded to hurt feelings but did not give any details. Max wrote his own vague letter to Ehrenfest referring to this same problem. Except for indicating tensions in the Born household, he, too, talked around it.

> Hedi is in [Bolzano], I have sent your letter on to her, she will really be pleased about it, as I am myself about my part. . . . I had meant to write to you earlier, for at your last visit you became extremely upset state and could not get better. That made me very sad. Meanwhile [the situation] has improved and when you return, you will find peace and comfort with us.

After describing his research for the next three pages. Born ended by asking Ehrenfest why he felt low.

> Why then a hang-dog mood? Can't you save up beautiful impressions as joyful possessions? You see I can still do that in spite of my occasional feelings of insufficiency. For example, a drive through the Dolomites or a steamer trip on Lake Garda.

By this time, Born was aware that he would need such beautiful impressions. The problem did not really concern Ehrenfest. Rather, Hedi was not alone in Bolzano. She had found her own solution to her husband's unavailability. She had replaced him with another man, Gustav Herglotz.

Herglotz—or Gusti, as Hedi referred to him—had been a professor in the Mathematical Institute since 1925. He and Born already knew each other. During Born's student days in Göttingen, he was a lecturer and had assisted Hilbert and Minkowski in the electron seminar that had so enthralled Born. When Born later struggled to define the concept of rigid bodies in relativity, Herglotz wrote a couple of excellent papers that cleared up some of Born's problems. Although Born felt a bit envious at the time, he genuinely admired Herglotz's mathematical ability.

As a student at the University of Vienna in 1899, Herglotz had been best friends with Paul Ehrenfest. This relationship made Ehrenfest the perfect confidante for Hedi who lost no time in establishing a closer bond. Writing him upon her return from Bolzano, she gushed girlishly and rapturously.

> Yesterday evening I came back via Munich where for three days I walked around in a forbidden inner feeling for life. There are such times—you have had it right now in Switzerland—where one is consumed from the inside out by joy—by everything the dear God has created so well: by a marvelous shin-

> ing early morning or by a beautiful Roger v.d. Weyden picture (there is a marvelous Annunciation in Munich)—consumed also by joy (and gratefulness) *that* one is able to enjoy oneself *so* much.

Hedi had undergone quite a turn around since she went off quietly by herself in the spring. Years later, she confided to a friend that when Max became too involved in quantum physics and neglected her, she found someone else.

Born began to collapse emotionally. When Ehrenfest wrote to him that Oppenheimer might visit Göttingen, he panicked.

> I shiver at the thought he should turn up again. I like him personally. I'm sorry for him—his nervous disability. I don't want to hurt him, but his presence destroyed the last remnants of my scientific capabilities and also ruined my young people. Couldn't you tell him gently that he should not extend his visit here for too long?

Fortunately, Oppenheimer never arrived, but Born's nerves still rattled. One weekend he and Hedi, among others, boarded the train to go skiing. Born anxiously announced that he could not go, abruptly got off the train, and went home. Around the same time seven-year-old Gustav found his father sobbing in the bathroom. It was one of the deepest impressions of Gustav's life.

In December 1928, Born wrote to Einstein and to the university Kurator: he had overtaxed his system. He needed a three-week leave because of "sleeplessness and other nervous symptoms." A few weeks later, Norbert Wiener, Born's former MIT collaborator, received a postcard thanking him for his New Year greetings. "I am here in Konstanz in a sanatorium because it has not gone well for me," Born wrote. "I have overstressed my nerves."

Born's breakdown had been building for more than a year. With difficulties at home, a few weeks rest was not going to cure him.

The winter of 1928–1929 was so severe that from Born's sanatorium on the German side of Lake Constance, one could walk across the frozen surface to Switzerland. His plan to sit outside and restore his health through an open-air cure was impossible. For the first few days, he stayed alone in his room. As he tried to rest, he kept turning over worries in his head: Would he ever be able to work again? What would happen to Hedi and the children?

When Max felt stronger, he joined the other patients in the lounge; it was to be a short visit. The main topic of conversation among the middle-class clientele—

"greasy businessmen, manufacturers, military officers," he later described them—was the problem caused by the Jews and the promise that Adolf Hitler offered Germany. The breadth of support for Hitler shocked Born. He retreated to his room, the benefits from his first days of rest completely lost. The only pleasure during his three weeks at the sanatorium was a visit from Irene, who came over from her boarding school at nearby Salem. At the end of the stay, he asked the university for an extension. His nerves still bad, he was not yet ready to face academic life in Göttingen.

Born traveled to Königsfeld in the Black Forest and continued his convalescence at the aptly named Pension Waldsruhe (Forest Calm). Walking through the old, picturesque town after a day of skiing in the forest, he heard a beautiful Bach fugue flowing out from the church. Captivated, he went in. As he sat alone in the eighteenth-century nave, listening to the organ music, oblivious to the cold, he wondered who the mysterious and masterful organist was; a letter from Hedi provided the answer. Albert Schweitzer, the renowned physician, was in Königsfeld to visit his wife, who had tuberculosis. Born returned to the church regularly, hoping for another musical encounter. One day, hearing Bach again, he climbed into the loft and found Schweitzer sitting before the organ. Schweitzer welcomed him and, recognizing his name, invited Born to his home. For the next several days, the two men walked in the woods and talked about religion, art, and politics. The discussions and company helped Born regain enough equilibrium to return to Göttingen at the end of February.

Born's assistants, Léon Rosenfeld and Walter Heitler, continued to lecture for him. Still not strong, he only took part in seminars. The rest of the time he stayed home, working on two books that he had begun some years earlier, one on quantum mechanics and another on optics. He thought this would be less stressful than research and provide extra income. Students who came to Göttingen to work with Born were disappointed in his near absence, but Franck, Courant, distinguished visitors, and a host of talented assistants kept the atmosphere lively. Delbrück's "whirlpool" of past years had quieted down—the quantum revolution had taken hold—but Göttingen still held the promise of discovery.

Born wrote to Bohr that he was better but still mentally and emotionally exhausted. He knew that it would be some time before he could work again. In truth, he was not much better at all. The new IEB representative W. E. Tisdale visited in mid May, and as the two of them chatted over coffee at Born's home, Tisdale witnessed "a nervous wreck," with Born complaining about his inability to work, his lack of money, and his sleeplessness—basically, "despairing of his future." Born brought up a wish to go to Caltech for a semester, but Tisdale thought that Bohr's plan to visit there made that idea improbable. One noted quantum physicist in Pasadena seemed to be enough at a time. Two weeks later,

Born wrote to the Kurator saying that his health still suffered, and he wanted a week's leave to visit Henry Goldman, who was vacationing in Baden.

Gossip about Hedi and Herglotz was swirling around town. When Ehrenfest wrote, inquiring about Göttingen news, Hedi responded that she was "aware of being news herself." Much of Göttingen's science community knew of "her friendship with Herglotz," and although she professed indifference to the gossip, she pulled back from socializing.

That July, when Ehrenfest came to Göttingen to lecture, as he did most years, he found himself entwined in a peculiar set of entanglements: Herglotz was one of his best friends, and Born was a close colleague. In the past, Ehrenfest had often arrived with an entourage—his wife, his daughter, an assistant or two, and the Ceylonese parrot whom he had taught to say, "But, gentlemen, that is not physics." On this visit, however, he came alone. He stayed with Herglotz and joined him and Hedi for supper, the three talking into the evening. Hedi still lived with the family, but the Born house was just a few minutes' walk from Herglotz's on Hoherweg.

Their conversation revolved mainly around Hedi and Gusti, who after a year of seeing each other were trying to sort out the jealousies and doubts inherent in love affairs. Before leaving, Ehrenfest gave Hedi some advice: to "create clear external conditions."

Hedi took the advice to heart. Shortly after Ehrenfest's departure, she told Max that she wanted to separate and marry Herglotz. Max would not hear of it. He felt he would endure anything rather than let his children experience the loss of either parent.

Max asked the university for a two-month leave and in the first week of August took off with the children to Waldeck, a rustic village to the southwest. Hedi stayed at home, enjoying Herglotz's company. In a lighthearted letter to Ehrenfest, she put the return address as *Am Hohen Luft*—on high air, a play on the name of Herglotz's street. By the twelfth of the month, Max and the children were back home; Waldeck's peaceful rolling hills and Eder Lake had offered no solution; so he left again and traveled alone to a hotel in a Swiss alpine village, but solace eluded him there as well.

The bright, lively, indulged twenty-year-old Hedi Ehrenberg that Max had married in 1913 had evolved into a very complex person by 1929. She was "always fighting between her altruistic side and her egomaniacal one," Irene later said of her. Which would win was unclear—the volunteer who helped the poor and wanted a simple life, or the grande dame who flitted about and pampered herself? One week she would talk about retreating to a primitive hut in the mountains with a dog. The next would find her at an expensive sanatorium in order to calm her nerves, the simple life all but forgotten. She did not recognize the contradictions and labeled others' reactions to her high-handedness as "oversensitive." Hedi's relationship with her children reflected her self-indulgence.

A friend of the Borns once remarked that what the children lacked was "nest warmth" from their mother. Elizabeth and Gretchen Heller, sisters hired to care for them, compensated for some of this. The sisters ate supper with them, helped with homework, and generally supervised. Sometimes, the children went to the Heller house, only a few blocks away. With five children, Frau and Herr Heller, and his aromatic pipe smoke, it was the three Borns' special home. In the corner Frau Heller had filled a cabinet with toys just for them.

Of course, Hedi felt she gave them warmth. "We are firmly attached to one another (freely!)," she reported to Ehrenfest, explaining how wonderful Gustav was, how she wanted Gritli to confide in her more, and how Irene was like a best friend. But Hedi, who considered Irene difficult, sent her away to boarding school when she was twelve years old so that she did not have to deal with her. Fortunately for Irene, she liked Salem, whose founder and Max's old friend, Kurt Hahn, challenged students to discover themselves. In the old converted castle near Lake Constance, she was out from under Hedi's control. When Irene received her mother's letters—which were usually highly emotional—she simply placed them in a drawer, unopened.

Gritli stayed at home and went to a nearby girl's Gymnasium. She was more shy and insecure than Irene, and Hedi, being able to intimidate her, found their relationship less difficult. At age seventeen, Gritli asked to go to Salem for her last two years of school.

It was different with Gustav. Hedi adored her sweet little boy and kept him close to her.

Max overlooked Hedi's incongruities and any problems she had with the girls. He needed her and depended on her. Having grown up without a mother, he could not bear another loss. His colleagues did not appreciate this and among themselves frequently sniped at Hedi, Einstein included. Some years later, Theodore von Kármán lamented to him that "Mrs. Born does not like me." (In fact, she had always held von Kármán responsible for careless errors that had detracted from the first paper on which he and Born had collaborated.) Einstein looked at him, paused for a moment, and said, "How do you think I feel? She likes me."

Hedi's vitality and self-assurance had been charming when the Borns and Einsteins were in Berlin but became less appealing in maturity.

After Max returned from his unsettled stay in Switzerland, the Borns wrapped themselves in silence. None of their turmoil escaped to the outside world. Hedi did not even write to Ehrenfest.

At the end of September, Hedi went off to a sanatorium to rest. She spent several weeks "lying in simple vegetative loneliness" at Braunlage, a health resort east of Göttingen in the Harz Mountains that became a frequent retreat for her in the ensuing years. She had ended her relationship with Herglotz. She never explained her decision but it was not one that gave her a sense of pleasure or even relief. She continued to question whether she had made the right choice.

In her first letter to Ehrenfest, which was, for her, unusually short, she described her state.

> I am now in the Harz at the sanatorium to learn again that life is worth living. I live completely by myself, which is good for me. And I can not tell you anything about the last couple of months. My life ended up in the absurd. You can't be at one place and be a whole and at the other place just a half. One is only 1 and not $1^1/2$. . . . I write to you with little joy.

By the end of her stay, after having battled with her emotions, she began to feel that she could reach a place of peace and even thought about working—writing, learning stenography, or perhaps refreshing her language skills. While waiting for that to happen, she wished for a desert or cloister where she could temporarily disappear.

Herglotz belonged to the past, but Max did not belong to a harmonious future, at least from Hedi's perspective. Max, she complained, reduced her to a formula and did not understand her or what forced her "to live a more and more lonely life." She did not see how he could be happy and act as though everything was alright while she was miserable.

Max was not happy but, as he himself later admitted, he could not show his deeper feelings. With little intuitive insight into Hedi, he tried gently wooing her with tender love notes. Wanting to be understood, not appeased, she was furious at not getting more from him. After referring to "my husband" in a letter to Ehrenfest, she exclaimed, "I hate that description—even in the best of marriages, it is something almost frivolously challenging: *my—my*—from a human being. So I'm going to say Born or 'B' as the women did in Goethe's time." She wanted to buy a dog and name it Anger.

A somewhat undefined part of this situation was Herglotz, who was still a part of the community and science faculty. He lived close by, shared many of the same friends, and participated in university events. He and Max reached some kind of understanding that was never described. As time went on, Hedi wrote in her diary that Max had gone to Herglotz's house; or when writing to

Hedi, Max occasionally made a remark about a card from Herglotz or doing a favor for him. There seemed to be a recognition and acceptance of what had been. Herglotz returned to the category of "friend."

This sense of normalcy did not transfer to the Borns themselves. Hedi converted two rooms in the attic into a hideaway and largely isolated herself from the family. Max turned to finishing a major book on quantum mechanics.

The book—a full exposition of quantum theory using matrix mechanics—had been in the works for three years, its completion crippled by his illness and coauthor Pascual Jordan's move to Hamburg. Frustrated and impatient with Jordan's lack of productivity, he just wanted to finish it and start a book on wave mechanics. By the beginning of 1930, he had completed the book.

Born was still on a reduced teaching schedule, but Tisdale, who was in Göttingen checking on Fellows, reported back to the IEB that he was much better. Born had even begun to accept lecture requests again. One of the first was an invitation to Paris, and he and Hedi planned a trip there in March, followed by a vacation in Italy.

Tisdale's other piece of news in his January report was the marriage of Franck's American assistant, Joe Mayer, to Maria Göppert, "Born's best pupil." Maria was beautiful, charming, and brilliant. Joe was a well-to-do, somewhat brash American. Nicknamed the "young kittens" by the faculty, they had gotten to know each other when Joe rented a room from Frau Göppert one year earlier.

Maria's father, who recently had died, had been a pediatrician in the university clinic, and Maria, an only child, had grown up in the company of the Borns, Courants, and Francks. When Maria started her university studies, she planned to be a mathematician, but Born spotted her one day at the university and invited her to sit in on his seminar. It was the beginning of Maria's physics career—which culminated with a Nobel Prize—and the beginning of a close personal relationship between the two. Not only did Maria join in Born family activities, such as swimming at the Göttingen pool or skiing in the Harz, she and Max were often seen bicycling off together. To Victor Weisskopf, who became an advanced physics student in Göttingen in 1928 and was himself smitten by Maria's charm, it was not an ordinary professor-student relationship.

Hints to the nature of their relationship in Göttingen are few. Later, when they had gone their separate ways, Max's letters to her opened with "Dearest Maria" and closed with "in old love." From the beginning of their correspondence, they used *Du*, the only student with whom Born was ever so casual. He confided in her about his marriage. He wistfully told her in one time of distress, "What I wish is a long, close get-together with you." That she was an essential support for Born during the troubled days that began in the fall of 1928 is certain. Just as Hedi had Ehrenfest as her confidante, Born had Maria, and she was an excellent choice. Although young, she was sympathetic and loyal. Later, when comparing the lives of Born and Franck in Göttingen, Maria said of Born, "he was not a happy man. . . . He had a much more difficult life." And yet, she professed not to know the source of the problems.

At the time of the engagement, Hedi described Maria's balanced, unruffled nature to Ehrenfest and gave no hint of suspecting anything or of being jealous. "The dear God gave [her] the gift to be neither sorry for herself nor ecstatically happy. The kitten licks herself clean with grace."

Born and Jordan's long-awaited *Elementare Quantenmechanik* came out at the beginning of 1930 and immediately received a scathing review by Wolfgang Pauli in *Naturwissenschaften.* It was pure Pauli—to the point and acerbic. The book's introduction defined *Elementare* as making "use of elementary means, and elementary means are purely algebraic; the use of differential equations for example will be avoided as consistently as possible." Pauli attacked this approach, criticizing the one-sided picture given to a small circle of readers (of which he said he was one) and the limitations imposed on the theory by restriction to the matrix method. The book did not even give Schrödinger's wave equation, he said. How, he asked, could the book present elementary quantum mechanics when

> many results of quantum theory cannot be derived at all with the elementary methods defined above, while the others can be derived only by inconvenient and indirect methods? . . . The restriction to algebraic methods also often inhibits insight into the range and the inner logic of the theory.

His scathing conclusion: "The layout of the book is excellent with regard to printing and paper."

Unusual for Born, it seems he completely misread the needs and interests of the physics community. He acknowledged that the criticism had merit, and to his former assistant Léon Rosenfeld, who had helped with the book, he made light of Pauli's "abuse": "I make nothing of it. In any case, I learned a lot from [writing] it." But in a letter to Fritz London thanking him for a more balanced review, Born made Pauli a central topic: "Naturally the book is, as you rightly remark, 'a torso', but that actually has external reasons, first my illness and second Jordan's exit from Göttingen. Pauli knew that exactly and it is not very friendly of him to write in such a tone."

The book's introduction states that Born planned to write a companion volume on Schrödinger's wave theory. His comment to London suggests that he made this decision partly in order to limit it to the work done with Jordan. In any case, he never followed up. His mathematician friend Hermann Weyl wrote it before he could do so. "I've just gotten Weyl's book," Born told Rosenfeld. "And I'm quite discouraged because I see that in forty pages he has put so beautifully all the things I was going to put in a book of two hundred pages." Quantum physics, which had given Born such a sense of achievement, took a toll on his confidence when he could least afford it.

Born had written Rosenfeld from Paris, where he and Hedi were alone together for the first time in almost two years. It was the initial stop on a five-week trip, and they delighted in seeing the sights—the Casino de Paris, Mistinguett, "Queen of the Paris Music Hall," the Louvre, and the theater. While Max gave six lectures at the Institut Poincaré, Hedi toured more sights. In minute and cramped handwriting, she recorded the highlights in her new toy, a tiny "Bobby-Kalender," $2^1/2 \times 2$ inches, the first of her day calendars that eventually contained decades' worth of scribbled thoughts and impressions.

After eleven days in Paris, they took a train to Avignon, where the Mistral—the cold wind from the northwest—quickly pushed them on to Corsica. They backpacked around an unspoiled landscape that was very nearly a paradise; and for the last five days of the trip, "drowned" in Catholicism in Florence—three services in three different churches, and "above all old music!" For Hedi the trip allowed her to "build a wall and put it in front of the past"—her continuing efforts to distance herself from memories of Herglotz.

Once home, Hedi started work on a play. Delving into writing, philosophy, and literature was partly soul searching, but she kept her work hidden from friends in Göttingen. Feeling disapproved of by those such as Ingrid Franck, she maintained a distance, as she continued to do with Max, in spite of their trip.

Max's enthusiasm for work, which had reawakened a bit before going to Paris, was gone. Whether from disappointment over Hedi's persistent remoteness or the failure of *Elementare Quantenmechanik* or both, he was academically fallow. Instead of the six or seven articles he had published in most years since 1909, 1930 saw only two: a one-page note and an article coauthored with Franck. At the beginning of June, he admitted to Hermann Weyl, Hilbert's successor and an old friend, that he could no longer keep up with new research and was merely satisfied to understand the most important questions and answers. Lacking the self-confidence to be more than simply a specialist, Born thought his future lay in deciphering important questions for young people. He hoped that he could lean on Weyl "a little" until his strength returned. A year and a half had passed since Born had gone away to improve his nerves.

At the end of summer 1930, the Born family took off for Ehrwald in the Tirol for their annual vacation. Regardless of their financial situation, these vacations—usually in awe-inspiring locales with inexpensive, rustic accommodations—were a must. For six weeks, their home was a primitive one-room apartment over the town bakery. The heat from the bakery below warmed their room during a ten-day stretch of rain and cold, and the dough machine woke them at 4:30 in the morning, but they had a good time with long hikes, bicycle trips, and overnights in mountain huts.

After the children returned to Göttingen, Max and Hedi met the Ladenburgs in Munich. (Hedi saw Herglotz on the street there, but they did not speak.) The foursome saw Bertolt Brecht and Kurt Weil's *The Threepenny Opera*, which had premiered in Berlin in 1928 and was still all the rage. At a time when Berlin's cabarets were flamboyant and decadent, the opera showed an entertaining but seedier, immoral side of life—prostitutes, murderers, thieves, and beggars—juxtaposed to the self-satisfied bourgeoisie. The anticapitalist satire stratified society into two levels and showed the power of capitalism manipulating both; in fact, a different kind of power was infiltrating German society.

While the Borns had vacationed in the serenity of the Austrian Alps, Germany's cities had echoed with the sound of heels clicking in syncopation with *heil Hitler* and an outstretched arm. With national elections scheduled for September, the floundering economy provided an audience for hate groups like the Nazis. Right-wing street violence against Jews and Communists was commonplace.

At the start of 1930, the Nazis had only twelve representatives in the Reichstag, less than 3 percent of the seats. With a strong economy in the mid 1920s, they had remained an impotent parliamentary postscript; but by 1930, more than 3 million Germans were unemployed. Adding further pressure to the tenuous political situation was the Young Plan, a new international agreement that instituted a less onerous reparations schedule for Germany but retained the clause that maintained German guilt for the Great War—a clause hated by the German populace.

The result was that the National Socialists, who had garnered 800,000 votes in 1928, drew 6.5 million in the 1930 election. These votes translated into 107 seats in parliament, or 18 percent, second only to the Social Democrats at 143. The Communists were third with 77. No consensus among the three was possible. The political structure was splintered, and the social structure appeared to be in similar danger.

Writing to Ehrenfest, who was lecturing in America, Hedi railed against Göttingen's part in the election results.

"In Göttingen, 10,000 Nationalists! That's significant: It is every fourth one or really every second *adult*! . . . The Nazis have a program: Get rid of all Jewish professors—no Jewish doctors. Basis: missing empathy in the German people's soul!!!"

The documented percentage was 38 rather than Hedi's 50, but the spirit of her concern was justified. Compared to the national average of 20 percent, Göttingen was filled with Nazis or at least their sympathizers. Gritli came face to face with the incipient rancor. In the schoolyard, her fourteen-year-old National Socialist schoolmates sang, "We are walking over corpses."

Equally distressed by the "wicked" conditions in Germany, Born told Ehrenfest that "the future looks dark. Perhaps we will pack off to America one day." If Hitler were to take power, Born said he wanted Ehrenfest to get them out.

Following the election, things were quiet for Born. He was gradually becoming more engaged in work and running the institute. He gave a series of lectures in Berlin and kept up scientific correspondence; but in a letter to Harvard professor Percy Bridgman about some ideas on lattice theory, he said his health kept him from undertaking the calculations. He had improved, but still not enough to do real research. Instead, he worked on a book. Having published *Elementare Quantenmechanik*—hardly a stress-free endeavor—he now struggled with one on optics. No modern textbook in Germany presented a complete electromagnetic theory of light, and he wanted to fill that gap.

Born had never done research in the field, but optics had interested him since his early Göttingen days when he took a course with Professor Woldemar Voigt. His attitude toward Voigt's course had initially echoed the verse carved into Born's desk by an earlier sufferer.

> *Twelve o'clock will soon be striking, then I can have a breather,*
> *Settle thine accounts with Heaven, Voigt,*
> *For thy time has come.*

Through the intercession of fellow classmate Max von Laue, however, who dragged Born to the lectures, he had finally realized their merit. He had used these class notes as the basis for his own lectures, adding new research findings to them.

Born's fundamental understanding of physics was so broad that his lack of direct experience in the field was no drawback. He drafted two students to transcribe his dictation as well as Victor Weisskopf to write a chapter, which the perfectionist Born completely rewrote. Gently, Born was starting to come back to research.

The relationship between Hedi and Max continued to be distant. Hedi spent most of her time in the attic rooms, where she had a bed, writing table, new electric gramophone, and her mementos. One was a picture of her forty-three-year-old cousin Franz Rosenzweig, the philosopher-theologian who had recently died of the horrific paralytic disease amyotrophic lateral sclerosis. She kept it, she told Ehrenfest, to remind herself of "somebody who voluntarily endured his pain out of an inner faith." Her comment seemed to suggest that she felt she was doing the same.

Born left her notes when she was tucked away upstairs. One December night around 11 o'clock, he wrote,

> I have to send you a greeting because I must leave the house early tomorrow. I want you to know that I think of you a lot when I'm alone in the evening. It is beautiful to think of you. I see your noble face—that is so beautiful if only it didn't look so sad.

Two days after Christmas, Born traveled alone to Arosa, Switzerland, to celebrate the New Year with Carl Still, his institute benefactor. Then he went to lecture in Zurich and met Einstein's son, Hans Albert, arriving back home in early January. Born and Einstein had maintained their friendship in spite of Einstein's frequent trips abroad. Born wrote to Einstein, who was in the United States, saying that the son had the same wonderful laugh as the father.

Born began 1931 preoccupied with his impending one-year term as Dean of the Mathematics-Natural Science Faculty, a position that rotated among the professors. For ten years, he had maneuvered to escape the responsibility of running the faculty meetings and dealing with bureaucratic duties. His duties would not begin until September, but he was so anxious about the additional stress that by February he was already trying to organize a semester's leave at its conclusion (which was still more than one and a half years away). He pursued an offer that Robert Millikan at Caltech had made a few years earlier, but found that with the Depression in America destroying financial support for guest lecturers, he could not get an invitation.

Hedi was not specifically part of this anxiety. She had left Göttingen for a few months. Carl Still had been impressed with her household accounting system and asked her to look over the accounts for his estate at Rogätz on the Elbe. He suspected that his farm managers were doctoring the books. The offer fit perfectly with Hedi's wish to get away and stand on her own. "Merely being a housewife," she told Ehrenfest, "is more senseless the less it is based on 'being a wife.'"

Rogätz, about eighty miles northeast of Göttingen, was a mélange of a castle renovated by the Stills, a large farm with some 17,000 animals, and an old medieval tower built for knights. From her balcony, Hedi could see livestock lying around and tugboats moving slowly on the Elbe. In the daytime, a choir of nightingales entertained her; in the night, an orchestra of frogs. For several weeks, she rose at 4 or 5 A.M. and joined the farm workers in the courtyard, observing their daily routine, chatting, and looking through the accounts. Her diligence and intuitive head for business were rewarded. She uncovered the embezzlement scheme suspected by Still: On several occasions, the managers had recorded the purchase price of bulls as four times higher than the actual price, then skimmed the difference for themselves. Hedi was proud of her success. Her "quiet hope" was to make some kind of regular job out of it.

After returning from the farm in June, Hedi went off with the children to a small apartment in Selva Gardena, a village 5,100 feet up in the South Tirolean Alps. Max stayed in Göttingen. He had new company: Heisenberg had recently sent him two of his protégés, student Carl Friedrich von Weizsäcker and new Ph.D. Edward Teller. Teller's actual assistantship was with a professor in physical chemistry, Arnold Eucken, but Born persuaded him to write a chapter for his optics book.

Like Wigner and von Neumann, Teller was a Hungarian Jew who had come to Germany to escape a repressive, anti-Semitic university system. He stayed in Göttingen for the next two years. Von Weizsäcker came from an old Prussian family, his father being a senior member in the foreign office. Born's book on relativity had fascinated von Weizsäcker when he was a Gymnasium student, but he now disliked what he called Born's formalistic approach. After one semester, he went back to Leipzig and Heisenberg.

Two old friends kept Born company as well. Maria Göppert-Mayer and husband Joe traveled from America and spent the summer with her mother. Joe worked with Born on crystals and had some success investigating lattice energy and compressibility, research that harkened back to Born's work with Landé during the war.

Born still enjoyed a close relationship with Maria. When he went away on vacation, he wrote to Joe about the crystal research and added a postscript to Maria: "The next time I will write to you very personally." She was still a crutch; but their relationship respected their marriages. When she returned the next spring to write an updated overview of lattice dynamics with Born, his very genuine letter to Joe apologized for enticing Maria away and urged Joe to follow soon.

Hedi delighted in having four weeks with the children without Max. Every day they picked edelweiss in the meadows, explored villages, or hiked up into the Dolomites to rocky passes—"stone oceans," she called them. Carried away by a sense of closeness, Hedi decided to tell Irene and Gritli about her life. She confided to the two, now seventeen and fifteen, that she had married their father because he was desperate to be married and he was suitable. She had not really loved him. Hedi, lost in her expansiveness, felt that the growing inner confidence and friendship with the girls made it possible to share this without their loving her less. Irene, at least, felt burdened.

Max arrived at the beginning of August and had two weeks of hiking and relaxing with the children before they left for home. He and Hedi stayed in Selva for another two weeks, until they went south to Bolzano for five days. Hedi scanned through the foreign registry list and discovered that Herglotz was there too, but she did not see him. Then she and Max went on to Munich for a few days of museums, theater, and an evening with Fritz Haber before returning to Göttingen.

On the day he returned, Born began his tenure as dean of the Mathematics-Natural Science Faculty. His immediate departure for the Faraday Celebration in London did not give any opportunity for problems to arise, but his earlier apprehensions were well placed. A split in the faculty and student body, driven by anti-Semitism, was growing.

In 1928, the Circle of Friends and Supporters of the German Inquiry Office—the name adopted by the students who had searched through library records for the religious affiliation of professors—had published its first book, *The Jewish Influence in German Universities*, Vol. 1, *University of Göttingen*. Born was one of sixty-two listed in seven categories ranging from "Mosaic" to "Mosaic born and later baptized" to "Jewish influenced." The Circle quietly circulated the volume throughout the university, using it as propaganda for indoctrinating students. With unemployment in the millions, the environment was perfect for the simple question, Where are you going to get a job? The Circle now had the means to show that Jews and converts, who were about 1 percent of the total population, constituted 19 percent of lecturers and 7 percent of professors. A majority of the students were ripe for the argument that barring Jews from being professors, doctors, and lawyers would improve their own opportunities.

The effort was sufficiently successful that in January 1930, a committee of the mathematics-natural science faculty felt compelled to call on the university rector to "reject all attempts to bring political matters into university life. . . . Students with pushy political views or with political uniforms should be forbidden to have access to university premises." The faculty maintained that it did not want political views distorting scientific work and affecting the image of the university. The rector agreed, but to no avail; students persistently agitated. By 1931, Nazi students attained a majority in the student congress. What had been a cohesive unit was disintegrating. Communists placed leaflets in the physics library, and National Socialists spread anti-Semitic rumors in the mathematics institute.

As dean, Born had to monitor the students' activities. Two months into his term, they planned a torch march to celebrate the Reich's fiftieth anniversary. To guard against inflammatory political rhetoric, he asked the rector to approve all speeches by students. The celebration occurred without mishap, but there was a constant, simmering tension.

Confrontations within the faculty were more problematic. The Ministry of Education's austerity reforms called for a 50 percent decrease in the budget of the university's six agriculture institutes (which had seen a large decline in students), 5 percent decreases in other departments, and a dismissal of some nonpermanent employees, including four faculty assistants. Born scheduled these issues for discussion at the December 5 faculty meeting, and twenty-five professors showed up, rather than the usual eight or nine. When he asked each department to hand in a list of the effects of the budget cuts, emotions flared—the professors in the agricultural institutes were particularly perturbed—and Born had to table the item and adjourn the meeting.

During the next week, tensions subsided enough to celebrate the arrival of Lord Rutherford from the Cavendish Laboratory in Cambridge. At five o'clock in the large lecture hall in the physics institute, Born addressed the standing-room-only crowd of 700 (300 more than the seating capacity) and presented Rutherford with an honorary degree. The introductory remarks by both men acknowledged the wisdom of King George II of England, Elector of Hanover, in founding the Göttingen Society of Science 200 years earlier. Rutherford, winner of the 1908 Nobel Prize in Chemistry for his work on radioactivity, then lectured on radioactive decay and gamma-ray spectra.

That evening, which ended with a small dinner party at Hotel zur Krone, was one of the few diversions in the year's tumultuous end. Born's only other pleasure was the musical quartet he had finally managed to organize. His relationship with Hedi was still unsettled. Just after Christmas she wrote to Ehrenfest, asking "Did I miss the right path at the crucial point 2¹/2 years ago?" Herglotz was still in the back of her mind.

The 1932 New Year began with a faculty meeting to resolve the impact of the budget cuts. During the holidays, Born had conferred with trusted colleagues such as Franck and Courant and developed a plan to save the four assistant positions. He calculated that a voluntary contribution of 1 percent from professors' salaries would finance them.

Born also sent the agricultural institutes a resolution expressing feelings of close association and regret. The words were little reassurance, however, for those who worried that economic forces were shifting financial support toward industry and away from agriculture, ultimately diminishing their importance and influence. Perhaps it was because Born himself was so dependent on assistants (although their loss was not specific to him) or because historically, the agriculturalists had opposed appointing Jews and increasing the number of science chairs. Whatever the reason, he was oblivious to their fears.

Born began the faculty meeting by presenting his proposal for the voluntary contribution from professors. The agriculturalists immediately reacted with fury. They had also worked over the holidays and had come up with their own budgetary plan. They estimated that spreading out their 50 percent budget decrease across all mathematical-science institutes led to only an additional 2.5

percent decrease for each one. The meeting maintained a hostile tone until twenty-three faculty members—none from the agricultural faculty—voted to approve Born's proposal. It passed, while that of the agriculturalists was rejected. A compromise allowed the dissenting group to receive a share of the funds that were collected from professors' salaries and a concert by Nina Courant. The gesture did not appease the professors in the agricultural institutes. Shortly after, Born expressed a fraction of his frustration to Fritz London. "I wish sometimes that I were young and free like you and could work peacefully in Rome instead of attending to the Dean's business."

Hedi, who had little sympathy for Max's plight, was involved in her own "contest of wills" with Irene over the holidays. The scenes were sufficiently unpleasant for Hedi to want to escape to "a log cabin in the midst of farmers and cows" and listen to the crunch of the bovines walking through dewy grass early in the morning. She wanted to be there with Herglotz, "still the one and only whom I miss."

Just a few weeks later, an event occurred that made Hedi begin to change her attitude. She walked past a house called Loufried a few blocks away and saw an older woman in a fur cape standing by her garden gate. Hedi was strongly struck by her; something about this woman, perhaps her radiant eyes, made Hedi want to meet her. She knew the woman's name, Lou Andreas-Salomé, and also knew that she lived in isolation and practiced psychoanalysis. Hedi went to a bookstore and bought Salomé's books on Rainer-Maria Rilke and Friedrich Nietzsche. A few days later, Hedi wrote her an introductory letter describing her feelings of isolation and interest in knowing more about psychoanalysis. Lou invited Hedi to visit her.

By the age of seventy, Lou Andreas-Salomé had lived a life that was part fantasy, part scandal, and part mystery. She now dressed in gypsylike clothing and wore dozens of long, ropey chains that cascaded from her neck over an ample bosom; but when young she had been a seductress. Beautiful, intelligent, charming, the daughter of a Russian general, she had left home at age seventeen and by age twenty had bewitched Nietzsche, who wanted to marry her. Years later, the same fate befell Rilke. She rejected both men—Nietzsche after a kiss, Rilke after a three-year passionate affair. By the time of her affair with Rilke, she had been married to her husband, Friedrich Carl Andreas (a professor of West Asian languages) for ten years, a marriage she said she never consummated. In 1911, she met Sigmund Freud, became part of his psychoanalytic circle in Vienna, and, as she told Hedi, had an affair with him. All the while, she wrote novels as well as accounts of her relationships with Nietzsche and Rilke.

Hedi was enthralled. In their first conversation, Lou talked about the pleasures of aging, of being a woman, and of having lovers. Hedi perceived in her an inner sense of independence and security, "a wonderful, shameless openness that calls everything by its name! That 'confidence in life'—quite similar to Einstein's."

The day after the visit, Hedi wrote an adulatory letter to Lou in which she described feelings of renewed freshness and security. Ehrenfest was the only person Hedi told about this new relationship, and she swore him to secrecy. One of the tidbits she passed along to him was Lou's quote, "I am a completely immoral person, I do not know erotic faithfulness."

Hedi savored Lou's anecdotes. The first she recorded in her daybook was,

> In a sanatorium in Tegel when she was with Freud, there was gossip that every spring and fall, she traveled with "friends," usually a new one. When one "well-meaning" person told her about it with indignation, she said, "No, no, that is not true at all. I offer myself at all times of the year."

At the end of February, Max started traveling back and forth to the Technische Hochschule (Technical College) in Berlin to give lectures on matter and radiation. The first of them, at least, was a grand affair with an audience of 1,000 people. While on the train to give his fifth lecture, he felt the right side of his face grow stiff. Suddenly, he could not speak. When he arrived in Berlin, he gave the lecture with the assistance of loudspeakers and returned by night train to Göttingen.

The doctors first diagnosed rheumatism but finally determined that it was Bell's palsy, a paralysis of the facial nerve. Three weeks later, he felt no better: He suffered from a stiff face and a bad headache, and the pain that kept him from sleeping at night also kept him from working during the day. When he did lecture, he had to rest afterward.

Hedi and Gustav went to Karlsruhe to visit her brother Kurt, a trip that coincided with May 8—the nineteenth anniversary of Hedi and Max's engagement. He always proclaimed loving thoughts and wishes for her happiness, but this was a special day that enkindled more.

> I praise the day nineteen years ago and, in spite of all difficulties and clashes, have become a lucky man through you. . . . I love you with a really deep, strong passion and, without you, am no complete man. You first made me a person. But it is not so much gratitude that I feel for you but a real enchantment that one calls only love.

He acknowledged Hedi's distance and spoke of their mutual agreement that he was free to be with other women. It seems that they had come to this arrangement after Hedi had withdrawn to her attic rooms. He explained that he usually felt uncomfortable with other women, although he seemed pleased to tell her they were attracted to him and tried to entice him, contrary to when he was young. He hinted that on occasion they may have succeeded; but what he really wanted was "a joint life, brewed together out of the most contradictory elements and yet harmonious through self-discipline and good intentions to show understanding for the beloved other."

As Max continued to have eating and speaking difficulties, Hedi softened, becoming more supportive and sympathetic than she had been in years. She wrote worriedly to Ehrenfest,

> You can imagine how much the thought disturbs me that his face could remain distorted. One, because such a wonderful head would be destroyed—and then because it would mean for him the loss of so much which I want for him and which he deserves—a loss for which he is much too young. But since yesterday's consultation my hopes for a recovery are rising again.

She may or may not have recognized her change in attitude toward her husband, but she now saw herself as "slowly climbing out of the dark valley of life and becoming healthier." She thanked Lou for this—"Since Einstein in Berlin during the war, no other person has had this liberating effect on me."

At the end of May, the family celebrated Irene's eighteenth birthday with a dance and ninepins bowling. For the rest of the summer they stayed in Göttingen with no plans for the usual monthlong outing in the Alps. Hedi visited Lou once or twice a week, two nephews came to visit, and the children were in and out of the house. Max busied himself with helping students with their research. He had improved, but his face, now with a lopsided look, was partially paralyzed. The end of both the term and his tenure as dean gave him the opportunity to go alone for treatment at an unusual resort in Bavaria.

Castle Elmau near Garmisch was a huge "old-fashioned" *Schloss* built sixteen years earlier by Dr. Johannes Müller, a minister who was as unconventional in his Christian faith as the castle's setting in the Wetterstein Mountains was breathtaking. Müller believed that one experienced Christ through individuality and creativity, a foundation for a Christian philosophy that, he preached, supported the healing of body and mind.

Elmau was a rather curious choice for the nonreligious Born, but he settled in quite contentedly. He took pleasure in the quiet of his room, the peace at the edge of the woods where he sat and read, the swimming, hiking, and evening performances of *Hansel and Gretel*, *Marriage de Figaro*, and Brahms sonatas provided by visiting artists. The windows in his room framed a beautiful mountain view, magnificent when clouds balled up and thunderstorms rolled in with crackling and flashing. He received electrical stimulus for his facial paralysis, which improved, and on some evenings attended group discussions held by Dr. Müller. He began to feel refreshed.

Hedi, astonished that Max tolerated "the pious mystic," as she called Müller, teased him for sending such insights as "You should live cheerfully, and you should not worry unnecessarily." Max agreed that the philosophy was shallow—"as what life philosophy would not be?"—but he defended the minister's personal appeal.

To Ehrenfest, she wondered how "Born," as she still referred to him, would react if she sent him similar banal thoughts about physics. She hinted that another feature of *Schloss* Elmau was the real point of interest—"The mysticism includes in its teaching the need for pretty young female assistants, and that is why you can amuse yourself there without mysticism." The *Schloss* employed "helpers" rather than maids, who were to be treated with the same respect as the guests. For this privilege, they worked long hours and were paid little. Having discussed this arrangement with Hedi before his departure, Max reassured her that the "helpers" were not especially pretty—although they looked nice—and were always stressed from too much work.

Hedi's intuition was not far off. Max had fantasized that Elmau might hold "a little erotic adventure"—after all, they did have an understanding; but he found instead, as he admitted to her, an enriching friendship that was neither adventuresome nor erotic. His new friend was Frau Knopf, a beautiful young pianist from Jena, who had recently experienced a difficult divorce. They had dinner together and sometimes went for walks. As for anything else, she was serious, reserved, and old-fashioned, to Max's obvious disappointment. He wanted Hedi to meet her, and he had invited her to Göttingen, enticing her with their two grand pianos. Whether anyone else interested him is unclear, but four months later, he acknowledged to Hedi that he had had three "adventures"—place and companion left unspecified.

As Max relaxed under the spell of Elmau, Hedi was increasingly seduced by Lou's philosophy. Even without formal therapy, Hedi's conversations with Lou had awakened feelings about her relationship with her mother and comparisons with Irene. Lou wanted to analyze Hedi so that Hedi could then analyze Irene. Max asked Hedi to keep a healthy skepticism about the process; his cousin, Cläre Kauffmann, had been analyzed twenty years earlier, and Max had not been impressed with the outcome.

Outside of Elmau, the world was less and less peaceful. Max had heard the "disgraceful political results" of the July 1932 elections but tried not to think about the Nazis' claim to 230 seats in the Reichstag (37.3 percent). Half of the voters in Göttingen supported the Nazis, and they were now a majority on the town council. The general situation was volatile. Around the country, several hundred people had died in street fighting. When Max heard that a new institute at Princeton had offered Einstein a position, he could not blame his friend if he accepted; but Max did not question his own situation.

Hedi came to Elmau for a week—she was not particularly enamored—then they departed for another week in Bolzano, an auto tour around Lake Garda, a steamer trip to Malcesina, then north to Merano. For Max, it was a happy two weeks, a special moment of reunion, one he had waited years for; Hedi felt the same, but when she happened to catch sight of Herglotz in Bolzano, she noted it in her diary. She now thought of herself as living two lives—"a visible one and a hidden one. An active busy one—and a lazy dreamy one full of wishes." As Lou told her, "A life without conflict is really a poor life, not especially admirable."

Max went to see the girls at their boarding school. Hedi rested alone in Merano for another three weeks. When she left, she brought something home that she had wanted for some time—a dog, an eight-month-old Airedale named Trixi that she found in a pound.

Irene had been at Salem for six years and had thrived, her spirit strong and undiminished. She had only one more semester until graduation. Gritli, having been there for only one term, liked it. Life was Spartan. Rain, sunshine, or snow, the day started with an early run in the park, then a shower, and finally breakfast. During his visit, Max watched Gritli play volleyball, had tea with Irene, and met their friends. To him, they were "splendid and charming," and he was very proud.

One of Hedi's concerns that fall was Ehrenfest. He was depressed, as was often the case, but this time was worse. Though he ascribed it to feeling inadequate in his job, Hedi energetically counseled that he could not reduce his feelings to a simple formula. The job was only a symptom. She agreed with Freud's idea that neuroses developed in childhood and believed that the root of Ehrenfest's problems stemmed from his youth and "hinder all later experiences of happiness and unhappiness."

Hedi had not taken up Lou's offer of analysis, but they had developed a therapeutic relationship. Hedi went to talk with Lou at least once a week and was now writing down her dreams. She was changing. In a letter to Ehrenfest, she even forgot and referred to Max as "my husband"—for the first time in years.

December was always a special month in the Born household. Besides the holidays, it was both Max and Hedi's birthdays. This year it was even more special:

it was Max's fiftieth. He expressed displeasure at this midpoint, musing that he would rather be thirty without a celebration than fifty with one; but he realized that he was already older than his father when he died and that, with the world so lovely, he wanted to reach sixty.

The assistants from the physics institute planned a surprise party for him, but Frau Göppert, Maria's mother, accidentally let it slip. The celebration, held at the Borns' apartment, was, nonetheless, unforgettable—a unique mixture of family and leading scientists "as could only happen in Göttingen in 1932," as one guest remarked. The Göttingen Society for Comical Physics—Born's assistants—published a "special edition" of the *Göttinger Nachrichten* with contributions from physicist friends far and wide. Some appeared in person. Heisenberg came from Leipzig and entertained with selections from Bach and Beethoven. He also brought some kind of marzipan "tallies" that represented the uncertainty principle.

Edward Teller was Master of Ceremonies and left a memorable impression on eighteen-year-old Irene. Years later, she described him as "a personification of Mephisto—a dark swarthy face with black fathomless and somehow intensely luminous eyes under thick jutting eyebrows joined in the middle." In fact, he announced one participant, Walter Heitler, as the "Hell-born" who spouted comic verse. One bit of doggerel he patterned on the German version of "Mack the Knife" from *The Threepenny Opera*, and teased Born about his love of mathematics.

And the matrix has an index
and the index has a prime
and one reads and reads it again
still one cannot understand.

One guest, who later reminisced about the evening, described it as "a uniquely beautiful celebration, especially in contrast to what came later."

NINE

SEEING HOW EXPENDABLE YOU ARE

ON JANUARY 31, 1933, BLAZING TORCHES LIT UP GÖTTINGEN'S NIGHT SKY AS THE NAZIS' celebratory procession wound through the old town. The throng proudly wore the political uniforms formerly banned by the Weimar government and the university. The previous day, Adolf Hitler had been appointed Chancellor of the German Reich. From that followed an avalanche of manufactured crises, propaganda, and edicts. The first round erupted with the Reichstag fire on February 27, which Hitler blamed on the Communists, and ended the next day with an emergency decree suspending civil liberties—necessary, according to the Nazis, to protect the country against a communist takeover.

In the run-up to the general election that Hitler set for March 5, the Nazi propaganda machine took over, controlling every element, even down to stripping the posters of opposition parties from the cylindrical billboards on the sidewalks of Göttingen. On election day, the Nazis polled 44 percent of the national vote—not quite the majority that Hitler sought—but, by forming a coalition with the right-wing German National People's Party, which had 8 percent, he achieved a majority in the Reichstag. Three days later in Göttingen, citizens and brown-shirted youths celebrated in the market square with a ceremonial raising of the Nazi flag atop the Rathaus.

On March 23, the Reichstag met to vote on the Enabling Act, five simple clauses of two or three sentences each that gave Hitler dictatorial power to pass legislation and change the constitution. To guarantee passage, Hitler's brown-shirted storm troops, the SA, and his personal body guard, the SS, arrested all the elected Communist and Socialist Party representatives that they could find, then surrounded the voting chamber in a show of intimidation. Of the 538 representatives present inside, 441 voted for the Enabling Act, officially the "Law for the Removal of the Distress of People and Reich." Only 94 leftist Social Democrats—those who had not fled or been arrested—voted against it.

Born sensed impending doom. On a train from Göttingen, he and his former student, Friedrich Hund, passed through a village with Nazi flags flying high from flagpoles around the station. He remarked to Hund that he would leave Germany when the flags flew throughout the country.

The Nazi power grab and the swift vengeance against his friend Kurt Hahn certainly made this prospect more likely. A year earlier, five of Hitler's storm troopers had trampled a young man to death. When they were tried and sentenced to death, Hitler sent them a congratulatory letter. Hahn, the founder of Irene's boarding school, sent his own letter to Salem alumni urging them to separate from Hitler. Hahn did not stop there; he spoke out publicly against fascism just before Hitler rose to power. By the end of March, the Nazis had him in "protective custody," one of thousands of opponents of the regime who were snatched up and thrown into makeshift camps. A few months later, he got out somehow and made it to England.

The Borns, like most Jews in Germany, struggled nervously to carry on with their lives. Hedi contended with gall bladder problems and the flu. Max finished his book on optics. It was not a propitious moment for books by Jewish authors; *Optik*, having almost no sales, would have to wait for another day—another era, in fact—to be appreciated.

That was day-to-day life. Below the surface, the Borns worried—Max about the possibility of leaving, and Hedi about Gustav. She dreamt about his going to war and felt desperate enough to consider sending him to the United States. Max, Hedi, and Gustav took a three-week break at the health lodge Braunlage to try to escape from the daily stress of Göttingen. Irene came from Salem and joined them after passing the Arbitur. In the tradition of her father's education and in accord with his wishes, she finished the classical course in Latin and Greek—though she would have preferred to study mathematics.

The night the Borns returned to Göttingen, Hitler's storm troopers led a march through town, throwing rocks at the windows of Jewish-owned stores. Youths in the group chased and beat up any Jews they spotted. Threatening and intimidating incidents multiplied. A ringing phone woke up the Borns one night, and when Hedi answered, she heard violent shouting, "Out with the Jews, may Judah perish," followed by the Nazis' anthem—the Horst Wessel Song—"Millions, full of hope, look up at the Swastika; the day breaks for freedom and for bread." When their friend, Maria Stein, was walking to the Borns' one day, a group of about fifty boys wearing the Hitler youth uniform marched by her, singing "And when Jewish blood sprays from the knife, then all will be twice as good."

After the government started searching people's houses for communist literature, an American student in the mathematics institute got scared and hid a copy of Karl Marx's *Das Kapital* in a drawer under his shirts. The Borns went further. They burned their copy in the basement furnace, along with any other material they thought might endanger them. Afterward, their housekeeper, Fräulein Schütz, warned them to be more careful. She had heard gossip that neighbors had seen their activities in the basement. With denunciations starting to fill concentration camps, the Borns worried that someone might turn them in.

On Saturday, April 1, the SA barred the entrance to Jewish stores. The new minister of propaganda, Joseph Goebbels, had ordered a national boycott to intimidate foreign Jews who were agitating against Nazi policies. Huge placards posted in Berlin streets read, "The Jews of the entire world want to destroy Germany. German people defend yourselves. Do not buy from Jews!" The world protested, and Goebbels abandoned the boycott after one day.

Four days later, Hedi recorded in her diary that all Jewish teachers in Baden had been put on leave. The purge had begun, even though new civil service laws for the whole country legalizing the suspension of "non-Aryan" employees were not issued for two more days. Professors were civil servants, since the state governments ran the universities. Unlike professions such as law and medicine, the Ministry of Education in the states had easy and direct control over them; and students, now well indoctrinated by the Nazis, wanted instant action. By the next year, 1,145 university teachers would be suspended.

Hedi left immediately for Zurich to confer with friends about emigrating. They had been urging the Borns to leave before it was too late, before the borders closed. They promised to help, even if it meant smuggling them out of the country.

A week later, Max called her to come home immediately. While she was away, he, Franck, and Courant had tried to work out a plan. Franck pushed for all of them to resign their positions in solidarity with those who had been suspended. Born and Courant resisted; they wanted to stay at their posts to try to save the institutes.

Franck followed through on his idea alone. On April 17, Easter Sunday, he sent a short letter of resignation to the ministry saying that the decision was "an inner necessity because of the attitude of the government against German Jewry." His longer letter to Göttingen's rector explained, "We Germans of Jewish descent are being treated as aliens and enemies of the Fatherland. It is demanded that our children grow up in the awareness that they will never be allowed to prove themselves as Germans." On Monday, the *Göttingen Zeitung* printed an excerpt from the letter, which was quickly followed by reports in national and international papers. The Borns spent those days in endless discussions, trying to decide what to do next.

The reaction in Göttingen was swift—first rumors, then retribution. The rumors were that "certain colleagues"—namely, Born and Courant—had encouraged Franck to resign. The initial retribution was a letter in the newspaper signed by forty-two professors. "We are unanimously agreed that the form of the above declaration of resignation is equivalent to an act of sabotage, and we hope that the government will therefore accelerate the realization of the necessary cleansing measures." Twelve out of the forty-two signers came from the agricultural institutes—six of their Ordinarius professors and six of their assistants. Their loyalty was rewarded a few months later when the Ministry of Education gave the department a new chair for agricultural politics. The names on the letter opened Born's eyes. He better understood the agriculturalists' animosity toward him when he was dean.

Apart from his closest friends, Franck received no support. Other "friends" melted away. On the day his letter of resignation ran in the press, his secretary waited in vain for the phone to ring. Not one person called.

A former assistant who did visit with both Born and Franck at this time was Pascual Jordan. He was extremely upset about Franck's resignation and the possibility of Born being suspended. He had always leaned firmly to the right, but his sympathies for Nazi politics, which were strong, did not include harm to his former mentors. After the trip, he told his mother that, if he had belonged to the Nazi Party, he could have perhaps prevented these misfortunes; he became a party member a few days later.

The Ministry had to act quickly if it wanted to enforce the new civil service law and minimize the disruption at the beginning of the semester. It sent questionnaires to university officials, including student associations, to determine who should be suspended. A particularly influential group within Göttingen's student organization was a large, rabid gang from the natural science institutes. They had been some of the earliest Nazi Party members at the university and now took revenge on Born and Courant by smearing them as both Communists and fronts for wealthy Jews. A few months later, these students wrote a lengthy diatribe condemning the Mathematics Institute under Courant as a cesspool of Liberalism, a collection place for the worst Marxists, the assistants there being either Jews or slaves to Jews. They labeled Born's institute "Communist-infested." Richard Courant later attributed this hatred to exclusion: First-rate students became part of the professors' intimate circle that Hilbert had initiated long ago; second-rate students, who were left out of this milieu, became easy prey for the Nazis. The Ministry formulated the initial list of suspensions based on input from the various groups.

On Tuesday evening, April 25, 1933, a newspaper editor, who was a family friend, told the Borns what to expect on the next day's front page. Hedi Born's diary entry read, "Max suspended by the government." Born, Courant, and four other professors were the university's first victims under the new "Law for the Restoration of the Professional Civil Service."

Even though Born had anticipated the suspension, he was devastated.

> All I had built up in Göttingen, during twelve years' hard work, was shattered. It seemed to me like the end of the world. I went for a walk in the woods, in despair, brooding on how to save my family. When I came home nothing seemed to have changed. Hedi kept her head and showed no sign of desperation.

The night after the dismissal, their phone rang. When Hedi answered, she heard a chorus singing the Lutheran hymn "Now Let Us All Thank God." Born had made enemies, and they made sure he was ousted.

The six Jewish professors were in academic limbo, suspended with pay for the next three months, but barred from stepping inside the classroom. Born told von Laue that he was on the list because of "supposed political agitation."

Born was deeply pessimistic about the situation, far more than Franck and Courant. They had decided to stay in Göttingen for the time being—Franck to look for a job in industry, Courant to fight for his position on the faculty. Born knew how difficult it would be to leave Göttingen. Hedi and eleven-year-old Gustav were natives, and he had lived there much of his adult life; but because many of his colleagues had believed the rumors spread by the students, and because almost none had spoken out about the suspensions, it was unbearable for him. He told Ehrenfest, "I shudder when I imagine standing in front of students who, for whatever reason, have thrown me out, or living among colleagues who were able to live with this so easily."

Born wrote to Max Reich, who had succeeded him as dean, to say that he needed to take a leave to recover from an asthma attack and promised to return immediately if needed. He enclosed a note from his doctor and another from his landlord in the Italian South Tirol where, anticipating their annual summer vacation, Hedi had already rented the same small apartment they had enjoyed two summers earlier. Max received a salary and was technically still a faculty member. Since he had no idea what the future held, he felt he had to ask the university for permission.

Max, Hedi, and Gustav received their exit visas at the beginning of May. The plan was for Gritli to stay at Salem and Irene to live with the Weyls in Göttingen. Max made an overnight trip to Berlin to discuss the situation with Ehrenfest (who had traveled from Holland) and Schrödinger, and when he returned, he began to pack. The evening of May 9 was one of sad farewells, and the next day, with a couple of suitcases in hand, the three boarded the train to Italy. They left almost everything behind, from cherished mementos to all of Max's physics work at the institute, with no idea of when they would return. The Borns were but three of some 25,000 Jews who fled Germany that spring. Another 300,000 of the 500,000 German Jews (less than 1 percent of the entire German population) were to follow.

Born had had a goal when he came to the university twelve years earlier—"to bring Göttingen physics to further heights." He had secured Franck's appointment, raised research funds, and recognized outstanding young talent: the eight future Nobel Prize winners Werner Heisenberg, Wolfgang Pauli, Maria Göppert-Mayer, Max Delbrück, Enrico Fermi, Eugene Wigner, Linus Pauling, and Gerhard Herzberg; the equally superlative intellects of Pascual Jordan, Friedrich Hund, J. Robert Oppenheimer, Victor Weisskopf, Edward Teller, Fritz London, Yakov Frenkel, and many more. During his tenure, Göttingen became an international center for physics that was instrumental in the quantum revolution. On May 10, as the train pulled out of the station, Born said a final good-bye.

On the night of the Borns' departure, student associations had organized massive ceremonies throughout the country to burn seditious literature. With bands playing and flags waving, students lined up to toss the works of Thomas Mann, Berthold Brecht, Upton Sinclair, Ernest Hemingway, Paul Tillich, and dozens of others, onto huge blazing pyres, cleansing Germany of the "un-German spirit." As the Borns rode into Munich, they saw the flames rising from the bonfire.

In Göttingen, crowds gathered in the auditorium to listen to rousing speeches. After trumpets blew and the SS band played a march, crowds of townspeople carrying torches followed the swastika through the old city to the newly renamed Adolf Hitler Park. Irene Born watched as they surged toward a ten-foot-high pile of books that had a sign sticking out of the top reading "Lenin." One of the marchers set the pile on fire. Afterward, as Frau Ilse Neumann-Graul recorded, the crowd drank beer and celebrated. Frau Neumann had an extra reason to rejoice. The Nazi government had just chosen her husband to be the university's rector. The friendly, seemingly apolitical Neumann, who in the past had exchanged dinner invitations with the Borns, had been a Nazi all along.

It took the Borns a few days to travel from this surreal world to one that felt a million miles away. The direct route through Austria to the South Tirol required a visa that cost 1,000 marks for each of them. The Austrians had imposed the fee to discourage Nazis from entering the country and softening it up for annexation. So the Borns had to travel through Switzerland, on the Munich-Zurich-Milan-Bolzano route, about 700 miles in all, then take a tram from Bolzano into the Alps.

In Selva Gardena in mid May at 5,000 feet, it still snowed every day. The stores in the alpine village were closed. It was lonely, quiet, and remote. The Borns quickly settled into the four-room apartment they had rented from the local woodcarver, Signor Perathoner. When Gustav was not bringing in firewood for the oven and fireplace and similar chores, he chatted with the Signor and watched in fascination as exquisite animals came to life from blocks of wood. Hedi cleaned and cooked (something she had never done much before); and Max, recovering from a severe asthma attack, helped a little and read. A week after arriving, he described the primitive but peaceful routine to Ehrenfest.

> This is a completely new and wonderful experience—the high mountains in the spring. I feel like thanking the powers in Berlin for making this possible. Down in the valley the meadows are already green. The higher you get, the more colorful the flowers are: gentian, yellow anemone, and many others. There are patches of snow that become thicker until everything is white. When the sun is out, it's wonderful, but even when there are clouds and fog, the rocks of the Dolomites are of a deep and serious beauty. In the afternoon the three of us go for walks and pick wild flowers in the meadow. Gustav rolls and tumbles down the slope and he is very cheerful. We forget all the evil we have experienced and take a deep breath at night and breathe the good air. In the morning I walk for a couple of hours and have time to ponder the meaning of what we've been through. Anger and bitterness slowly disappear except at night. They come back in nightmares and in long hours of wakefulness. It's clear to me that it's really not bad to be thrown off an accustomed track and to have to start something new. It's also useful to see how expendable you are. And finally I think of you both and I'm completely not worried about us. Not many people have such good friends. We're going to find a place somewhere.

In spite of his reassurances to Ehrenfest, the outside news from the German press, which gave a hazy, distorted view of the upheavals there, made Born extremely insecure. So far from the world, suspended with no job, no future means of support—no country—he worried that he would be forgotten. Letters from friends such as Franck, who wrote to him as "Maxel," as he had in the early days, kept him in touch with what was happening and provided support; but even letters to and from Germany were a concern. Born and his friends sometimes used initials and innuendo to share information, because they strongly suspected that the mail was no longer private.

Born's correspondence circle included Werner Heisenberg, and Born sent an excerpt from one of Heisenberg's letters to Ehrenfest. He described the content as the "well-formulated attitude of well-meaning German colleagues"—an attitude with which Ehrenfest was familiar. Heisenberg apparently was shocked to learn that Born did not want to return to Göttingen. Little realizing the nature of Born's predicament, Heisenberg hoped he would wait until the fall to make a final decision.

Together with many other German intellectuals at the time, Heisenberg clung to a future that did not affect their lives adversely. He believed things would settle down. "Surely," Heisenberg wrote, "with time the ugly will be separated from the beautiful . . . in the new political situation there are those who are well worth the wait." Planck had even spoken to Hitler, who told him that the government would do nothing more to affect physics negatively. Under the influence of this illusion, Heisenberg envisioned Born, Courant, and Franck returning to a Göttingen unaffected by the political upheaval; surely some would suffer, but only a few—certainly not Born and Franck, and probably not Courant. Heisenberg shared his optimism—if not the dismissive attitude toward those affected in the junior faculty—with many of Born's friends, including Planck and von Laue. According to them, the right thing to do was simply wait until the situation sorted itself out.

Born responded to Heisenberg in a lengthy and emotional letter, telling of his pain and anger, of his having been "born and raised German" with no Jewish identity and of his anguish at the loss of his home. He described his inner conflicts to Heisenberg, saying that if he had not been suspended, he might perhaps have tried to stay in spite of Franck. He acknowledged that one inside voice wanted things to improve and another feared no one could really do anything. "But now I have been put out and experienced in my own soul what so many others have shared. Now I belong to this and feel their bitter feelings. . . . You see how complicated it all is and how the various motives work against each other."

Born said he would see about postponing the decision about another position until the fall. He thought a visit from Heisenberg would be helpful so that they could talk it over. He tried to lure him by describing the beauty of the Alps, with no success.

Expert at seeking jobs for his students, Born began a letter-writing effort on his behalf, starting with his two closest friends outside Germany—Maria Göppert-Mayer at Johns Hopkins University and Rudi Ladenburg at Princeton. Ladenburg followed up with a letter to Niels Bohr, who was visiting Caltech, asking if he would talk to some of Born's friends at Berkeley and the University of Chicago. If there were a vacancy, Ladenburg intimated that he could cover some of the salary from private sources.

At the end of May 1933, Born received a letter from his former student, Alfred Landé, who had taken a position at Ohio State University several years earlier. Landé asked him to come there for a semester. Born accepted immediately, on the condition that he did not receive a permanent offer from someplace else. He was overjoyed. No longer fearing oblivion, he began to sleep less fitfully.

Soon offers began pouring in, and Born felt humbled: they came from Léon Brillouin in Paris, Yakov Frenkel in Leningrad, and others in Belgrade, Brussels, and Constantinople—as well as from Einstein, who was safely in Oxford and shortly to leave permanently for America. He suggested a possibility in Jerusalem.

The latter proposal gave Born pause. He did not want to go to Palestine, but worried that Einstein might take offense if he were unenthusiastic. In a somewhat circumscribed explanation, he responded that he now felt acutely aware of being Jewish, and that "oppression and injustice provoke me to anger and resistance"; but Hedi and the children (who were raised as Lutherans) had never been aware of being Jewish—"or 'non-Aryan' (to use the delightful technical term)," he noted sarcastically. Not only would such a change be difficult for them, but above all, he wanted the children to grow up in a Western country.

Born addressed the issue directly with Ehrenfest and said that Hedi completely rejected the Palestine option and that he agreed.

> Among all the other possibilities this one would please me the least. You know, Hedi and my children, they are really Jews for today's German laws but they have never thought about it before. They are missing and, I also, *every* single emotional connection to Judaism—its ways and its laws. Down deep I believe myself to be a liberal western European with a strong cultural identity. With Hedi (and the children) the German emotional component is even stronger.

He cited the political instability in Palestine and his lack of familiarity with Hebrew and Jewish literature and art, which he contrasted with his closeness to German language, poetry, and art. He saw beauty in Zionism and appreciated it for its emphasis on harboring persecuted Jews, but he was not a nationalist. Were he single, he would take Einstein's suggestion to heart, but his goal was "to create a life for my wife and children so that their inherited or acquired emotions for their cultural environment are not crushed more than is unavoidable."

Born had told Einstein that what he truly wanted was a position in England. Having spent a semester at the Cavendish Laboratory in Cambridge, where he had learned from J. J. Thomson as a novice physicist, he had had the idea to return for a semester or two to study nuclear physics, but he had never found the time. Now he had all the time in the world, but no idea if it could happen.

A couple of days later, Cambridge University answered Born's wish. Mr. M. H. A. Newman, secretary of the Faculty of the Board of Mathematics, St. John's

College, wrote to offer him a lectureship in mathematics for three years. This was just what Born had hoped for, and he knew that a number of people were responsible, not least of all Einstein, Rutherford, and Ehrenfest. Curiously, however, he did not accept the offer; rather, he explained to Newman that he needed time to consider further, since he had received offers elsewhere.

To Ehrenfest, Born confided that he was very pleased, but felt guilty that he might be taking a position from a deserving Englishman. He communicated the same to his friend Ralph Fowler at Cambridge, who wrote that this particular position was "a very special one with terms which could not have been offered to anyone else." If Born did not accept it, it would not be offered to anyone else. Even so, Born debated between Cambridge and Brussels. Ehrenfest favored Cambridge, warning him about the ugly émigré atmosphere that could envelop Brussels and affect Gustav.

Before deciding, Max and Hedi left their sanctuary in Selva and traveled to Zurich. It was a trip with many agendas: consulting with friends, attending a physics congress, meeting a Cambridge physicist to discuss the lectureship offer, and helping to develop a committee for assisting exiled scientists. After six weeks of isolation, business was refreshing. Born enjoyed listening to brilliant lectures at the congress on nuclear physics and seeing old friends; but his true interest lay in "The Emergency Committee of German Scientists in Exile." In the spring, a few professors in Zurich had formed a committee to help find positions for scientists excluded from their jobs in Germany. The group had invited him to be on the executive board. As Born already knew, it would be a frustrating task, especially finding positions for those who did not have an international reputation.

Scores of scientists were anxiously trying to leave Germany, but opportunities were few. In April, at the outset of the suspensions, Rudi Ladenburg wrote pessimistically about positions in the United States. "There is little one can do here—the Depression is so strong that many capable people wait for appointments. . . . There are naturally no new positions." Once Einstein had reached America, he echoed this sentiment to Born.

> Conditions are very difficult here, as the universities, which in the main have to live from hand to mouth on a combination of private contribution and diminishing capital, have to struggle for their existence, and for this reason many capable young local people are unemployed.

The British had the same constrictions. With their own countrymen waiting for appointments, they had little freedom to give opportunities to foreigners. The solution of the Emergency Committee was to relocate the 300 German scholars

Gretchen Kauffmann Born and Gustav Born with Max in 1883. (All photos from the Born family private collection.)

Three-year-old Max, the winter before his mother's death.

Max (far right) with cousins (from left) Helene Schäfer and Hans Schäfer, sister Käthe, and cousin Luise Mugdan, all grandchildren of Salomon and Marie Kauffmann.

Gymnasium student Max at age 16.

Max, sporting a mustache and wearing the uniform of the Kaiser's bodyguards, the Lieb-Kürassier Regiment No. 1, with friends and relatives in the garden at Kleinburg, 1907. Standing from left, friend Käthe Eppenstein, Max Born, cousin Alice Kauffmann, friend Charlot Strasser, and cousin Cläre Kauffmann. Aunt Luise Kauffmann, seated.

Six-year-old Hedi Ehrenberg with her childhood nanny Minna Oberdieck in Göttingen, 1897. Minna later took care of Gustav when he was a newborn.

Hedi and her parents on the steps of the home of their Rosenzweig cousins in Kassel, July 1915. From left, Georg Rosenzweig, Adele Rosenzweig, their son Franz, Hedi Born, Hedi's uncle Richard Ehrenberg, Hedi's mother Helene Ehrenberg, Hedi's father Victor Ehrenberg (back to camera). Kneeling in front, the wife of Hedi's brother Rudi, who was also named Helene Ehrenberg. Franz Rosenzweig became an important philosopher and a colleague of Martin Buber's.

Max and Hedi in the garden of his sister's house in Grünau, just before their wedding in August 1913.

Hedi with Gritli on her right and Irene on her left in Frankfurt, 1919.

Born, seated, with colleagues during the Bohr Festspiele in Göttingen, June 1922. Standing from left, Swedish physicist Carl Oseen, Niels Bohr, James Franck, and Swedish physicist Oskar Klein, who was working with Bohr in Copenhagen.

The Born children in their Göttingen backyard, circa 1924. From left, Gritli, Gustav, and Irene.

Gustav on his father's shoulders, in Göttingen, circa 1926.

Born's student Maria Göppert in the mid-1920s. She won the 1963 Nobel prize in physics.

Gustav on the shoulders of James Franck who is standing next to Richard Courant, Göttingen, circa 1926.

Born with physicist Paul Ehrenfest in the late 1920s.

Born with colleagues in Bangalore, India, 1935. From left, Prof. MacAlpine, Sir Mizra Ismael (Dewan of Mysore), Sir C.V. Raman (director of the Indian Institute of Science), Max Born, Prof. E. P. Metcalfe.

The Borns with friends from Germany in the South Tyrol, summer 1933. From left, Max Planck, Hermann Weyl, Anny Schrödinger, Hedi Born, and Max Born.

Gritli and Gustav with Trixi in their backyard in Edinburgh, 1937.

Born with Gustav in Edinburgh in 1942 when Gustav was a medical student at the University of Edinburgh.

Born's close friends Albert Einstein and Rudolf Ladenburg at the Institute for Advanced Study, Princeton, circa the mid-1940s.

Born, right, with Werner Heisenberg in front of Born's Edinburgh house in November 1947.

Born with Irene in the backyard of his house in Bad Pyrmont, 1957.

Family gathering at Caius College, Cambridge, to celebrate Max Born's seventieth birthday on December 11, 1952. First row, from the left: Irene Newton-John, Hedi Born, Max Born, Käthe Bo[illegible] Koenigsberger, Renate Born, Maurice Pryce. Second row: Gustav Born, Ann Born, Susi Born, Gritli Pryce, Brinley Newton-John, Kurt Ehrenberg, Otto Koenigsberger.

he town square in front of the Rathaus in Göttingen on the occasion of Max Born, James Franck, nd Richard Courant receiving the medieval honor of the "freedom of the city" in June 1953. Max nd Hedi are in the first row on the far right. Richard and Nina Courant are further down in the ixth and seventh seats past Hedi.

Max Born at age 60.

Max Born, feeding the birds on a path in Bad Pyrmont, circa 1960.

that it saw affected to positions around the world—a few here (Zurich, Brussels, Geneva, London) and a few there (New York, Istanbul, Belgrade)—so as not to burden unduly any one place.

Burden—and not merely the financial one—was very much on the minds of all the organizations that were being set up to help refugees. The Emergency Committee took pains to point out, without being explicit as to the exact reason, that in Zurich and elsewhere, trying to get positions for too many Jews in any one place would result in no positions. Germany did not hold a monopoly on anti-Semitism.

The Rockefeller Foundation was aware of similar pressures. By May, it had appropriated $140,000—and continually increased the sum—for grants to institutions wishing to offer positions to displaced scholars, and immediately faced the problem of placing Jews. The University of Pittsburgh wanted Otto Stern—Born's former assistant in Frankfurt, who had already completed the research that would earn him a Nobel Prize. Stern wanted to bring two Jewish assistants with him. Tisdale reported,

> Prof. Stern wants to bring [Immanuel] Estermann with him as an experimental man, and probably another displaced German Jew on the theoretical side. I drew attention to the fact that too many in one place might cause criticism, and S. indicated that while he is not desirous of having all of his associates Jews, he feels he must have some of them.

Later, a summary of refugee problems prepared by the foundation expanded on these concerns.

> Fears have been expressed, too, that this influx of foreign scholars will start a wave of anti-Semitism. Quite apart from this is the point of view of the younger faculty members in some institutions who see men from abroad receiving appointments with the assistance of funds which are not available for the employment of native scholars.

Those involved in the placement efforts—Jews and non-Jews—were acutely sensitive to the potential resentment about local jobs lost to Jewish refugees and to the potential anti-Semitic backlash; yet hundreds, soon to be thousands, of people were gravely in need.

From the start of the upheaval, Born considered a number of the younger Jewish generation—in particular Walter Heitler, Lothar Nordheim, Fritz London, and Edward Teller, to be "under his wing." He had already arranged a position at the University of Bristol for Heitler and thought that Teller would receive the IEB fellowship that he, Franck, and Eucken had proposed. Consequently, just after his arrival in Selva, Born had been very disturbed to receive a letter from the IEB rejecting Teller's application. It was for purely technical reasons—Born and Franck, who had guaranteed the position, were no longer in Göttingen.

Born was concerned for Teller, writing to Ehrenfest, "Teller is an unusually talented person who deserves every support, and he is already weighted down enough by being a Hungarian Jew." When Born heard that British scientists had also begun an organization to help refugees, he wrote to his old friend Frederick Lindemann in Oxford, who was planning a trip to Germany to search for talent, telling him "to take Teller's name to heart." Just before Born departed for Zurich, Lindemann wrote to say that he had met with Teller in Göttingen and arranged a two-year grant for him with Frederick Donnan, professor of physical chemistry in London. Lindemann also found a place for Fritz London in Oxford. Like Born's own options, the positions were temporary. Of the assistants close to Born, only Nordheim was still anxiously chasing uncertainties—a possibility in Paris, another in Delhi, maybe a Rockefeller fellowship—until he finally received a firm offer from Paris.

Born's trip to Zurich was fruitful. By the time he left there, he had decided to accept the Cambridge offer. Beginning October 1, he would be a Lecturer in Mathematics at Cambridge for three years. His responsibilities would be light—one course in each of the three terms. He would have a position at the Cavendish Laboratory, which was under the direction of Lord Rutherford. Born's tone instantly changed from anxiety to energy. The only problem was the salary of £600 ($2,800), which was only about half of his previous level in Göttingen. A little later, a Rockefeller Foundation grant increased it by 50 percent.

Born immediately notified Kurator Valentiner of his acceptance. He was still an official member of the faculty because Valentiner—who hoped that Born would be able to return to the university—had manipulated an employment law in Born's favor. Under the civil service laws issued in April, frontline soldiers from the Great War were excluded from dismissal. Even though Born had traveled irregularly to the front on his wartime duties at the Artillerie-Prüfungs-Kommission, Valentiner exploited the exclusion in Born's favor. Born had not wanted this applied to him and told Valentiner, "I agree with Franck's opinion and do not want special treatment. The reasons that Franck gave apply for me as a German Jew, and I hope you respect this opinion and share it." Valentiner, though, ignored his wishes. How Born now changed his status of "suspended with pay" to something else—resignation, sabbatical, or retirement—affected whether the government imposed oppressive emigration taxes or canceled his retirement rights.

So, despite his suspension and departure, the Nazis continued to have some control over Born's life. The government seemed content to sit back and watch as Jewish professors were wrenched between the compassion of a few university personnel such as Valentiner and the malice of groups such as the student association. The government was glad to be rid of many who left and unsure of what to do with those who stayed.

Yet more and more, Born appreciated his good luck at being thrown out of Göttingen. To David Hilbert, he wrote that all the Göttingen business (presumably administrative) had disrupted his work. To Ehrenfest, he altered this sentiment to read "the cursed 'business' in Göttingen used up my strength so that I could not carry out my physics." Whether or not the "business" in the second instance referred to Hedi and Herglotz, the crisis of exile had created a closer union between Hedi and Max.

Without these pressures, Max now felt he could do research. He always said that theoretical physicists were fortunate to need only pencil and paper for their work. With neither library nor mathematical tables at his disposal, he had to find the right topic; he reached back to his roots and what had stirred his imagination almost thirty years earlier: the electromagnetic mass of the electron. In his habilitation thesis, his attempts to develop a model of the electron as a rigid body had foundered. Sitting on the porch in the Italian Alps, he started to think through the problem again. He wanted to apply quantum theory to the electromagnetic field and create a new field theory. Others had tried and arrived at unsatisfactory results. He decided to investigate if Maxwell's equations for the electromagnetic field and quantum theory were compatible.

Born even had a couple of assistants. In June, two students—one from Britain and one from South Africa—had found their way to Selva Gardena and were studying with him. He held the lectures at "Selva University," as they called it, on a bench outside the house or in the nearby woods, and the two young men helped him with research.

Life in the Val Gardena had become even more refreshing by the end of July. Irene and Gritli arrived bringing their dog, Trixi. As in years past, the family trekked up to surrounding ridges and stayed overnight in mountain huts. Tourist season in the Alps meant that others were close behind. The family welcomed an assortment of friends and relatives who trickled through for the next two months, sometimes carrying candid letters that were too dangerous to send by mail. The scientific community was well represented: Hermann Weyl, Wolfgang Pauli and his sister, Frau Anny Schrödinger, the Plancks, and Percy Bridgman from Harvard. Later came Joe Mayer (Maria's husband), Arnold Eucken from Göttingen, Hermann Mark from Vienna, and Frederick Lindemann with an offer to Born from Oxford. "Nearly a mathematical physics congress" was Born's assessment. The arrival of pianist Artur Schnabel added some balance on the artistic side. As for relatives, Max traveled to the Brenner Pass to meet his brother, Wolfgang.

Wolfgang had become a painter and had had some success. In Munich in 1921, he had met Thomas Mann, who asked him to do the illustrations for the first edition of *Death in Venice*. After completing a Ph.D. in art history, he had been working as an art critic for newspapers in Vienna. The brothers did not often see each other, but nonetheless there was a close sense of family. When Wolfgang needed help, Max responded.

One August day, the Borns had unexpected visitors—eight boy scouts, one of whom, eighteen-year-old Otto Still, was the favorite son of Born's Göttingen benefactor, Carl. The youths had a delightful day in Selva before moving on. Twelve days later, Born received devastating news. The boys were sliding down the slippery wet grass on their thick leather backpacks high up on a ridge; going at great speed, Otto lost control and went over the edge of the cliff, tumbling down 650 feet and breaking his neck.

Born had the sorry task of informing the Stills. He wrote to Courant in Göttingen, who traveled to Recklinghausen to break the news to them as gently as possible. A few days later, Born met the other Still son, Carl Friedel, at a little chapel down the mountain in Brixen to help send the body back home. Born began his letter to his friend Still, "I am so horrified and shattered by the terrible misfortune that I cannot find words to express it."

At the time of the accident, Hedi was resting in Bolzano before returning to Göttingen to pack up the house. Max did not tell her immediately because he did not want to destroy her peace, which was smashed when she heard the news.

Hedi left Selva at the beginning of September to pack up the house in Göttingen. Irene had already gone back there before heading to a housekeeping school in the Hartz Mountains. (Rebelling against domestication, she would find a way to spend her days happily planting trees with foresters.) Gritli was about to leave for her last year at Salem. Max stayed behind in Selva with Gustav and Trixi because friends advised him not to return to Germany. Whether Born would be in actual danger or rather did not feel emotionally composed enough to face the situation is unclear. Franck and Courant, who were both greatly disliked by the student association, were still living there. In any case, Hedi went alone. Max did not see his home again for many years.

On a circuitous route to Göttingen, Hedi made a sad farewell tour, stopping off in Munich to visit Heisenberg and his mother, then to Recklinghausen and the distraught Stills, and to Berlin to say to good-bye to Frau Blaschko.

Once in Göttingen, she dealt with problem after problem. The government would not allow money to be taken out of the country. The travel agent lost her passport. The financial issues associated with Max's position in Göttingen were still ambiguous. The authorities were trying to hinder property being taken out of Germany. Everywhere the atmosphere was poisonous. Max wrote to an associate, "You will understand that she is not far from breaking down." Ironically, when staying overnight in nearby Braunschweig, she watched as the actors in Beethoven's *Fidelio*—an opera in which the hero challenges tyranny—gave the Nazi salute at the end of the performance.

Hedi packed up the apartment (sending furniture and silver to Cambridge) and at the last minute, rented it. She learned that the Ministry of Education, at the behest of the Kurator, had allowed Max a three-year foreign leave to go to Cambridge, a step that resolved some concerns about future retirement benefits and avoided the emigration tax. Valentiner thanked the Ministry for its willingness to keep Born connected to the university, saying that many others were also thankful for this declaration.

On September 26, Hedi went for a short stay at a health clinic to recover from the ordeal of moving. She found a letter there from a friend and made two entries in her diary. The note in the back of it read, "As I came to the clinic on Wednesday—a letter from [Erich] Rosenberg: Ehrenfest has shot his sick (mentally retarded) child and then himself.—Friend—This year gives and takes too much." In a unique gesture, she encircled the entry "Ehrenfest shot himself and sick child" recorded on the actual date in the calendar section. She did not leave behind any other reaction to the loss. Max too buried his feelings, giving the tragedy only a sentence in his memoirs, written several years later—"This tragedy was a horrible shock for us."

For years, Ehrenfest had vacillated between high energy and deep depression, even telling Hedi in the past of a death wish. In winter and spring 1933, he had been in a major depression. He told Hedi he felt a sense of failure concerning his ability in physics—a recurring theme. He was despondent about politics in Germany and in despair about many friends who had to leave the country or were unemployed. On July 16, he had written that he was suffering the worst crisis of his life. He was going away to an isolated place and would write something definite in a couple of weeks. That was the last they heard from him. Ehrenfest's death was a particularly painful and final farewell.

At the end of her stay in Germany, Hedi said good-bye to friends and family. She recorded in her diary that Gustav Herglotz sent her a love letter. On October 8, she boarded a train for England.

Max, Gustav, and Trixi departed by train from a wet and cold Selva. On a night train from Zurich to Calais, with sleepers unavailable, Max sat up in a crowded compartment, Gustav dozed, and Trixi slept with her head on Max's lap. Max worried about what awaited his family. That night was one he would never forget; but as he wrote years later, "The future was not so dark as it then appeared to me."

TEN

TALKING OF DESPERATE MATTERS

WHEN THE BORNS ARRIVED IN CAMBRIDGE, POSTERS THROUGHOUT THE TOWN READ "THE Man Born to Be Hanged." The movie title caused the Borns a double take and then a laugh. Max knew the town well from twenty-six years earlier, when he was torn between mathematics and physics. Then a new Ph.D. escaping from misery in Göttingen (and Klein's wrath about his dissertation), Max had gone to see what physicists in Cambridge were doing; now he was escaping to Cambridge with his direction in question, again.

Hedi had seen the town during a recent house-hunting trip, but for Gustav it was brand new. When they arrived in Dover, the twelve year old, who knew no English, sobbed as his beloved Airedale Trixi was placed in a cage and lifted by crane from the deck of the ship to the dock, bound for a six-month quarantine. That and his first ghastly meal of English peas did not endear him to his new home, but he proved resilient.

Born slid back into the life of a scholar. Officially known as Mr. Born—few faculty held the highest rank of professor—he was happy to be with friends and have a workload light enough to accommodate his research. He gave three lectures a week, held an hour seminar with Paul Dirac and Ralph Fowler, and attended evening physics clubs every week or two—the scholarly Delta Phi Club and the more informal Kapitza Club, which gave him plenty of time for the new field theory he had begun that summer.

By the time Born had left Selva, he had finished his first article, "On the Quantum Theory of the Electromagnetic Field." His research program was the ambitious goal of determining the structure of particles, a mystery that had first fascinated him thirty years earlier in the electron seminar in Göttingen. At that time, a researcher only had Maxwell's equations of the electromagnetic field to work with, which are linear, and when the field is generated by a charged point particle, such as an electron, it goes to infinity at the origin—an unreasonable result.

The structure of particles was still unknown, Born's fascination had not ceased, but now the puzzle was more complex. Researchers had to consider quantum effects. In his initial report, Born acknowledged quantum theory but focused more on making Maxwell's equations nonlinear in order to introduce an electron with finite mass. When he presented his theory in a seminar, the young Polish physicist Leopold Infeld questioned his calculation of the electron's mass. At first, the two found no meeting of the minds—but by the next day, Born had a new collaborator with whom he worked well. Within a couple of months, Born was writing to friends such as Franck that his nonlinear theory followed from "the postulate of general relativity." Maxwell's theory, he was pleased to see, did not.

The initial feeling of success restored Born's confidence. Infeld observed an "enthusiastic attitude, the vividness of his mind, the impulsiveness with which he grasped and rejected ideas." Born attributed this spirit to an "intense desire to discover something fundamental to regain my self-confidence after the loss of my job in Göttingen and to 'show the Germans what they had lost.'" In letters to Franck and Courant, he held on to the hope that his research would lead to a discovery. Along with a desire to prove the Germans incorrect, he needed to impress the British so they would keep him. His appointment was only for three years. He did not know what the future held.

After only a month in Cambridge, Born took off to lecture at the University of Geneva. He, Louis de Broglie, and James Franck were invited to speak to physicists and mathematicians on quantum theory. Articles in Swiss papers described Born with an energetic face, fit body, and graying hair. Silhouetted against a blackboard covered with equations and graphs and speaking German, he led the audience through quantum mechanics as well as his new field theory. The group gave Born and Franck standing ovations in recognition of their achievements and exile.

It was the first time Born and Franck had seen each other since Max had left Göttingen. During the summer, they had assured each other that the separation would not affect their friendship.

The Born family had made it through a hellish year of rejection and loss, but the convulsions of 1933 were not over. In November, the world—and Born—learned that Paul Dirac, Erwin Schrödinger, and Werner Heisenberg would receive the Nobel Prize. The 1933 prize awarded jointly to Dirac and Schrödinger cited their contributions as "the discovery of new productive forms of atomic theory." The reserved 1932 prize for Heisenberg alone cited "the creation of quantum mechanics, the application of which has, inter alia, led to the discovery of the allotropic forms of hydrogen."

As in other pivotal moments in his life, Born left little trace of his reaction to the news. The following spring, when several Nobel Prize winners, including Heisenberg, entered the Cavendish Laboratory to applause, Born said with emotion, "I should be there." Twelve years later, he wrote simply to his son, "I was deeply hurt."

Born wrote a letter of congratulations to Heisenberg and on November 25 received a letter of apology and thanks in return.

> Dear Herr Born, If I have not written to you for such a long time, and have not thanked you for your congratulations, it was partly because of my rather bad conscience with respect to you. The fact that I am to receive the Nobel Prize alone, for work done in Göttingen in collaboration—you, Jordan and I—this fact depresses me and I hardly know what to write to you. I am, of course, glad that our common efforts are now appreciated, and I enjoy the recollection of the beautiful time of collaboration. I also believe that all good physicists know how great was your and Jordan's contribution to the structure of quantum mechanics—and this remains unchanged by a wrong decision from outside. Yet I myself can do nothing but thank you again for all the fine collaboration, and feel a little ashamed. With kind regards, Yours, W. Heisenberg

Heisenberg was sincerely embarrassed by the Swedes' omission of Born and thankful to him for his contributions to quantum mechanics. He knew what Born had done for him. If Born had not recognized the significance of Heisenberg's insight and set out the complete formulation—while Heisenberg was hiking in the Alps—Dirac would have done so first. Without Born, there would be no "Heisenberg matrices."

Born later said that only the three collaborators could disentangle the separate contributions to the final formulation of quantum mechanics. Yet earlier in the summer of 1933, he had explained his part to Ehrenfest.

> At the beginning of quantum mechanics I had the experience that bold things occur to me (e.g., the matrices, the confidence relationships, the perturbation calculations, the transformation $\mathbf{SUS}^{-1}$, etc., the shares of Heisenberg, Jordan, and me are rather equal, the formulas about 90% mine).

One of those "bold things"—the fundamental commutation law of quantum mechanics, $\mathbf{pq} - \mathbf{qp} = h/2\pi i\mathbf{I}$, the basis for the entire formulation, which Born was the first person to write down—he rated as his most important contribution to physics. The statistical interpretation of the wave function was the second most important. He did not consider his and Jordan's development of the principal features of quantum mechanics trivial, either. Although he described their joint work as "only a formal step beyond Heisenberg's" breakthrough, he noted that Dirac had received his Nobel Prize for developing an equivalent formalism.

At that time, the Royal Swedish Academy of Sciences kept its records confidential so Born never knew why he was not included. One reason was, however, that his physics colleagues rarely nominated him. In the four years preceding the 1932 and 1933 awards, Heisenberg and Schrödinger received multiple nominations while Born had only one. Multiple nominations applied pressure but they were not, by themselves, decisive. Dirac received none before 1933 and only two in that year. More important than the number was Born's old friend, Carl Oseen.

From 1922 until 1944, Oseen was one of five members of the academy's committee on physics, the first of three levels in the internal nomination process. After considering names submitted by a large but select group of physicists, this committee made recommendations to the ten-person Physics Section who voted on them and forwarded the results, usually with no changes, to the full Royal Swedish Academy for the final vote. Oseen was an excellent theoretical physicist, a classicist in outlook, and a shrewd manipulator. In 1929, he became chairman of the committee and wrote up the evaluations of theoretical research for the other members.

Physicists began to nominate Heisenberg and Schrödinger for a prize in the late 1920s, sometimes separately, sometimes jointly. In 1929, Oseen's first report on quantum theory argued that it was not yet sufficiently systematic "from a logical point of view" to merit consideration. Instead, that year the committee under his leadership recommended Louis de Broglie for the wave nature of matter. This theory had experimental proof behind it.

The next year, when Born was nominated in conjunction with Heisenberg, Oseen offered a number of reasons to deny them a prize: the need to recognize Schrödinger equally; the difficulty of matrix mathematics, as exemplified in Born and Jordan's *Elementare Quantenmechanik*; and the importance of Jordan's contribution. If the prize was for the systematic development of the theory, he said, then Jordan must be included—and Jordan had not been nominated. Most of all, the tone of Oseen's remarks suggested that he did not particularly value the matrix formulation of quantum theory. Quantum jumps, as opposed to Schrödinger's wave formulation, were not to his liking.

For the next couple of years, Oseen maintained that quantum theory was revolutionary, but not a "discovery or invention" as envisioned by Alfred Nobel. Also, unsure that quantum theory was final because it did not include relativistic effects, Oseen found various arguments for not recommending its originators.

Pressure within the academy to recognize this theory was building. Suddenly in fall 1933, Oseen recommended that Heisenberg, as the originator of the matrix theory, receive the 1932 reserved prize, and Dirac and Schrödinger, as fundamental contributors to Heisenberg and de Broglie's work respectively, share the 1933 prize. Earlier in a preliminary report, Oseen had dismissed Dirac but, somewhat inexplicably, he now justified his inclusion. "Dirac is in the forefront of research and realized Heisenberg's bold theory. As opposed to Born and Jordan, he worked independently." Although some of Dirac's work paralleled "his German colleagues," he had arrived "at a higher level of creativity," as the influence of "Heisenberg's [theory] waned." With no nomination for Born that year, Oseen avoided recognizing him and the matrix formulation of Heisenberg's idea.

The next year, Born and Pauli were nominated to share a prize, Born for both lattice theory and quantum physics and Pauli for the exclusion principle. Oseen sidestepped both nominations. He described Born's contributions to lattice theory as "predominantly deductive," and saw them as coincident with those of Debye, Madelung, Ewald, and, not least of all, Oseen himself. As for Born's work on quantum theory, Oseen described him as the originator of its basic formulation (his earlier report had said a mathematical clarification with no physical extension) and again raised the need to include Jordan. Oseen dismissed Born's statistical interpretation of the wave function, which he considered Born's greatest contribution, because some "like Planck have rejected" it. He remarked that he "would have expected him [Born] to use matrix mechanics [to analyze collisions], but he doesn't"—implying that not even Born wanted to use matrices.

Oseen could view Born's dearth of nominations as support for his judgment. This lack was perhaps based on Born's contributions fading from view during the previous five years. His breakdown had removed him from scientific activity; the Copenhagen Interpretation had swallowed up his ideas; and Heisenberg—who was charming, competitive, and had clearly completed the last step—did little to credit Born's earlier creative work upon which he built or Born's mathematical formulation. Heisenberg lauded Bohr's influence instead.

Weyl, who was in Princeton, wrote that Einstein thought Born's part in the founding of quantum theory had not been sufficiently appreciated. Weyl probably did not know that Einstein had proposed only Schrödinger for the 1933 prize. (In 1928, Einstein suggested dividing a prize between Heisenberg, Born, and Jordan but thought that Heisenberg was the more deserving.) Weyl counseled, "Don't make yourself sick, dear Born, that the golden Nobel Prize passed you over! Heisenberg and Schrödinger really got there first."

Born moved on and reflected on life in a Christmas letter to Hilbert. Generally, he said, things were going well, but he spoke wistfully of being homesick. His mood shifted as he reported to Hilbert on his new field theory and his collaboration with Infeld, his enthusiasm about the theory's beauty and agreement with general relativity obvious. He was particularly pleased to tell Hilbert,

> The leading spirit comes from your old lectures on the calculus of variations of 30 years ago and even your more beautiful optimism. . . . One must believe that through quiet thought one can solve the secrets of nature . . . —but it must not be forgotten that no human fantasy can guess the truth without hints from experimentation.

Nothing diverted Born more from his feelings of loss than finding truth and beauty in the abstract world, especially in relation to his long quest in understanding the structure of matter—except, perhaps, the welfare of his family, and they were doing fine. The girls were still in Germany, and Gustav, who had the greatest challenges, had adjusted nicely, making friends and bringing home good school reports. "Forest poor" England made it difficult for them to recreate a German Christmas but they found a small tree.

Born did not tell Hilbert that Hedi was unhappy, but Courant, who stayed with them briefly, wrote to Franck that she was "somewhat fragile." The Heller sisters, who had looked after the children in Göttingen and had come to help in England, had gone home. An English girl was now with them, but if she were anything like the others, she was not destined to stay long. Hedi coped with the household and constant guests—old friends and relatives from Germany as well as new friends and colleagues from Cambridge, all with the requisite teas and dinners. The entertaining and her sensitivity to small cultural missteps bothered her, but mainly she was homesick. She began dreaming of and writing to "G"—Gustav Herglotz.

The Borns spent the end of the year with old friends. Fritz Haber visited the day after Christmas and reminisced about the war for the whole afternoon. The Schrödingers, who had moved to Oxford because Erwin hated the Nazis, celebrated New Year's Eve quietly at the Borns with a candlelight dinner. Later that evening, Kurt Hahn came by. Since escaping from the Nazis, he had been in England and was setting up another boarding school. This one, also based in an old castle like Salem, was in Scotland and became Gordonstoun.

January 29 brought the first news of the 1934 New Year: at sixty-five, Fritz Haber had died from a heart attack. A month earlier, Hedi had glimpsed a bitterness and unhappiness behind his lively reminiscing, a "great actor" she called him, a superpatriot destroyed by Germany's total rejection of him as a Jew. Planck almost wished that Haber could have died the previous year, to save him from this tremendous pain. In Germany, Max von Laue wrote two obituaries for Haber and endured the government's reprimand.

Planck's letters to Born had lost their optimism from the summer, when he had asked those who were dismissed to have patience. Now he told Born that he could only talk of the difficulties face to face. Planck's new caution, however, was not universal; Courant had turned down two opportunities for work in America and still waited for one in Germany. Teller's former mentor in Göttingen, physical chemist Arnold Eucken, whom Born considered "one of the best" of the lot, hoped that soon the individual, regardless of origin, would again be accepted. Eucken believed that if the right person were appointed Reich Minister of Education, German science would not sink and there would be "confidence in the future." He was glad that Born was waiting to see how things turned out. In fact, Valentiner was the one pushing to keep Born connected to the university, although emotionally Born may not have wanted to face a complete break.

Optimism in Göttingen flew in the face of evidence. Ludwig Bieberbach, professor of mathematics at the University of Berlin and a former Göttingen student, wanted to purge the influence of a certain "nationality, blood, and race upon the creative process." The former Einstein attacker Philipp Lenard sought the same "purity" for "German" physics. "In reality, as with everything that man creates, science is determined by race or by blood," he wrote. The theories of relativity and quantum physics were labeled "Jewish science." Most German physicists did not subscribe to this ideology, but the "Reich" spirit still infected them.

In the spring, Born received a letter from his former assistant Friedrich Hund. Springer Publishing had chosen Hund to be coeditor for the series on the structure of matter that Born and Franck had started years earlier. The other editor was the physical chemist Hermann Mark. Mark had been dismissed from IG Farben in 1932 for being half Jewish and was now back in his native Austria teaching at the University of Vienna. Like Heisenberg's letter of the previous summer, Hund's letter showed a regard for dismissed scientists that varied according to their stature.

> The acceptance of Mark made me somewhat uncertain at first because Mark, if I am correctly informed, does not meet all the requirements which would be expected or demanded from a university teacher in the German Reich. Since he has been accepted, I have gotten over it. The editors who live in the German Reich will have to see to it that the series can stay connected to physics in the German Universities. Obviously, first-rate physicists like you and Franck are always welcome as authors even if you live abroad. I pray for your understanding of this point of view. Because of it, I am against the book by Stobbe.

The letter infuriated Born—similarly Courant, who was visiting Born when he received it. Nonetheless, when Born saw Hund at a scientific congress in London a few months later, he treated him kindly "in spite of that stupid letter." Other Germans at the congress he found disgusting and avoided.

Hund manifested what Hedi had once characterized to Paul Ehrenfest as "a lack of empathy in the German soul." He was not a Nazi, but he was willing to tow the party line. His letter was satiated with the language of the new order. He did not live in "Germany" but in the "German Reich." Mark did not meet the new Reich "requirements." "First-rate" Jews—only physicists of world renown—were acceptable. Martin Stobbe, a former assistant to Born who now lived in England, did not qualify.

Stobbe belonged to the group that Born described to Einstein—"Almost every week some unfortunate wretch approaches me personally, and every day I receive letters from people left stranded." Although Stobbe's story was similar to that of many others—no country, no job, no means of support—he was different: he was not Jewish; his father was a Prussian general; he had left Germany solely because he could not tolerate the Nazis. Born felt responsible for Stobbe because, at Born's suggestion, he had given up a fellowship in Bristol in favor of Walter Heitler, a Jew. Born supported him through all kinds of means, and Stobbe eventually found a short-term position in Paris.

All of this passed Hedi by; she was enduring her own problems. She had gone to St. Evelyn's nursing home to recover from a bad case of flu mixed with a bad case of nerves. The housekeeping, entertaining, and her increasing dislike of British social conventions had wrung her out. She checked into the nursing home at the end of March, just a few days after Gritli and Irene had finally arrived from Germany after six months. On Easter Sunday, she wrote Max a note, apologizing that she could not give him an Easter egg and sending him a few dear words, namely, that she loved him through and through. A week later, she was on a boat crossing the channel for a stay at the Braunlage sanatorium in the Harz Mountains, her solution to regain her health. She was away for three months.

Max kept the household going. Irene and Gritli, ages twenty and eighteen, adjusted to England quickly and began taking dressmaking classes with the aim of turning professional; Gritli took sculpting classes as well. Twelve-year-old Gustav continued doing well, studying hard and becoming a flutist. When the maid gave notice, the girls tried to run the house but with only partial success. Anny Schrödinger, "the perfect and nice German housekeeper" as Born called her, came over from Oxford to help out, returning home when husband Erwin came down with chickenpox.

Max's sister, Käthe, came from Berlin for a couple of weeks, just happy to breathe some "pure air." She was there to check on her children. So far, she and her twenty-six-year-old son Otto had managed to get three of the other four children out of Germany and into schools in England. Until the Nazi takeover, the children had not even known about their Jewish heritage. Georg, who had died two years earlier, and Käthe had converted and raised them as Lutherans.

The next visitor was Werner Heisenberg, who came to lecture in the Cavendish Laboratory, and Max reported to Hedi that everyone was pleased. He did not give further personal details about Heisenberg's visit because he could not put them in a letter to Germany; but he later told her a tale. While in the garden one afternoon, Heisenberg offered Born a research position in Germany. He had been directed to do so by the German Physical Society, which had permission from the Nazi government to make the offer. Confused and homesick enough to be interested, Born inquired about bringing the family. When Heisenberg said no, they were not included, Born ended the conversation, amazed and offended that Heisenberg could be so insensitive as to make such an offer. They did not speak about it again.

After only a month in Cambridge, the pretty Born girls were caught in a whirlwind of social engagements. Like the town itself in the spring, they were in full bloom, and it was "raining invitations for balls and concerts." Max wrote to Hedi,

> May Week approaches . . . Irene's friend John has invited the three of us to the Caius concert, where one wears evening dress; then there is dinner in the hall. I am also going to the B Minor Mass of Bach in King's Chapel because Gritli plays in the violins. Furthermore Matthew Landau invited the girls to the Trinity Ball. And we want to watch the boat races.

Within two weeks of this letter, Born was telling Franck that the girls had found some "good friends; by the way both splendid boys (or did you imagine they were girls? No, my good friend)."

Hedi seemed to be improving. After a long hiatus, she was working on her novel again, but Max remained concerned. He thought she was gripped by a "clinical and fanatical homesickness." His letters to her were newsy and loving, filled with advice not to stimulate her nerves. When she left the sanatorium to stay with friends in Göttingen, he said the same thing—"not to visit anyone!" and "naturally refrain from G.'s [Herglotz]."

His other request of her was not to "speak too much [to him] of the Hainberg. I already have more than enough longing for it!" He missed his home and the German landscape and walking in the forests. He wrote to Franck, who was also depressed by the situation, "there is a crack inside me that will never heal."

Hedi arrived in Göttingen on May 10 and left at the beginning of July. She did not tell Max about the Hainberg, but she did see Herglotz. Later she cut those months out of her diary.

The realities of exile were sinking in. Hedi began to realize that she would have to adjust to a future in England; Max, that the position in Cambridge really was only temporary. When, in 1934, the Cambridge Board made him the Stokes Lecturer, an honorary title that brought neither extra duties nor financial benefits, he hoped that it was a sign that people liked him and that his appointment would become permanent. But within months, Rutherford and Fowler both told him that as a foreigner he had no chance of a permanent position. The most they could do was extend his position for a year.

Britain saw itself as a "clearinghouse" for permanent emigration to someplace else. One British refugee organization stated, "If German scientists with senior qualifications compete for these junior posts it will be regarded by British graduates as unfair competition, . . . an abuse of the displaced scholar's circumstances and would be an unfair act against younger British competitors." This concern, together with the fear expressed by both Jewish and non-Jewish refugee organizations, that a concentration of Jewish refugees would increase anti-Semitism, severely limited the number of permanent positions at British universities.

Born relied on Frederick Lindemann for advice. Lindemann was a contemporary and old acquaintance of Born's who, because his father was German, had studied in Berlin and spoke the language fluently. He was wealthy and also a snob who loved the British aristocracy. His intellectual arrogance and sarcasm at times caused even his friends to dislike him. Born, however, admired his "energy, wit; sometimes his judgment, not so much in regard to scientific facts and theories but to personalities." He did not seem to mind his friend's other traits.

Lindemann advised Born that opportunities in America were probably better than in England. Courant, who was now settled in New York, disagreed, however. Courant thought Born could probably turn to his former benefactor, Henry Goldman, but only in an emergency because he was already supporting Franck and the art historian Erwin Panofsky. For Born and thousands of others, options were few.

Eighteen months after Hitler had issued the 1933 civil service laws, the League of Nations' High Commission for Refugees (Jewish and Other) Coming from Germany estimated that out of the 65,000 German refugees, 1,300 were academics, 5,500 were professionals, and 7,000 were students. Only half of the academics had found positions, and the situation was worse in the other two categories. Overall, only 25,000 refugees had been settled—most of them in Palestine.

Born was the physics representative on two refugee organizations that coordinated positions. He answered daily inquiries about the merits of candidates, suggested names, and wrote to acquaintances in distant countries about job possibilities. Franck, Courant, and Ladenburg, all in America, sent him money to distribute to penniless friends stranded in one or another of Europe's major cities. At times, he was overwhelmed because the need for money and positions seemed never ending. As temporary appointments ended, the same names came around.

Born had one possibility for himself. C. V. Raman, India's only Nobel Prize winner in physics and the first Indian director of the Indian Institute of Science, had invited him to come to Bangalore. Raman wanted to build a premier organization and thought that his greatest need was "a first-rate Mathematical Physicist who could guide and inspire the work . . . on the theoretical side." Born's reply noted a number of problems—separation from his children, loss of the research environment in Cambridge, lack of familiarity with India, concerns about the effect of a tropical climate on his health. Nevertheless, he expressed interest in lecturing there for a few months and seeing what it was like. They finally agreed to a six-month visit the following year, from October 1935 through March 1936.

At the end of 1934, the German government imposed a tax of 10,000 Reichsmarks on Born's savings there, about a third of the total. Born protested to the government, and Hedi went to Göttingen to ask Valentiner for help. She also asked Planck to intercede. He agreed but did not think that the Minister of Science would be receptive to him.

To get the tax canceled, Valentiner submitted affidavits to the Ministry affirming that Born's stay in England was in the best interest of Germany. The German ambassador in London, the vice chancellor of Cambridge University, and the president of the Prussian Academy of Science all attested to his scientific reputation. For added weight, Valentiner invoked the name of Hedi's famous grandfather—"Prof. von Ihering, one of the most important German lawyers of the past century"—to assure the Ministry of the Borns' strong connection with Germany. The Ministry wanted something in exchange—for Born to resign his chair in Göttingen of his own free will. If Born did this, the Ministry would allow him to retire with possible pension benefits and would cancel the tax. Valentiner wrote to the Minister, "I know that Born belongs to those Jews conscious of their race and reserved enough to be willing to give up a position." Born submitted his retirement request.

Born's old friend, Hans Bolza, one of the housemates long ago in *El Bokarebo*, sent him a postcard from Lisbon. Bolza painted a picture of a continent in the grip of demagogues and "hopeless disorientation." The Portuguese were about to hold their first "election," a word he placed in quotes because candidates from only one party were on the ballot. There was a frightening right-wing campaign of anti-Semitism in France, and, of course, Nazi Germany and fascist Italy. This was his Christmas greeting.

Surrounded by uncertainty, Born asked an important favor of Lindemann.

> [Hedi and his daughters could] find their way of living somehow, starting with the little capital left. But my boy whom you know, would not have the means of completing his schooling and studying. He is very gifted and clever, interested and eager to learn. It is a sad thought for me to imagine his uncertain future, and it troubles me sometimes so much that it distracts me from my work. If you could tell me that you would take care of him and help him to get a good education to start his own life, it would give me a great deal of comfort and peace. I hope that the case may never occur; but I feel sometimes rather tired and old.

Lindemann assured him that he would do everything he could to help, but hoped that Born would "not envisage all these painful possibilities."

Gustav was becoming "quite a little Englishman." As Born proudly related to Lindemann, "He has got a Foundation Scholarship of the Perse School, the only one of the 40 boys who took the examination." Gustav and his father enjoyed each other's company—everything from occasional piano lessons to building a huge kite that they flew on the Perse playing fields, which backed onto their house. Gustav's schoolmates so admired it that they started a kite club and made him president.

The girls were doing equally well. Irene was engaged to Brinley Newton-John, the young man who had escorted her during May Week. Gritli was seeing one of her father's assistants, Maurice Pryce. Max and Hedi liked both of them very much.

The young suitors, each man extremely smart and capable, came from different worlds. Brin was the son of a Welsh woodworking teacher who was also a puritanical Presbyterian. Through the church, Brin had spent his childhood summers with families in France and Germany and spoke French and German like a native. He studied German literature at Cambridge. Maurice had a British father and a French mother and was bilingual. He had learned German, the language of choice in the Born household, when studying in Heidelberg. He had a degree in physics from Cambridge and was working on his dissertation under Born, who considered him very talented. The two girls were happily planning futures away from Germany.

All thrived except for Hedi, who, unhappy and sick, spent as much time as she could in Göttingen. In 1934, she was there for five and a half months. She went again in March 1935 and again destroyed the pages from her diary.

Max also traveled to Germany that spring—in his case, to see his sister and numerous other relatives in Berlin. It was his first time there since his departure almost two years earlier. He reported to Franck that Germany looked splendid—the Nazis had put the vast number of unemployed to work on municipal projects—and that his family and physicist friends talked of "desperate matters."

When he returned from Germany, he brought his nephew Otto Königsberger back with him for a temporary stay. Otto, an architect whose successful career in Berlin ended with the Nazi takeover, was to illustrate Max's new book, *The Restless Universe.*

Born was writing a popular science book primarily to make money. From the outset, the family had faced a significant income reduction from the Göttingen level. Increasing the financial pressures, the Rockefeller Foundation decreased Born's grant by £100 each year.

The Restless Universe had a loftier purpose as well—"that a deep insight into the workshop of nature was the first step towards a rational philosophy and to worldly wisdom." For Born, it was one step towards a hopeful future.

He began the book, "It is odd to think that there is a word for something which, strictly speaking, does not exist, namely rest." He used the restlessness and chaos of air molecules, molecular beams, electrons, waves and particles, to encompass all of basic physics. Along the side of each page, seven drawings come alive when the pages are rifled—among them, colliding gas molecules and an electron moving around its nucleus. Inspired by doodles of Gustav and his schoolmates, they were Otto's contribution to the successful book.

A book titled *Atomic Physics* was also in progress. It was based on the seven lectures Born had given at the Technische Hochschule when he was stricken with Bell's palsy. With eight editions from 1935 to 1969, it went on to become a standard text in physics.

Born had stopped collaborating with Leopold Infeld on field theory to work on the books. The two had been struggling with the second part of their research program anyway, which was to quantize their field equations. Maxwell's classical theory did not hold in the interior of the atom, and they had to introduce quantum principles to have a consistent theory of how matter interacts with a field. They assumed that the particles' mass is derived from the field energy as opposed to the more usual assumption that particles provide the charge. They ran into difficulties but maintained this assumption and continued to change their quantization methods. With each shift, Born hoped they would reach something fundamental, but they kept hitting obstacles. Halting research to write books gave him time to think about where the research was going.

In June, Born went to a conference in Paris. After having been in Göttingen for two and a half months (another financial drain on the family), Hedi met him there. Shortly after they returned to Cambridge, Max received a friendly letter from Courant asking about the children and inviting Gustav to visit, but not mentioning Hedi; nor did a card sent by Courant and Franck about two weeks later. The omission needled Born. "You did not mention Hedi," he replied. "Presumably, you think that she is not here. She has already been back a long time and runs the house. She is also coming with me to India."

When Hedi had been in Göttingen, Courant and Born both heard rumors that she planned to stay. Now the old friends exchanged a few letters with Born trying to find out who was saying what. They dropped the issue when Courant asserted that he was not the source of the gossip. He pointed out, however, that Hedi had said repeatedly that she did not intend to live in Cambridge.

Courant had inadvertently hit a sensitive nerve. Max had been sufficiently distressed about Hedi's goings-on to consider divorce. Two months after her return, however, he wrote to Maria Göppert-Mayer that things had improved; they had become closer. He had made his own emotional accommodations with her relationship with Herglotz, even agreeing to her wish to spend two weeks with Herglotz in the South Tirol before sailing for India in September. Clearly, he wanted her to stay with him.

The Borns made arrangements for the children and got ready for their trip. Irene was to be in London at a dressmaking academy along with Gritli, who would study window decoration. Gustav was to stay with a family in Cambridge.

Just before leaving, Born received a letter from Germany. It was an official thank you for his service in Göttingen, a conclusion to the resignation of his chair. The certificate of appreciation, emblazoned at the top with the words *In the Name of the Reich*, read as follows.

> According to your request, I relieve you of your duty at the Mathematics-Natural Science Faculty at the University of Göttingen at the end of March 1935.
>
> I thank you for your academic work and your service to the Reich.
>
> *Berchtesgaden, 23 July 1935*
> *The Führer and Reich Chancellor*
> *(signed) Adolf Hitler*

The Ministry had a choice of routine certificates signed by low-level functionaries and more courteous ones signed by important officials. Born was one of the few Jewish professors who received one from the führer (although signed by autopen). Presumably this "honor" reflected Valentiner's intervention.

The beginning of Max and Hedi's two-week sea journey on the steamer *Staffordshire* was a mix of wondrous nights filled with shooting stars and daytimes with blue seas, flying fish, and warm ocean swims. That changed when they neared the Suez Canal. On the fifth day, they awoke at 5 A.M. to a sunrise breaking over palm trees in the desert. Once they were through the canal and into the Red Sea, dreadfully torrid nights followed sizzling days. The air did not move, the heat was sickening, and the boat swayed constantly. The only way to survive was to remain absolutely still. In the "cooler" evenings, Hedi played ninepins to take her mind off her discomfort. She was pleased when they steamed into Cochin, and she could finally step down on the red earth of this bustling city on India's southwest coast.

A friend from the journey, a Captain Harvey who was the tutor for the children of the Maharajah of Travancore, invited the Borns to spend the night in Cochin. They postponed their departure to Bangalore for a day. Riding through the city in a rickshaw, they saw sacred bulls roaming the streets and eating the vegetables in the bazaar while goats ambled nearby. They went to the old Jewish quarter and saw the four-hundred-year-old synagogue, famous for its floor of hand-painted blue tiles from Canton, China. Their first night in India introduced them to mosquito netting and the small lizards that ran up and down the walls. The next day, Joseph, their new butler who had come from Bangalore, met them at the train station and settled them in for the twenty-hour trip.

Bangalore, in the state of Mysore, is on a plateau 3,000 feet above sea level. Daytime temperatures in April, the hottest month, reach ninety-two degrees; in January, the coldest month, seventy-eight degrees. From the train, Hedi saw terraced rice fields, water buffaloes, and palm leaf huts—a "magical countryside," she said. Max described "cloudless skies, fragrant air, and lovely climate." But they were far from home. There was no contact by phone; letters took nearly three weeks; newspapers arrived in batches; and Bangalore's first airline connections were still two years away.

In Bangalore, the Borns' new staff greeted them in front of their "bungalow," a sprawling house surrounded by a large garden filled with exotic flowers. The wife of the vice chancellor of Mysore University had arranged everything—house, furnishings, and servants. Mrs. Raman welcomed them. Her husband was away, and the Borns did not meet him for another couple of days. When they did, he captivated them, a prince out of "'1001 Nights,' . . . young and slim with a sparkling, intelligent face, wearing a fine white muslin turban with gold braid on a dark head." At this first meeting, they talked for two hours about art, physics, and politics.

The Borns were enchanted with everything and adapted quickly, usually rising around 6:30, drinking tea, and either taking a walk or playing tennis on their private court. Hedi attended to social obligations and drove around in a second-hand Chevrolet, the first and only car the family ever owned. Max strolled to the Indian Institute of Science where he lectured on his new field theory and held a seminar on optics twice a week as well as a colloquium. He worked with five students, two of whom he ranked as very good; and he had time for research. With monkeys looking over his shoulder, he often worked on the institute's veranda. He wrote a couple of papers about his field theory using a classical treatment but knew that to couple matter with the electromagnetic field in any meaningful way required quantum principles. He still could not get the two theories to reconcile.

Hedi quickly shed her British malaise and was "lively, interesting, and in a better mood," Max wrote to Maria. In Cambridge, Hedi had wanted to escape from a large house with no maids, financial obligations, two lively daughters with boyfriends, an inadequate income with increasing education costs, and no relaxation. In Bangalore, she found a beautiful bungalow with a wonderful garden and a service staff that was inexpensive and so capable that the house ran itself. The move had improved their finances, and she enjoyed the socializing, especially with Raman and his wife, who lived minutes away.

A month after they arrived, a British professor of electrical engineering named Aston came to teach at the institute. He, his wife, and child stayed a few days with the Borns until a house became available. They impressed Hedi as "tired, old, miserable people," but in the months following, she dutifully accompanied Mrs. Aston on shopping trips and socialized with her. Mrs. Aston told Hedi that English colleagues had made her husband accept the Bangalore position for the purpose of "clearing up" Raman's institute. The Astons were a minor inconvenience. Hedi continued to find the country magical.

In October, Max and Hedi traveled eighty miles south to Mysore. The maharaja had invited them to the Dasara Celebrations, a Hindu religious festival. For four days, it was a fairy tale with a bejeweled and bedecked maharajah carried in a sedan chair, splendid state elephants, and an enormous palace with a special room for the Durbar where the maharaja received his subjects. The Borns stayed in an elegant tent with all the conveniences, one of many in the park surrounding Government House. There were torchlit processions in the evenings, receptions, polo—a time from another world. Hedi kept reminding herself that this was not a dream or a play but real life.

By the beginning of November, Born was considering a permanent position in Bangalore, although he was ambivalent. The mass poverty upset him. "The state of the lives of the underclass is still distressingly primitive and there are villages where the people have windowless huts, 20 to a room, in filth and neglect," he told Maria. He worried about Gustav and his education. Raman wanted him to stay, however, and Born, like many other refugees, had no clear alternative.

The situation in Europe continued to decline: Italy had just invaded Abyssinia and Germany passed the catastrophic Nuremberg Laws on Citizenship and Race, which defined citizens of the Reich as those who had "German or kindred blood." A decree issued two months later defined Jews as those with either three Jewish grandparents or two Jewish grandparents if certain other conditions applied, such as practicing Judaism or being married to a Jew. Otherwise, those with two Jewish grandparents—and therefore presumably two German ones were "individuals of mixed Jewish blood" and subjects of the Reich. The addendum, Protection of German Blood and German Honor, placed further restrictions: Jews could not marry a national of German or kindred blood, have a relationship outside marriage with such a national, employ female household help who was such a national, vote, hold public office, or hoist the national flag.

Born's current side of the globe offered few possibilities for others. The National Tsing Hua University in Peiping wanted to take a scholar but had to wait until the political situation with the Japanese cleared up. India was a delicate situation for different reasons.

> If a man comes here, who is not excellent or fails, he brings about enormous damage. The few German Jews who are here all stand out prominently, observed by the English with unfriendliness and envy, by the Indians with mistrust. Many could perhaps find jobs here and achieve. But failure can ruin everything. Especially here at this Institute, which is visited daily by politicians, journalists, etc. and plays a great role in the public.

Born kept some hope that the dewan of the state of Mysore, Sir Mirza Ismael, who was a clever and progressive man, could help.

Born corresponded with Fowler about his situation in Cambridge and learned that the faculty board could propose Born for a readership, but uncertainties remained. Fowler counseled Born to take the position in India if offered, and Born wrote to Lord Rutherford that he was "rather inclined" to do so. Although he did not want to leave Cambridge, his income in India would let him help relatives in Germany as well as provide security for his own family if anything happened to him.

To keep Born, Raman needed approval for a new chair of theoretical physics from two groups: the Council of the Institute with members from across India, and the Institute Senate. One member of the Senate was Aston, the Englishman with the task of "clearing up" the institute. Raman thought that the Senate, which did not want Europeans, would allow Born to have the chair, but only Born. Otherwise, it would go to an Indian.

Raman scheduled the Senate meeting for the first of February and insisted that Born be present. Born complied, although when the time came, he was uncomfortable listening to Raman praise his scientific and scholarly attributes.

When Raman finished, Aston stood and addressed the group. He came directly to the point. He did not understand, he said, why the Senate wanted to give a chair to a man "who was rejected by his own country, a renegade, and therefore a second-rate scientist unfit to be part of the faculty, much less to be the head of the department of physics."

Born and Raman were stupefied. Years later, Born wrote that he went home and cried, completely shattered. He had convinced himself to stay in India because it was the only place that offered some security; now, together with the apparent loss of this solution, he was humiliated and slandered.

That afternoon the Borns had a visitor. Mrs. Aston came to make peace "after the evil Senate meeting," Hedi wrote in her diary. The next day, she brought Hedi flowers. The day after that, Aston and Born talked over the matter. In the evening, the two couples went to the movies together to see *On the Wings of Song*. Neither Max nor Hedi ever explained Aston's behavior or their own cordiality.

Bangalore's princely state of Mysore was not under direct control of Britain but was bound to it by treaty. Although the state managed its own domestic affairs, the British still pulled many strings. Born learned that before arriving in India, Aston had visited representatives of the powerful Tata family, who ruled a major Indian conglomerate. With the Tatas' economic success closely tied to Britain, their initial support for Raman as director of the institute shifted when the British reported on disgruntled British scientists at the institute. Once there, Aston had quietly collected complaints about Raman and roused latent discontent among the staff. Raman's nature made this job easy: he indulged in such practices as posting invitations to council meetings for certain members so late that they could not attend. Despite these kinds of political games, Raman nevertheless was trying to shake up a sleepy institute and transform it into an international center.

Deliberations on Born's new position fit right into the middle of what he described to Lord Rutherford as "a terribly complicated system of intrigues, of real politics, connected with the jealousy between North- and South-India, of Bengalis and Mysorians, between low caste and Brahmins, and last [but] not least connected with English-Indian relationships."

The Senate meeting, however, had not ruled out Born for the position. The Council of the Institute was to make the final decision just before the Borns' scheduled departure in mid March. Although Born no longer wanted an offer, he waited for the answer. At the same time, Hedi developed a strong interest in staying.

Shortly after reconciling with the Astons, Hedi went across the street to discuss something with Mrs. Raman and found her sitting on the veranda with a white-bearded monk. Hedi asked her neighbor a question and departed. The monk, Swami Avinasanda of the Ramakrishna mission in Bangalore, caught up with her in the garden and said that he must see her. Two days later, he came to the Borns' bungalow at 4 P.M. For two and a half hours, they sat on the veranda and talked about "the individual." He told her, "A chrysanthemum cannot grow on a rose tree." For Hedi it was as though they were sister and brother. She noted in her diary, "It was destiny that we should meet."

Hedi spent the next day reading Romain Rolland's biography of Swami Vivekananda. Vivekananda's master, the Hindu mystic Ramakrishna, had dedicated his life to contemplating God and warned of the loss of religion's transformational abilities when consciousness of God fades. Vivekananda used this consciousness of God to serve the needs of man and inspire others to worship God in this way. Disciples of Vivekananda had established the ashram in Bangalore.

Hedi spent most afternoons with Swami Avinasanda, who soon became simply Anda. His message was that she needed to expand her consciousness. He told her, "You are like a blossom on which a stone is lying. So many weaknesses. You will overcome them. Expansion of the soul through service, suffering, loving others."

One Thursday evening at the end of February, she and Max had a long discussion. The crux of the issue was that she did not want to return to Cambridge; she was thinking about leaving him. Max was stunned. He thought that during the months in Bangalore, they had become tied to each other. To him, it had been a beautiful time.

For the next several days, Max, Hedi, and Anda discussed the situation together for hours at a time. One late afternoon, she was not feeling well and went to bed. Max and Anda continued the discussion sitting on her bed. She later compared their attitude to two frogs who fall into a pail of milk. In this story, the pessimist, accepting that he could not get out, drowned; the optimist, swimming and kicking to wait for help, finally jumped out of the pail via the butter he had churned. Presumably, Max was the pessimist; Anda the optimist.

A week later, Max and Hedi painfully talked through the situation—just the two of them. What Max thought of Anda and how he fit into the picture was not recorded, but at the end, they had reached some decisions.

Hedi drove to see Anda at the ashram then went to the temple. She performed the Hindu ceremony of *pujah* in which she created a spiritual connection with a deity. Although she still wanted to stay in Bangalore, she had decided that she would not do so without Max. Her anxiety over the council's deliberations prevented her from sleeping nonetheless.

The days started ticking off to their departure date of March 18, 1936, and the council's decision. On March 14, a farewell tea was given in Born's honor at which the physics students presented him with a scroll that thanked him for the "rare combination of clarity of expression and depth of thought characteristic only of a master mind" and his "sweet and affable manner that endeared you to our hearts." It expressed the hope that "soon you will be on the staff of the Indian Institute of Science."

On March 16, the council met all day to resolve the job issue. The results came in the evening. Hedi wrote in her diary, "The dreams of the promised land are buried. God help me to find the good side of this decision." The drama was not quite at an end. The council had not voted against Born but rather postponed its decision until the following meeting in July. Raman advised Born to accept any reasonable offer from Cambridge.

The next day, the Borns continued packing and went to tea with the council, Hedi reluctantly. (Her comment: "It makes you sick.") On their final day, they said good-bye to Anda, had lunch with the Astons, and tea with the head of the Botanical Gardens. In the evening, they took the train to Cochin and thence the steamer home. For Hedi, it was "a heavy farewell to India." She still hoped to return and felt distant from a life in England. Max must have wondered to what kind of home they were going. He had few possibilities in an increasingly chaotic and restricted world.

At the Suez Canal, Max's nephew Otto Königsberger joined them for a day to talk about jobs. Otto had migrated to Egypt after making sure that his three sisters, younger brother, and cousin Renate, Wolfgang's daughter, were safely in England. (Children under eighteen were relatively easy to get out. Käthe was still in Germany.) He worked in Egypt as an assistant to Jakob Burchardt, a famous archaeologist and historian. When the dewan of Mysore had asked Born about a non-English architect to help with an extensive building program for the state, he recommended Otto. Otto eventually got the job, the start of a career in India that culminated in his becoming the country's chief architect and city planner under Nehru.

The Borns reached Marseilles on April 3, traveled overland to Paris, and finally arrived in Cambridge. Irene and Gustav, in good shape and eager to see them, waited at the train station. The family settled in. Max quickly applied for the readership that Fowler had put on hold while the India debacle played out, an action that caused Hedi further stress. To her it meant a definite farewell to India. She talked to the children about her wish to leave England.

Two weeks later, the readership application was rejected because of a missed deadline. The faculty extended Max's position for a year and planned to reapply for the readership, but there were no guarantees. Hedi's hopes of India buoyed again. Max wrote to Maria,

> Hedi has an almost sick aversion to England and the English overcome it in no way. Yes, it is intensified through correspondence with an Indian, a clever Brahmin and Hindu monk. In no way does she want to stay in England. The house life does not fulfill her, she hates the socializing, and she strives to find her own life's calling. All that, dear, is only for *you*—speak to no one about it. She has really definite plans and she will carry them out. Then I will be alone for a long time. But I understand the dissatisfaction in her life too well to place obstacles in her way.

Hedi planned to stay until September, when they were to move into a new house, "then she will try her way." Where she was going was not made clear. Under the definitions stated by the Nuremberg Laws, she was Jewish—two Jewish grandparents and married to a Jew; so Germany with Herglotz, to whom she still wrote, was not an option. India and Anda seemed a likely destination.

Max had been through enough to be philosophical about Hedi, but the situation was ripe for taking stock. He looked hard at himself and saw a man in his midfifties, homeless, and exiled. He felt that he had little standing or influence in the world of physics. His only glimmer of hope, an invitation from Bohr to visit Copenhagen, was perhaps an "acknowledgment" of the value of his recent research on field theory. He regretted that he had not been offered a position in America. Since he shared some of Hedi's attitude toward England, he also had not completely given up on the council and the idea of a peaceful existence in India.

In the midst of Hedi's decisions, Born received a long letter from Peter Kapitza, a Russian physicist who had been a noted experimentalist in the Cavendish Laboratory from 1921 to 1934. He now lived in Russia. Perhaps, in retribution for the escape of George Gamow, a rising star in physics, the Stalinist government had seized Kapitza on a visit home in 1934 and put him under house arrest. Kapitza was the head of a new institute in Moscow and had the idea, as he put it to Born, "of playing a nasty trick on you and making your state of mind a little more perturbed." He wanted Born to start a school of theoretical physics in his institute.

Born had not been in Russia since he steamed down the Volga in 1928. He was intrigued. He saw some advantages, which he enumerated for Kapitza.

> An independent position, collaboration with you, enough comfort even for my bourgeois customs, a great number of gifted students, the enormously interesting structure of the socialistic state and perhaps a little collaboration at its development, and last [but] not least the neighborhood of good friends, as you and your wife are! I am sure that my wife should prefer to live in Russia as she attributes very little importance to the comfort of the body and very much to that of the soul.

Likewise, he had concerns, foremost about Gustav and his schooling. The girls, planning to marry Englishmen, were already making their own way. Yet to leave Gustav behind would be very difficult for Born. The impossibility of learning a new language at his age and fears that living in Russia could mean never visiting Germany again weighted on him too. The Nazi government had such a hatred of Russia that Born felt they might imprison him and Hedi if they returned with Russian visas stamped in their passports. They continued to "love the old country in spite of all that has happened" and did not want to be cut off.

Putting little faith in governments, Born then predicted,

> But the attitude of governments are changing. There may be tomorrow a great "friendship" between countries which just have "hated" one another. Morally they are always behaving as children. May be that the Nazis discover soon a strong friendship in their hearts to the eastern neighbor, as they have already with the Poles.

Kapitza responded psychologically, telling Born that his wish to find a quiet spot such as India or America was "to get away from that history which treated you so badly. No hope Born, do not make yourself such illusions." Unless Born lived in a jungle, he could not escape from the politics that had driven him from his home. The only way he was going to get his fatherland back was for the Nazis to be driven out by socialism. Kaptiza's advice to Born was to decide which side he was on. Then "all the remaining arguments like the language difficulties etc. vanish as very small and insignificant."

Born did not quite buy Kapitza's arguments. Unsure that Germany would be destroyed, he rebutted, "History shows that in general the stupid, irrational, things happen, that nonsense is victorious and reason smashed down. Do not underestimate the strength of the Fascist countries. All depends on which side England will take."

Learning that having a Russian visa in his passport was not a problem if only for a visit, Born proposed continuing the jousting in person. He and Hedi would come to Russia from mid August to mid September. At the end of June, Max and Hedi began language lessons with the Russian wife of Rudolf Peierls, a refugee and one of Born's colleagues at the Cavendish Laboratory. Hedi had never been to Russia, and she was excited; Max, less so. For Hedi, who liked adventure, Russia was far preferable to Cambridge.

Then, on July 8, the physicist Charles Galton Darwin, grandson of the evolutionist, asked Born if he would be interested in succeeding him in the Tait Chair of Natural Philosophy at the University of Edinburgh—natural philosophy being the nineteenth-century Scottish designation for physics that was still in use. Darwin was on his way to a position in Cambridge as Master of Christ's College. A week later, Born wrote to Kapitza that he was accepting the Edinburgh offer.

Unbeknownst to Born, the Institut de Biologie Physico-Chimique in Paris was trying to contact him as well. They had written to a British refugee placement organization, which, not wanting to interfere with the Edinburgh offer, asked them to wait until other negotiations were finalized. The French recognized the delaying tactic and argued, but by the time the letters went back and forth, Edinburgh had become the Borns' future. Fortunately, as history played out, the charms of Paris, one of their favorite cities, did not have an opportunity to seduce them.

Hedi and Max went immediately to Edinburgh. It took them only one day to find their new home at 84 Grange Loan. Hedi's feeling of "endless joy for Max" expressed her own as well. They were away from Cambridge, they had an adequate income, and no longer were they guests in the country. With the appointment came the removal of the residential restriction in their passports. They were eligible to be naturalized citizens. Max returned to Cambridge, and Hedi stayed behind to organize work on their house. She had found an escape from Cambridge; there were no more thoughts of going off alone.

Several days later on the train back to Cambridge, a woman approached Hedi and asked if she were a nurse or in the Salvation Army. When Hedi asked why, the lady explained that she was a clairvoyant and saw Hedi wearing a nurse's uniform. Hedi was surprised. During her school days, she had wanted to be a nurse, but her parents had rejected the idea. The prediction intrigued her.

Born's new appointment in Edinburgh signified an honor that he could not appreciate at the time. He was one of only three refugee physicists who obtained a permanent position in Britain during the 1930s. Such recognition would have pleased and heartened him.

ELEVEN

WORSE THAN IMAGINATION

AFTER ARRIVING IN EDINBURGH, MAX BORN RECEIVED A LARGE YELLOW ENVELOPE addressed to "Professor Mac Born." Hearing this, his colleagues appointed him an honorary Scot and the forebear of the clan "MacBorn." He so enjoyed the malaprop that he framed the envelope and hung it on the wall in his department.

In winter 1936, three months after taking up residence, Born summed up his feelings about his new home in a letter to Franck.

> Dear James, . . . We like it here very much. Hedi is quite lively. Mostly we take long walks through the wild, secluded Pentland Hills on Saturday. But the enjoyments of the big city are also here, good concerts and theater. At last I have music at home—a good quartet with a few younger colleagues. The children are well. Irene works partly at home and partly at social services. Gritli is in Vienna, hard-working and contented. Gustav likes to go to his distinguished school. Unfortunately we are very worried about relatives and friends in Germany. . . . My brother in Vienna is naturally always in need. Well, one helps wherever one can. But I have heavy debts.
>
> My work goes quite well. I am always trying things at the limits of the known and unknown. . . . Well, you know my incurable optimism without which I cannot work.

The Borns quickly learned to love the town, the country, and the Scots. Open fields and hills were in abundance to the city's north and south. The forests in which they loved to roam were precious few, but the rolling moors covered with heather and moss soon became a frequent weekend destination.

Their first outing soon after arriving was to Arthur's Seat, one of Edinburgh's extinct volcanoes northeast of the city center in Holyrood Park. Only a ten-minute walk from the Borns' home on Grange Loan, this mass of rock and grass became a favored weekday stroll for Max and Gustav. With the dog Trixi, they hiked up undulating slopes that resembled a crouching lion to the "head," a peak 823 feet above sea level. Life's philosophies passed from father to son. The Borns had found stability in Scotland—a huge relief, especially after the changes they had recently witnessed in Germany.

In August 1936, the Nazis hosted the Olympic games in Berlin. Before the games began, they sanitized the city, removing anti-Semitic signs, moderating the racial tone of the newspapers, and removing particularly strident periodicals such as *Der Stürmer* from newsstands—all to demonstrate Nazi tolerance and German superiority to the world. (They largely succeeded: most foreign correspondents sent home glowing reports.) Jews in exile, like the Borns, Ladenburgs, Courants, and Francks, took advantage of the relatively safe interlude to visit family and to see one another.

Max, Hedi, and Gustav crossed over to Cologne at the end of July. Max visited relatives in Germany before proceeding to Karlsbad, a town in Czechoslovakia known for its curative thermal springs where, like his mother, he sought a cure for gallstones. Hedi and Gustav headed to Göttingen, where they climbed into the attic in Wilhelm Weberstrasse and sorted through boxes of personal family letters, household papers, and memorabilia. Except for what little they could pack, they destroyed everything. They did not want the Nazis picking through their personal papers. Max's professional correspondence at the institute had to be left there.

Hedi had not been in Göttingen for more than a year, and the dramatic change in the environment made her feel like "a foreign body." One day the postman said to her in English, "What are you still doing in the new realm?" She answered uncomfortably, "We have family here." He charged back, "Can't you take them out of the country, too? Where is your husband?" Hedi was disturbed, and it was no different in the surrounding towns when she and Gustav made short sightseeing trips. One day, they took a three-and-a-half-hour walk from the village of Dassel, just northwest of Göttingen, to Neuhaus. Every hamlet they wandered through fluttered with Nazi flags—a field of swastikas against a bright blue sky. That evening they took a bus to Höxter on the Weser River. A sign in the town chillingly announced: "Jews are not wanted in this place." They spent the night there in a hotel. As they went down the Weser on a steamer trip the next day, the same Nazi fervor smothered them.

Just before leaving to meet up with the rest of the family at her brother Kurt's house in Karlsruhe, Hedi received a "lovely farewell letter" from Herglotz and a photo. Her reaction to what was probably the close of the eight-year relationship was "God gave us a peaceful ending."

In his separate travels, Max was seeing a "new" Germany—modern roads, factories, and trains as well as workers who lived in labor camps but were gratified by wages that provided for their formerly hungry families. The "Aryan" population was beginning to prosper as the living conditions for the Jews, including numerous Borns, Kauffmanns, and Lipsteins, rapidly deteriorated. In Breslau, Max visited his mother's sister, Gertrude Schäfer, a widow who at seventy-seven was too frail to consider leaving. In Berlin, his stepmother, Bertha, also in her mid-seventies, was in the same predicament.

Max happened to see a copy of the Nazi newspaper *Der Stürmer* with its vicious harangues against Jews and could not sleep after reading it. Upon returning to Britain, he would write to Rutherford that he and Hedi wanted to become British citizens as soon as possible. With his permanent position in Edinburgh, they had to wait only two more years to complete the five needed for naturalization.

The Borns left Germany with another family member, Kurt's son Rolli. He was fifteen, the same age as Gustav, and had earlier spent summer holidays with the Borns in Göttingen and the Alps. Now he was coming to live with them in Edinburgh, leaving his family behind and entering a foreign world where he knew only four words: *corned beef*, *darling*, and *bugger*. Gustav told him that if he did not understand something, just to say "bugger."

Edinburgh was a welcome refuge, but the first winter tested the Borns' resolve. Dark gray days and constant bouts of cold and flu prompted a query to Rudi Peierls in Cambridge. Their doctor had said that their ill health partly resulted from lack of sunshine, so Born wanted to know about the possible dangers of using an ultraviolet lamp.

With no central heat, the Borns relied on fireplaces and stoves to warm their solid stone house—one "built for eternity," Hedi described it. Everything, even the staircase, was of stone. In bygone days, local quarries had supplied stone for houses, stores, office buildings, and churches that were now blackened with soot. Roads, fences, everything in the city was of stone, including the centuries-old majestic castle, symbol of Scottish strength and fortitude rising 300 feet above the center of the city on a volcanic outcropping.

The Borns' own "castle," even if cold, was homey, bright, full, and busy. Scientists and musicians were always coming and going. Cambridge and Göttingen friends visited and lectured. During the holidays, the family celebrated a real German Christmas. The big tree was decorated with brightly burning candles; an elaborate crèche was complete with moss that Hedi collected from the Pentland Hills; and the air lingered with the aroma of freshly baked German cookies. Gritli was in Vienna at art school, but everyone else was there, including Brin, whose beautiful tenor enriched the family's rendition of J. S. Bach's

Christmas Oratorio, for which Max played piano and Gustav flute. Hedi, Irene, and Brin sang, and Rolli, who was very homesick, turned pages.

World events struck increasingly discordant notes. The Spanish Civil War between the leftist government and fascist rebels had broken out the previous summer. Germany and Italy had recognized the rebels almost immediately. Born knew that this was merely a prelude to the major drama to come. Just after the first of the year, he wrote to Landé at Ohio State.

> One sees exactly in Spain how it will be. As slowly but certainly the earth devours Madrid, so it will go presumably in many places in Europe. . . . Now I have a splendid 15-year-old son and a couple of terrific nephews of the same age. You can imagine how troubled I am at the thought that they should be killed fighting for or against one of the stupid European "Idealisms."

Born wanted Landé's advice on how to get the boys into an American university. He asked him to reply to his institute on Drummond Street rather than home in order to keep the correspondence from Hedi, who saw the situation as less urgent than he did. Landé's response has not survived. In any event, neither Gustav nor his cousins went to the United States.

Born's new institute was joined to the Tait Chair in Natural Philosophy, named in honor of Peter Guthrie Tait, a scholar of quaternions (a type of complex number) and investigator of the aerodynamics of golf balls. The chair was established in 1922 for the purpose of teaching mathematical physics, and the holder was, in practice, head of the department of applied mathematics. Born's institute was located on the basement level of the physics department in what had once been the university infirmary (the concrete floors in the hallways still had troughs on either side for easy cleaning). It consisted of a lecture hall and one large room containing Born's desk and several long tables. At the beginning of his tenure, the tables were empty. No research fellows filled them.

The head of the departments of physics, Charles Barkla, and of mathematics, Edmund Whittaker, had been in their posts for almost twenty-five years. Barkla, an experimentalist, had received a Nobel Prize in 1917 for his work with characteristic X-radiation of the elements, and Whittaker was a highly respected mathematician. The physics was very respectable but not up to the standards of Cambridge, where most of Scotland's best theoretical physics students went, as Born discovered.

In the mornings, Born lectured to undergraduate students, a duty he shared with Robert Schlapp, a kind and excellent teacher whose reluctance to publish, even at Born's urging, kept him in the junior position of lecturer. With teaching and administrative chores, Born had no time for research. He also had to tackle a Scottish approach to physics that was foreign to him. The existing physics curriculum, based on rote fact and abstruse examples, was the antithesis of Born's, which emphasized new theories. In earlier days, he and his students in Göttingen had poked fun at this tradition by concocting examination questions: "On an elastic bridge stands an elephant of negligible mass; on his trunk sits a mosquito of mass *m*. Calculate the vibrations of the bridge when the elephant moves the mosquito by rotating his trunk."

The Scottish students struggled with Born's style as well as with his English. The precision of Cambridge English had diminished the German accent, but his rapid speech confounded the Scottish ear.

By the end of January 1937, after four months in Edinburgh, the family had settled down, yet Born had constant anxiety about war, their fate, and that of their relatives and friends on the continent—a stress relieved only slightly by finding positions and funds for refugee friends. The unalleviated tension was heightened by a sequence of losses: Hedi's muse Lou Andreas-Salome, Max's former benefactor Henry Goldman, Bertha Born, and Max's aunt, Gertrude Schäfer. Death had spared them the depravity that lay ahead.

When Born had met Goldman in London a year earlier, he could see that the catastrophe in Germany—a country Goldman always defended and believed in—had completely broken the old man. On the day before his fatal heart attack, Goldman had written to Born about positions for his half brother Wolfgang in the American art world. Wolfgang's job in Vienna as an art critic for a newspaper and radio had ended. He was unable to support himself and hoped for a job in the States. Goldman had nothing to offer; American universities had churned out too many art historians.

Amid losses were new beginnings. In April, Irene and Brin were married at the Rembrandt Hotel in London. Afterward, Max, Hedi, Brin, and Irene crossed the channel to France. Hedi went off to a little German village on the Rhine near the ruins of Ehrenfels castle to take a course in midwifery. Prodded both by her early ambitions to be a nurse and by the clairvoyant's vision on the train, she had enrolled in midwifery training at Edinburgh's Elsie Inglis Maternity Hospital for the fall and was taking an introductory course. The others went to Paris. Newlyweds Irene and Brin enjoyed the springtime glory of the City of Love. Max gave lectures at the Institut Poincaré and at the Free German Hochschule.

Gustav reached a milestone later that summer—his sixteenth birthday, for which he was in Fife across the Firth of Forth at an Officer Training Corps camp as required by his school. His father wrote a birthday greeting that he has treasured ever since.

> My best wishes on your birthday! . . . Tomorrow I will think of you. You are my best friend and comrade in spite of your youth. I am so pleased with your development and see in you a better copy of myself. Because of you I would

like to live really long and be by your side, not only with external support but also with exchanges of thoughts and feelings.

When Gustav returned from camp, the family went to Germany one more time. Again they split up, with Hedi—Gustav in tow—making for Göttingen and Max meeting Gritli and Rudi Ladenburg in Heidelberg. They found that the swastika had pushed out the gaiety and romance, but the picture of the knight in shining armor still guarded the entrance to Max's rooming house of thirty-five years earlier.

In Göttingen, Hedi and her son met much of the same. The statue in the market square of the girl carrying her geese was overshadowed by the Nazi flag that hung from the Rathaus. In June, the university had celebrated its founding by King George II of England (and Elector of Hanover) two hundred years earlier. The town's winding streets had resounded with marching SA troops, cheered by townspeople who lined them. At the close of the first evening's ceremonies, a chorus of several hundred sang a "Festival Cantata" written for the event; the paean to Germany ended with mystical, neopaganistic lines:

Who dies for you dies not in vain.
Thou bearest him into eternity.
We are mere hostages for thy life.
We are guarantors of thy glory.
Thou hast fed us with thy holy blood.
We were sworn to you ere we were born.
We are the mortal stuff of thy eternal life.

A physics faculty member at the event was Richard Becker, one of Born's first students in 1909 and now the successor to his chair. When the Ministry ordered him to Göttingen, Becker had written an apologetic letter to Born, asking forgiveness "for occupying this venerable position." Becker explained the move as the "least evil" of his options and thanked Born for the support to his academic start.

On his return to Edinburgh, Born found that he had two German research assistants. After a year, the long tables in his room at the institute were no longer empty. One of them, Walter Kellermann, a blond giant, was the son of a rabbi. The other, Klaus Fuchs, an intellectual type, was the son of a "confessional" pastor. Kellermann was a competent physicist, but Born considered Fuchs brilliant. A Communist who had escaped from the Nazis, he had finished his doctorate under Neville Mott at Bristol. Exhausting funds there, he was now in Edinburgh to do research with Born and work on a doctor of science degree.

Born described the twenty-five-year-old Fuchs as "weak in appearance but with a powerful brain, taciturn, with a veiled expression which disclosed nothing of his thoughts." Irene said that he looked like someone who needed a warm coat and a good meal. Fuchs was a frequent dinner guest and second violin in Born's quartet. On outings to the movies and on walks, the Born children and cousins braved long discourses on communism.

Having an excellent assistant did not mean that Born could begin research right away; he was still occupied with teaching and administrative duties. When Robin Schlapp offered to take over the latter, Born finally started working with Fuchs.

At almost forty-six, Hedi's life too was beginning to come together. Since the beginning of the year, she had regularly attended Sunday Quaker meetings and lectures at the Friends' House. After the swami awakened her spiritual side, she was drawn to the Quaker ideals of social justice. In the fall, the Edinburgh Meeting accepted her membership, and she began a round of conferences and speeches and wrote articles for their publications. After many years and false starts, her path to fulfillment was before her. She later wondered about her earlier life without God.

At the end of October, suitcase in hand, she reported for the midwifery internship at the Elsie Inglis Hospital. For the following two months, she lived and worked there almost nonstop watching deliveries—sometimes as many as seven in the morning and eight in the afternoon—going to lectures and taking rests as best she could. On her occasional afternoon or evening off, she met Max and Gustav for dinner and a movie. When she had a few hours at home, Max told her he was so happy to see her that he felt as though he "hopped around like Trixi."

In the course of that year, Max and Hedi's relationship had evolved into a "true loving marriage between old people." He thought it ironic that especially at that point, when they were close, they were so often apart. He understood the reason—she was "following God's direction"—and Max respected her faith. Religion, however, was an area rife with frictions. He told her not to proselytize him. Although he was neither an atheist nor a skeptic, he chafed at any form of group worship, even the silence of the Quakers. He needed to trust in God, but he wanted to do it his way, and "as I do not criticize the way of others I like to be left alone, too." God, he said, "does not leave me without direction but he talks to me differently."

Irene and Brin came for Christmas that year, and the family enjoyed a festive celebration on Christmas Eve. Hedi was back at her post on Christmas morning. That afternoon, she suddenly saw floaters in her right eye, and a week later, on New Year's Day, she was operated on for a detached retina. The doctors told her that it was unlikely she could work as a midwife. She was stoic about it.

The beginning of the year was difficult. Hedi stayed in bed for six weeks, after which her eye was functioning, though never normal. Then Gustav came down with whooping cough; and Max had gallstones.

Through it all, Born and Fuchs's research progressed. As many others had tried to do, Born wanted to merge quantum theory and relativity in order to further the theory of elementary particles. Questions about the electron's charge and mass continued to intrigue him.

He knew from cosmic rays that particles existed with energy many times that of their mass at rest. This led him to the conclusion that their velocities were almost at the speed of light. Since a small variation in their velocity would have a large impact on the momentum and energy, Born thought these two variables should have significance, independent of velocity. He wanted to extend classical mechanics to include this hypothesis. He called his new idea the Reciprocity Principle because he was looking for symmetry between the space-time coordinates and time in relativity with the momentum components and energy in quantum theory. Born had sent a couple of short articles to the *Proceedings of the Royal Society* but he was unsure what it all meant. He and Fuchs were only at the beginning; but he wrote to Courant that he was very curious.

All of this faded into the background as German aggression struck in earnest. On March 12, 1938, German troops crossed into Austria, and the next day Hitler announced its annexation. Prime Minister Neville Chamberlain, saying that Hitler's move could only have been prevented with armed force, did nothing more than to condemn it. The French government espoused similar thoughts.

Wolfgang Born had made it out of Vienna and, with Courant's help, found a position at Maryville College in St. Louis; but Gritli remained in Austria and wanted to stay until June to finish her art degree. Her parents were upset. She told them that she had no difficulties and was quite happy, but they were not comforted.

Born's letters to Francis Simon at Oxford were pessimistic about Chamberlain's appeasement policy. During their five years in Britain, Born and Simon had built up a close friendship on their common refugee status. Born confided his worry that Britain was naive about the Germans and making serious mistakes. He hoped it was not too late for "Lindemann's friend Churchill" to take over the government; but the idea that Winston Churchill, a marginalized voice, would provide salvation was premature. As Churchill ranted about the Nazis and foretold of ominous events, few joined him in challenging Chamberlain's course.

The Lindemann-Churchill friendship had begun in the early 1920s and centered on compatible personalities drawn together by a mutual interest in science and warfare. Lindemann quickly became part of Churchill's circle, providing data Churchill needed to argue for stronger air defenses. Lindemann had brought Simon to Oxford, and they were close colleagues there. The Simon-Lindemann-Churchill connection would eventually provide Born some outlet for his anxieties, but not until Churchill's dark prophecies had come to pass.

The Anschluss increased Born's stress and worry. His sciatica flared up and he had trouble moving. He wrote to Hedi, who was in London, that he was overtaken by "a tiredness and lack of interest," but she sensed his underlying anxiety. He obsessed over trifling matters such as whether to go to a conference in Paris—having trouble balancing difficult and draining lectures with the pleasure of seeing friends. Whenever his mind wandered to hiking in the German countryside with Hedi, he felt tortured. Not that he wanted to go back, but he knew that he could "never think of the home country without, at the same time, thinking of the sorrow and misfortune of the people whom one cannot help."

The Anschluss increased refugee pressures. Erwin Schrödinger, who had returned to his native Austria, was under surveillance in Graz by Nazi authorities. He wanted to get out. Max von Laue in Berlin felt likewise. Born and Simon became part of a clandestine messenger circle that arranged a position for Schrödinger in Dublin as well as one for von Laue. The Schrödingers escaped and went there, whereas von Laue changed his mind and stayed in Germany. Meanwhile, the Borns were visited by Max Planck, who received an honorary degree in Glasgow. Planck's thank-you note astutely avoided reference to any sensitive topics.

Max bought train tickets for Gritli to Edinburgh. In the beginning of July, forced to leave almost everything behind, she boarded an evening train in Vienna, carrying only her rucksack. She crossed into Germany and traveled through the country to Aachen on the Dutch border. She got off the train there and bought a ham sandwich from a vendor on the platform. Finding it moldy, she complained. An SS officer rushed over and yelled at her, demanding her name. Frightened, the twenty-two year old complied. When the officer said more calmly that he had a relative named Born, she responded with a nicety and took the opportunity to retreat quickly to the safety of the train. Her parents were relieved when she arrived in Edinburgh.

The week that Gritli left, delegates from thirty-two countries met in Evian, France, a resort town on the southern shore of Geneva's Lac Léman. Their purpose was to develop an emigration policy. Germany's annexation of Austria had exacerbated an already serious refugee crisis, adding 185,000 Jews to the 350,000 still in Germany. Only 150,000 Jews had fled Germany thus far, in large part because no country was willing to accept such a vast amount of refugees.

Ten days of meetings in Evian resulted in only one country, the Dominican Republic, increasing its quota of German and Austrian refugees. The United States, which had a strict quota system, promised only to fill all its allotted places, something it had not done for the previous five years. Great Britain thought it might have some space for resettlement in its East African colonies; France claimed it had reached a saturation point; Australia did not want to import a racial problem. Anti-Semitism and concerns about the economic depression restricted improvements in possibilities.

The German reaction to this? At the conference's conclusion, the *New York Times* quoted the official newspaper for the German foreign office, which wrote,

> Since in many foreign countries it was recently regarded as wholly incomprehensible why Germans did not wish to preserve in its population an element like the Jews . . . it appears astounding that countries seem in no way particularly anxious to make use of these elements themselves, now that the opportunity offers.

To "cleanse" the country, the Nazis had tried to push Jews out through violence and repression. The conference was a clear sign, however, that no one wanted the Reich's remaining 550,000 Jews nor the tens of thousands more that came with the Germans' march into Czechoslovakia's Sudetenland a few months later. The message from the conference to the Nazis was that their emigration strategy would fall short and that other governments had no commitment to protect the Jews.

The Nazis stepped up their level of ruthlessness. The Borns felt it even in Edinburgh. On September 2, Born wrote to Einstein that the Gestapo had just confiscated his house in Göttingen and his pension income. The pretext was his lecture at the Free German Hochschule in Paris the year before. The Gestapo statement, which Born was not privy to, read as follows.

> The Jew Born, by serving this institution [the Hochschule], the main goal of which is the fight against Germany, damaged the reputation of the Reich in a most serious way. There is a court case against Born for him to lose all rights. On the order of the Gestapo in Berlin of 18 June 1938, all of Born's possessions in Germany are to be held until confiscation. Part of this property is a special bank account.

Since summer 1937, the government had deposited 888 marks monthly into a special pension account at the Deutsche Bank for Born and had approved them through 1938. The money was restricted to use in the Reich, and the Borns used it to help relatives. In conjunction with the Quakers, Hedi also planned to transfer some of it to a refugee organization in Vienna. Hearing the depressing news, she immediately contacted the bank, and through careful questioning learned that the Gestapo's confiscation order had not yet made it there. She was able to transfer the monthly funds to Vienna until the end of the year. When the university closed the account in December, the balance was only 77.53 marks. Hedi had a small victory.

Hedi used the strength of her newfound faith to endure the worsening situation. Max had less to draw on. He could feel her strength, but he could not muster his own. His had always come from a belief in fundamental truth, followed without regard to tradition, interest or purpose, which carried its own value and redemption. He had found it earlier in his work. Now he felt his efforts were no longer sufficient to do so. Particularly troubling to him was that with "fanaticism versus fanaticism, belief versus belief, the distortion and repression of truth is raised to a political art."

Perhaps this was a reason why, for the first time, he used practical science—weapons designs—to quiet his anxieties rather basic research. One idea was an "air-torpedo" that struck incoming fighter planes. Born conceptualized "a small, extremely fast, unmanned airplane that, when it reached a certain height, automatically steers to the point of maximum (propeller) noise and uses" microphones on each wing to determine acoustical wavelengths. He wanted to pass it on to Lindemann but remained reticent. At the time of the Anschluss, Born's plea to Lindemann to support the creation of a Jewish state in Palestine had met with a perfunctory response.

Simon thought Born's idea a worthwhile suggestion that had a few kinks (like the new quiet German planes) and suggested that he contact people at the Air Ministry—but warned that they were not interested in long trials. He had recently brought Lindemann an idea about using liquid hydrogen as a long-distance fuel for airplanes and had it dismissed on a completely threadbare basis.

When not thinking of Britain's defense, Born contemplated escape from unsettling currents in England. He was disturbed by the complacency of the British middle class to the events on the continent as well as by the pro-Nazi propaganda of some newspapers and newsreels. His wondering whether the British feared bolshevism more than fascism led him to ask, "How long will we be able to stay in this country?" There was reality behind his perceptions, but any ideas of escape were merely fantasy.

October 1938 marked the Borns' five years of residency in Britain, and they applied for naturalization; approval was not perfunctory, and they did not know if they would be accepted. Coincidentally, at the end of that month, they received a standardized letter from the German consulate in Glasgow. By a new order of the Reich minister of the interior, passports of Jews were only valid when stamped with a *J*. The Borns had until November 15 to send in their passports to receive the stamp.

They did not comply. On November 15, the official Reich newspaper listed the five Borns among those who lost their German citizenship. The family was now stateless. A few weeks later, the University of Göttingen, at the request of the Ministry, revoked Born's doctoral degree.

The idea of the *J* had not originated with the Nazis. After the Anschluss, the Swiss government had more Austrian Jews in the country than it wanted. A Swiss official wrote to the Germans that "Switzerland, which has as little use for these Jews as has Germany, will herself take measures to protect Switzerland from being swamped by Jews." The solution was a three-centimeters high, red *J* stamped into the passports of the Reich's Jews.

On the night of November 9, the Nazis instigated *Kristallnacht*, "The Night of Broken Glass," a bloody, ruthless rampage against Jews. The murder of a Nazi embassy official in Paris by a distraught Polish refugee was the stimulus. Gangs destroyed thousands of businesses and burned down more than one thousand synagogues. The police arrested tens of thousands of Jews and sent them to concentration camps. The press outside of Germany reported the atrocities, but no country increased its quota to rescue any more Jews from Germany, Austria, and Czechoslovakia.

Born sent Courant a long letter about the plight of their mutual friends still in Germany. There was no discussion of work. It was too hard to do anything under the circumstances.

Kurt Hahn had given Born news about people such as his old friend Ernst Hellinger and cousin Fritz Jacobi, who had been sent to concentration camps. (Both were eventually released.) "In the camps, the poor people are killed by hunger and overexertion more or less quickly. If only they could be rescued!" Born wrote to Courant. Hedi's brother Kurt (Rolli's father) had just been expelled from Munich, losing everything; he and his wife, Marga, were living with his brother Rudi in Göttingen. Under the Nuremberg Laws, those with two German grandparents could stay if they did not practice Judaism and if their wives were "Aryan." Rudi's wife, Helene, was Catholic; Marga was Jewish. Rudi opted to stay; Kurt had to leave. Max applied to the British government as the financial guarantor for Kurt and Marga, so that they could obtain British visas.

Plenty of Borns, Kauffmanns, Jacobis, Lipsteins, and Ehrenbergs were still in Germany, including Max's sister, Käthe, in Berlin. As Max wrote to a friend, "In my family the catastrophe is horrific; with the best will in the world, I am unable to help." Käthe's five children and Renate Born had gotten out, and so had some of his Jacobi cousins, but not their mother, Aunt Selma. His cousin Hans Schäfer lived in Holland with his Dutch wife. Hans's sister, Helene Perlhöfter, had made it to London. Most who had emigrated from Germany were penniless, and to help them, Max earned extra money as a Ph.D. examiner for other universities.

Letters from colleagues who desperately needed assistance inundated Born. He did not confine his efforts to scientists. Alfred Wittenberg, a brilliant violinist with the Klinger Quartet in Berlin whom Born had known years earlier through his Lipstein relatives, sent him imploring letters. Born, unable to help, later wrote that "British musicians were not as magnanimous as the scientists and tried to keep competition out." Wittenberg disappeared.

Hedi helped friends and relatives too, but she also had further ideas. In December, she launched the Domestic Bureau, an agency that placed young Jewish women from the continent in domestic service jobs—the only kind for which visas were readily obtainable. Employers paid the girls in room and board. To complement this activity, the bureau set up a hostel for those who were out of work or had days off from their jobs. It also trained girls over the age of sixteen who arrived under the Care of the Children campaign and were not adopted.

Many of the women were depressed and in dire straits. In addition to the stress of immigration, they lived with loneliness, poverty, horrible memories of the Nazis, and anxiety about families left behind. Some had breakdowns, and a few tried to commit suicide. Hedi organized an administrative office in the Friends' House and for more than a year chaired the board—taking the lead in coordinating activities with other Scottish refugee organizations and raising funds for the 300 or so women whom the British allowed into the country.

At the start of 1939, the Borns made a concerted push to get relatives and friends out of Germany. Kurt and Marga made it to Edinburgh, as did Käthe, finally. Born arranged a physics lecture in Edinburgh to be given by his old friend Heinrich Rausch von Traubenberg, as a ploy to get his family out. (Heinrich was Estonian nobility, but his wife, Maria, was Jewish.) But the Nazis only let the two daughters go. Dorothee and Helen came to Edinburgh, and Hedi enrolled the teenagers in a boarding school. Heinrich and Maria stayed in Berlin, where he had a serious breakdown.

In spite of Born's worried mind, his little department was burgeoning with seven research students, whom he listed as "German-Aryan, German-Jewish, Irish, Indian, Chinese, Polish, and French." The long tables were almost full. As politics destroyed departments on the continent, Born thought his might become the biggest center of theoretical physics in Europe. The work gave him some diversion. Every morning, he stopped at each student's workplace to discuss problems, and no matter the area—stability of crystals, thermodynamics and melting, quantum physics—he would instantly change gears. Much of his versatility came from always having challenged himself from his days with Hilbert to start with the basics and think through the problem on his own without relying on the literature.

His own research with Fuchs still focused mainly on reciprocity, looking for the symmetry between relativity and quantum theory. As with his earlier work with Infeld, he hit an impasse with the mathematics. Much of his and Fuchs's time went to undertaking the transformation of the space-time coordinates and the momentum-energy components.

In March, the Royal Society, Britain's most prestigious scientific organization, elected Born a Fellow. He was the first of the recent refugees to be chosen. It was a more than fitting replacement for his dismissal from the Göttingen Academy of Sciences the previous year. In May, he went to London to be inducted. He did not return the following month for the society's lecture series given by German scientists, even though he was in Cambridge at the time. Throughout the "peaceful" aggression of the Nazis, scientists in the two countries had maintained contact. Born had no interest in meeting the foreign delegation.

The spring brought bright moments of happiness for the family. Gritli married Maurice Pryce in a small ceremony in Edinburgh. Afterward, the couple took off to the University of Liverpool where Maurice had a teaching position. Irene was pregnant with Hugh, who was born on July 3; and the Edinburgh sheriff called Max to his office to give him good news on the progress of his naturalization. It was not settled, but signs were positive.

In August, Max was relieved to learn that his naturalization papers had gone through. The undersecretary of state issued the official Certificate of Naturalization to him and Gustav on August 31, 1939. As a minor, Gustav came under his father's application. Hedi, whose papers were filed separately, did not receive a notice.

The next day, Friday, September 1, Germany attacked Poland, unleashing a devastating blitzkrieg on eight cities. The week before, Germany and Russia had entered into a nonaggression pact, as Max had predicted three years earlier. It was a time when one could usually be right by predicting the direst and least expected events, and Born had a penchant for seeing all events in the darkest light, including the resolve of his adopted home.

Not bothering to wait for a declaration from the Chamberlain government, Edinburgh started to prepare for war. The Scottish crown jewels went into a fireproof vault under the Crown Room in the castle. The one o'clock gun fired from the castle each day was silenced, thereby stopping the city's automatic time check.

Refugees began to adapt in their own modest way. Hedi wrote her diary and letters to other German refugees in English, a switch made by most. It was an emotional response at first, but as the government started inspecting letters, it became pragmatic. Letters in German, especially when they included mathematical equations, confounded the censors and led to a delay at the least. Friends could find the new language uncomfortable; Franck expressed his restraint to Born.

> I actually do not know why I am full of inhibitions to write to you. Certainly it is not lack of interest. Maybe the reason is that I do not like to use the English language in writing to you, or that I cannot convince myself that we always will be forced to communicate by letters only.

On Sunday, September 3, the Borns had a few people over in the late morning—Klaus Fuchs, Reinhold Fürth (a new associate who had recently arrived from Prague in one of the last group of émigrés) and the two Traubenberg girls, who were visiting from their boarding school. They were in the Borns' sitting room about to have tea and cakes when the air raid siren sounded. Everyone jumped up to stand in doorways or scoot under tables. Half an hour later, the all-clear signal rang. Friendly aircraft had been mistaken for the enemy. The Borns and guests returned to the sitting room to find that the youngest Traubenberg girl, Helen, had remained there and in her anxiety eaten all the cakes. Cakes would soon become scarce. That morning on the wireless, Chamberlain declared, "This country is at war with Germany."

Like the entire country, Edinburgh shifted quickly to a war footing. Gas masks were slung over shoulders; garages were converted to air raid shelters; and a blackout took effect—clear but quiet signs of war. The authorities evacuated city children to rural areas—thousands on the first day—and ordered all aliens to report to the police station for registration. All of Max's foreign research assistants, Hedi's brother and sister-in-law, and many friends went through the process. Fortunately, Hedi had been granted naturalization just days prior to the requirement.

A census taken in Edinburgh listed 1,401 aliens. Of these, 370 were "enemy" aliens—that is, of German descent, including German Jews. The government set up the East of Scotland Aliens Tribunal to distinguish between those who were in the country for business purposes and those who were there as refugees. By the end of the year, it had reviewed the registrations of some 800 aliens and given them one of three classifications: *A* for those detained because they were not considered reliable; *B* for those who were restricted, as by prohibition on radios or a 10 P.M. curfew, but not detained; and *C* for those free from all restrictions. The Borns' relatives and associates, as "refugees from Nazi oppression," fell into categories *B* and *C* and went about their usual business. Naturalized citizens such as the Borns were not classified, but received identity cards, as did all citizens.

On the clear afternoon of October 16, a German bomber swooped over Edinburgh's Forth Bridge and dropped two bombs, one for each of the navy cruisers docked alongside. The Scots, hearing explosions but inexplicably no air raid siren, went outside to watch, thinking it was a practice exercise. The Luftwaffe's first air raid on Britain did little damage.

Hedi saw it all while riding her bicycle. Most mornings and afternoons she rode the three miles to and from Niddrie Mains, a public housing project for Edinburgh's "poorest of the poor." With few possibilities of getting more girls out of Germany, she did less at the Domestic Bureau and more with her earlier interests of nursing, in spite of the doctor's prognosis after her eye operation.

Born continued his teaching—the number of undergraduates had not decreased. What he wanted to do was to work for the war effort, but government regulations specified that researchers had to be British-born. He sent a letter to Lindemann via Simon volunteering himself and his assistants. Lindemann had, in fact, pushed for refugee involvement but was constantly told, "The matter is under consideration."

With the exception of two air raid warnings, Edinburgh was quiet, along with the rest of Britain. It was the "phony war," a time of few noticeable hardships. Due to the blackout, no glow from streetlights or houses and restaurants brightened the night. Food was more expensive. Neither of Born's sons-in-law was called up, however, and Gustav, a second-year medical student at the University of Edinburgh, was told he could finish his studies. Everyone carried on as usual.

Gustav was following in the medical tradition of his grandfather and great-grandfather. When he finished the Edinburgh Academy in summer 1938, he had to decide, in the European tradition, what to study at university from the outset. History and music as well as the sciences drew his interests, although not mathematics. (He was in a singularly unique position to know that he was no mathematician.) He based his choice partly on his father's counsel. Max had said that a war was likely and if Gustav studied medicine, he would be in a position to save lives rather than take them. Gustav never regretted following his father's advice.

The quiet of the war thus far offered a chance for contemplation. Max borrowed a biography of Abraham Lincoln from Gustav's shelf to learn from history "how the knot in which we ourselves are tangled up, will be loosened." He was spellbound but, at the end, no further educated. The principle of suppression and racial separation that paralleled the human rights issues in Germany was only one factor. In spring 1940, he wrote to Gustav, who was visiting Irene in England.

> Then the slavery question was only part of the greater problem, union or secession, aristocratic Latifundium-economy or industrial state, etc. Just as today the conflict is not only caused by the Nazi theories of German racial supremacy but also by many other problems: the union of Europe versus nationalism, social equality, colonialism. It is so difficult to find a rational attitude.

He ended the letter with his persistent fear that Britain might not press the nation's resources sufficiently to withstand the Nazis, should that situation arise. Chamberlain's initial appeasement policy had planted this worry firmly in Born's mind.

Within weeks, the time for quiet contemplation ended. Germany's actions in April 1940, as it rolled over Norway and Denmark, signaled a close to the phony war. Watching these assaults, Born understood that there could be no peace with the German or Italian fascists. "Political ideologies are a much stronger driving force than a simple question of political equilibrium," he wrote to Courant in New York. At the same time, Born's pacifist nature struggled against the reality of the threat. He looked to Courant to help him sort out this dilemma, while chiding him about the neutrals' "fairy world where there is peace—yet." Born saw an idiocy in watching as one's neighbors were overwhelmed and waiting for one's own turn to come, rather than standing and fighting together.

One personal somber note in the letter was the emotional impact that the sudden death of their friend Otto Toeplitz had on him. It marked the first loss from the original Breslau-Göttingen group.

Unhampered by pacifist feelings, Courant answered that he felt greatly frustrated with the neutrality of America, whom he saw entering the war only when in "deadly danger." Britain was not going to get help from them any time soon. Yet he also knew that he had to be cautious about expressing his views. Not being a native-born American, he saw himself under "the suspicion of strong prejudice" and knew that immigrants—such as he—should not try to influence the situation.

Born continued his research. He thought his reciprocity ideas had at last allowed him to unify wave mechanics and relativity properly. In addition, he had discovered certain representations of the Lorentz group. The results were encouraging, especially with Pauli telling him, "I think you are on the right track." Max enjoyed telling Einstein about them. By the end of April, he had planned the next stages.

One evening in early May, the Borns invited Fuchs and Kellermann over to play bridge. It was to be the last time. The Germans attacked Holland and Belgium that night. The two countries mounted fierce opposition but did little to slow the German attack. A couple of days later, Hedi wrote in her diary that

Scotland's east coast had become a protected area. "All male Germans and Austrians, including refugees 'rounded up' and interned. Rolli, too."

Hedi's nephew Rolli was in his second university year at St. Andrews, now in the protected area. Police had knocked on his door at 5 A.M., put him on a bus, and transported him to Edinburgh. They locked him up in an old school with other "enemy aliens." Max discovered where he was and visited him, but all he could do was give him some pocket money. Born could not get him out. For Rolli, though, the feeling of not being forgotten was most important.

Fuchs and Kellermann were scooped up too. Born soon found that they, together with the majority of internees, were being held on the Isle of Man, a small island off the northwest coast of England in the Irish Sea. They lived in makeshift, segregated camps—usually hotels or rooming houses surrounded by barbed wire. Sometimes food and medical supplies were adequate, sometimes not.

This internment was the first of three that swallowed up increasing numbers of refugees as the Germans pushed closer to Paris. The policy resulted partly from rabble-rousing headlines such as "Intern the Lot" and partly from government fears of German treachery—the notorious Fifth Column so effective in the collapse of the Scandinavian countries and of Holland and Belgium. The government was taking no chances.

Max's niece, Renate Born, was also at St. Andrews in the protected area on the first day of the roundup. She slept late that morning then went to tea at a professor's house in the afternoon, knowing nothing of what was happening. When she walked in, the group looked at her as if she were a ghost. They assumed she had been interned. She was saved that fate, but when France fell, St. Andrews became off-limits to her. She ended up at the University of Glasgow with a scholarship from the International Student Service to continue her chemistry studies. Since Edinburgh was in the protected zone, she could no longer visit the Borns. Instead, she and Uncle Max occasionally met for tea in Peebles, a small town abutting the south side of the zone.

The authorities gave the Domestic Bureau three days to move scores of young women who worked in the protected zone to Glasgow. The staff and board members feared that the women would suffer from this hasty departure. Hedi and another committee member, Olive Ludlum, marched into the police station and delivered an ultimatum: If they stuck to the three-day limit, the bureau would deliver all the women and their luggage en masse to the police station. The two won extra time to move the distraught refugees. (Hedi was well known to the police: They regularly found her bicycle and called her to fetch it.)

The university had a break at the end of May, and Born went alone to the Scottish village of Bonskeid on Loch Tummel to stay at a YMCA hostel. He knew that his reciprocity research was at an end—not only because Fuchs was gone but also because he assumed an assault on Britain was coming. Having been optimistic about his success, he regretted abandoning work on his theory, and in a poignant gesture of finality, sent John von Neumann, who was at Princeton, his latest reprints and drafts "to use them when we are definitely prohibited [from] working."

Born had more compelling concerns on his mind, though. The well-being of his children plagued him. Both daughters were in the early stages of pregnancy, Gritli with her first and Irene with her second. They would have to stay in Britain with their husbands. Gustav had no such commitments, but he also had no good alternatives to Britain. The day after arriving at the hostel Born wrote to Simon.

> The last weeks have not soothed my nerves. But I feel that in this quiet valley I shall quickly recover. I wonder what you have felt during the last weeks. Although we have prophesied some of the events, facts have been worse than imagination. It worries me very much that so much life, wealth, happiness is wasted by the omission of proper foresight and precaution.

Foresight was something that Born thought the British had too little of and the Germans too much. Considering himself an astute observer of the "fiendish cleverness of the German militarists," he again wondered whether the British were really up to taking them on. This concern led him to write to Lindemann about the Germans' landing troops on the lonely Scottish lochs.

The British had incorporated the Highlands into the protected area, but Born wanted to emphasize their emptiness and the Germans' ability to land men in seaplanes. Lindemann's secretary responded that Born's proposal was passed to the appropriate authorities. Lindemann himself was too busy to answer. Churchill had just been named prime minister, and Lindemann, who had been a personal and scientific advisor to the former First Lord of the Admiralty Churchill, was part of his inner circle.

The peacefulness of the Highlands stimulated another of Born's ideas for Lindemann—a "dirigible bomb" that could be directed, as opposed to dropped in a free fall, and, therefore, destroy bridges over the Rhine and disrupt military supply lines. Nervous about bothering Lindemann again, he wanted to explain the idea to Simon and get his opinion.

> [It] could be directed from a plane at great height with help of electric waves. It looks [at] first sight impossible as no geographical direction can be fixed. But it would be done in this way: The bomb carries rudders, two sets

> in mutually perpendicular directions; one of these directions is visible as a line in the telescope of the aiming man in the plane who keeps a line in his field of sight coinciding with it. The motion of the telescope is coupled to two switches, which control the radio waves for the "parallel" and "normal" motion. In this way the motion is in principle completely determined by the observer. Whether it is possible to do is another question. But as Dymond has built a sending apparatus in his balloons for registering cosmic rays on the ship (as the balloons in Greenland were generally lost) I can see no obstacle for such a directing device.

Simon gave him the needed reassurance.

Lindemann's response is lost, but the British did not use Born's idea. Others thought of the same thing, though. In September 1943, high-altitude German aircraft dropped the so-called Fritz X, which a bombardier directed using radio signals, on the unsuspecting Italian battleship *Roma* and sunk it. The United States developed a similar type of bomb called the AZON, which it used to destroy the bridge over the River Kwai, cutting the major Japanese supply line that went from Burma to Thailand.

Born arrived home from his rest on June 13, 1940. Paris fell to the Germans the following day, and he put aside all thoughts of weapons. The British government initiated its final internment order, which included mostly Jewish refugees. Hedi's brother Kurt was one of them. Another, the chemist Fritz Meyer, a friend of Born's, committed suicide when he received his internment notice.

All totaled, the authorities rounded up 27,000 refugees—most of them Jewish—and kept them in camps. Of these, 7,500 were transported to Australia and Canada, the latter the destination of Fuchs, Kellermann, Rolli, and Käthe's son Helli. Fuchs sent Born a letter from the Isle of Man.

> [I want] to express in a few rushed words all I owe you during the time I have been with you. . . . I find it hard to cut away from a country which I have learned to love and until the last moment I hoped that, especially at such a time, [it] might not be necessary. But there is no point in deluding myself.

On June 17, Hedi wrote to Otto Oldenberg and Max wrote to Courant and Maria Göppert-Mayer. Faced with the prospect of an imminent invasion, the Borns were searching for a way out of Britain. Their emotions went beyond the fear of Britain falling, but rather to their fate as Jews. As Hedi wrote to Oldenberg,

> The Nazis may succeed in invading Great Britain—even Churchill admits this possibility. It would be terrible to fall into their hands again. I do not want to omit doing anything that would prevent this. . . . Nothing is known of the fate of the refugees living in Scandinavia, Holland, Belgium and France. We heard a rumor from Holland and Belgium that the refugees are being transported to Poland. This means—as you know—a cruel and slow death. You know that we wish to take part in any sorrow and sacrifice that must be born by the British. But the British are not threatened with this prospect of death in Poland. We still are hoping for the miracle that *must* come.

Hedi pleaded with Oldenberg to find Max a position in America. Max added a note saying that he feared they may not survive the coming weeks. His letter to Courant elaborated.

> I stand with the British as long as there is any possibility. But I must reckon with the possibility that a particular fate is in store for us, Lublin [Poland] or something equivalent. In this case, I shall not survive. But as nobody likes to go without any resistance, I wonder whether you and our friends, Einstein, Franck, etc. could find a way out for Hedi, Gustav, and myself. Gustav's fate worries me most. The girls are married to Englishmen and have to share their fate. We do not know yet whether this country will continue the fight. If you write that we shall live in a crazy world I am afraid that is an understatement—if we shall live at all.

If the facts were "worse than imagination" two weeks earlier, now they had become potentially fatal. Invasion, with German soldiers crawling over Edinburgh's rocky shoreline, was a nightmare. His letter to Maria Göppert-Mayer was more explicit. "If we have to submit too, I shall not survive; for I do not wish to be sent to Lublin."

Born knew what he would do. Göttingen friend and physician Maria Stein, who as an "enemy alien" was leaving Edinburgh, had said good-bye to the Borns that day. She gave Born half of her supply of sleeping pills—a deadly dose. She and Born were far from alone in contemplating this step.

On June 25, the first air raid alert of Britain's real war sounded in Edinburgh.

TWELVE

THERE ARE SO MANY IFS

THE WIRELESS NEWS BROADCAST ON FRIDAY NIGHT, JULY 5, 1940, SHOOK THE BORNS TO their roots. Former Germans, Austrians, and Italians in Scotland who had been naturalized since 1931 had to report to the nearest police station. The Borns pondered the obvious: Was an internment camp their next stop? When they registered at the police station on Monday, they remained free, but with a new line of red ink in their passports that marked them as special citizens. The police told the Borns that "in the meantime" there would be no restrictions on their freedom. They did not know if "in the meantime" meant days, weeks, or months.

One scenario the Borns foresaw was sitting in an internment camp surrounded by barbed wire, waiting to be seized by German storm troopers who would invade the beaches of Britain. Hedi and Max sketched out a long letter to Lindemann, telling him about the "nightmare of anyone who feels an object of Nazi persecution and vengeance." They asked for his advice, especially since Max had received a cable from Otto Oldenberg at Harvard proposing a U.S. lecture tour if he could obtain a visa.

They also worried about Gustav, who, having become thoroughly British, did not feel that he belonged "to an extra class of citizen." Max asked Francis Simon about getting him to the United States, but Simon said that Gustav, at nineteen, was too old. Only children under age sixteen could get into the country without a sponsor. Ladenburg wrote that he would join forces with Einstein and Oldenberg to get Gustav out if need be.

The day after the Borns wrote to Lindemann, German bombers attacked shipping in the English Channel and radar stations on England's south coast. This was intended as a prelude to a full-fledged attack. The Germans were amassing troop barges all along the French coast. They planned to knock out British air defenses then deploy troops across the channel.

Lindemann responded to Born's anxieties matter-of-factly: In wartime, for the government to keep track of recently naturalized citizens is reasonable, because some might be "unreliable." He saw no need to worry if one acted properly. "In these days," he chided them, "it does not do to be unduly sensitive." Lindemann's personal fear was being accused by his political enemies of helping German refugees. His good friend Simon already knew that Lindemann would not rescue him if he were interned.

A number of brave academics did, however, step in on behalf of the refugees, and by mid July, criticism of the internment policy forced the government to reconsider its orders. The policy gradually unraveled. Born stopped worrying both about it and Britain's ability to withstand the Nazi onslaught. He wrote to John von Neumann, "The situation that seemed hopeless after the breakdown of France has improved." In August, Otto Oldenberg wrote about a temporary academic position in the United States, and Born, after realizing the improbability of bringing Hedi and Gustav with him, asked that it be kept as an emergency backup.

Born turned his attention to getting Klaus Fuchs and Walter Kellermann and his nephews Rolli Ehrenberg and Helli Koenigsberger back from Canada. He asked Einstein to find the addresses of Fuchs and Kellermann and get books to them. He wrote to Bridgman and von Neumann too. Some of the refugees shipped to Canada were being released. Born thought Fuchs might be released under the "distinguished scientist" exception and worried that he would have no means of support. He told von Neumann, "Fuchs is really a rare personality and it would be a shame to let him go down after he has suffered by Nazi persecution more than any of us."

Cambridge physicist Ralph Fowler, now in Ottawa serving as science advisor to the British High Commission, was also on Born's list of contacts. Fowler could not do much; there was a new process. The Society for the Protection of Science and Learning, a refugee organization for academics, compiled a list of scientists who should be released. The society gave this list to a tribunal set up by the Royal Society, which, in turn, passed its edited version to the Home Office for action. Born learned that Fuchs and Kellermann were already on the tribunal's list.

At the beginning of August, German bombers assaulted British airfields, launching the Battle of Britain in earnest. Edinburgh was largely removed from this destruction, but had its share of air raid warnings. In mid August, the Borns sought peace in a cottage on a wooded slope that overlooked the Ullswater in the Lake District. For an entire month, Max, Hedi, Gustav, Gritli and Maurice Pryce, and Paul Dirac and his stepson, Gabor, rowed on the lake, biked, and picnicked. Dirac, with his long legs and lanky build, led the climbing on surrounding rocky cliffs (where Gustav discovered his fear of heights). In the evenings, they played bridge. There was to be no physics shop talk, but Gustav seemed to hear a lot of it. At night, they slept free of the worry of air raids and bombs.

By the start of September, the Germans began nighttime bombing of industrial cities. Incendiary bombs fell on London every night for almost two months. Maurice Pryce, who had to leave Ullswater early and return to Liverpool, reported bombs falling around the apartment. The Blitz killed 30,000 people and left a million and a half homeless. Even so, 60 percent of Londoners slept in their homes—not in air raid shelters.

The Borns were still in the Lake District on September 7 when the Home Guard in Edinburgh, as in other parts of the country, went on full alert against invasion. It was the last time that tide and moon conditions were favorable for the Germans to cross the channel; but, still fearing the threat of the Royal Air Force (RAF), they let the opportunity slip.

Bombers continued to attack the south daily, so after the holiday, Gritli, who was six months pregnant, stayed in the Lake District. Irene, also pregnant, came to Edinburgh with little Hugh to stay with Max's assistant, Reinhold Fürth, as paying guests. The RAF had called up Brin and the admiralty, Maurice.

Air raid alerts triggered by German bombers flying over the city on their way to Glasgow's industrial center sounded regularly in Edinburgh. Few bombs were dropped, but in late September, five hit a distillery containing a million gallons of whiskey, causing a great fire (as well as great sorrow in the townsfolk). By the end of October, after taking heavy losses in the air, the Germans cut back the attacks on Britain and turned their sights eastward. The RAF had proved more stubborn than the Nazis expected.

The Borns' sense of safety grew, as did the tedious and draining effects of war. Rationing, which had begun in January 1940, became more restrictive. Yet, with the government allocations based on nutritional requirements and with its subsidies for basic foods such as bread, cheese, and meat, many people actually ate better than they ever did. Sugar, fruits, ice cream, and most other nonessentials, however, were daydreams. Shortages were part of life because supplies came in convoys across hostile waters. Hedi reported that "after many weeks without, the whole country has been waiting for the 'onion' boat." When it arrived, she was delighted to get one large onion. The Borns were fortunate to receive occasional food packets from Ladenburg and Franck to supplement the rations.

The real war shrunk more than waistlines. The Indian, the Irish girl, the Pole, and the Czech in Born's institute all had departed. It now consisted of "a brilliant Chinese and two girls," as he put it. (Another girl came a little later.)

By fall, the British government had seen the shortsightedness of its policy on refugee scientists and allowed a limited number to work in weapons research.

After all, refugees Rudolph Peierls and Otto Frisch at the University of Birmingham had calculated just months earlier that only a few pounds of uranium 235 could create an atomic explosion. Simon, one of the best low-temperature experts in the world, was immediately pulled into "hush hush business," as he told Born. Born, however, had never been particularly interested in this "business" (radioactivity), thanks in part to a dull, disorganized course given by Johannes Stark during his student years. He was not pulled in. As he admitted to a friend, his ability was "not of the kind which is of great practical usefulness."

The loss of Fuchs stymied Born's research on reciprocity at first, but he found an assistant in the mathematics department to continue the work with him and was excited when new results showed no infinities occurring in his rigorously relativistic quantum mechanics. He had a new energy and optimism. His worst fears had not been realized; the British had stood firm against the German onslaught.

His interests expanded into other fields. Trying to improve melting theory, he generalized the thermodynamics of crystals but had no success. The work did lead, however, to a theory of lattice stability, which pleased him. When Fürth returned to the institute after authorities allowed him into the protected area, he and Born began analyzing why the properties of Fourier transformations hold in nonrelativistic theory but not in relativity. Rounding out his program, he worked on an explanation for the "new" Raman effect.

Raman had published an article earlier in 1940 describing new experiments on X-ray reflection in crystals (where X-ray photons bounce off the atoms within the crystal). He observed abnormal X-ray spots. Born set about trying to understand the phenomenon and developed a formula, the problems with which he laid out in a letter to crystallographer William Lawrence Bragg. (The title Sir Lawrence differentiated him from his equally notable father, Sir William, with whom he shared the 1915 Nobel Prize for their seminal discoveries in X-ray diffraction.) He wrote after seeing Bragg's article on Raman's research. Born explained that he had developed a formula similar to Bragg's but had rejected it because it captured only part of what Raman had observed experimentally and was theoretically unsatisfactory. In a letter five days later, Born apologized. He had just "found to my bewilderment that the article on 'Extra Spots of the Laue Photograph' is by your father, not by you." Born then wrote to Sir William, who passed some of Born's work on to another experimental physicist, Kathleen Lonsdale, at the Faraday Research Laboratory in London. Born and Lonsdale began an exchange, with Lonsdale correcting some of his assumptions.

Fuchs, Kellermann, and Rolli suddenly showed up in Edinburgh in mid January, almost like a New Year's present. (Helli Koenigsberger remained in Canada a bit longer.) Born wanted his researchers to stay but had no funds to support them, because the university had withdrawn all financial aid and the Carnegie Trust had decided not to give grants to "enemy aliens." He eventually found a teaching position for Kellermann. Fuchs worked in Born's institute until May, when he received an invitation from Peierls to come to Birmingham and participate in war research. Peierls did not specify the nature of the work, but Born and Fuchs both knew he meant atomic research.

Born had recently described his dislike of this research to Simon, while acknowledging that "it might be the only way out." As both men knew, *it* meant the atomic bomb. Simon responded with his own ambivalence, "If we have it long before the others and if it comes into the right hands and if it is managed by the right people also afterwards, then everything in the world may come to a happy end. But there are so many ifs."

Appealing to Fuchs's ideological beliefs, Born tried to dissuade him from joining Peierls. He later related to Gustav,

> When Fuchs went to Birmingham to join the uranium people, I had a serious talk with him. I told him that if they succeeded it would mean a new concentration of power in the hands of a few, and very likely the wrong ones, capitalists and nationalists; I warned him who confessed to be a communist that the result of this work would mean a strengthening of capitalism. He answered that it had to be done as they knew that the Germans [were doing] it.

Émigrés in Britain and America well knew the abilities of their former German colleagues. Just a few years earlier, they had all thrived together in the spirit of Göttingen, Munich, Berlin, and Copenhagen.

By June, Fuchs was in Birmingham, Born's warnings unheeded. A year later, he became a naturalized citizen and signed the British Oath of Secrecy, allowing him to work on the Tube Alloy (atomic bomb) project.

Born spent summer 1941 teaching cadets in a new air-training program, relieved to have any opportunity, however small, to fight the Nazis. Otherwise, he walked in the Pentlands with Hedi, went to the movies, played bridge with the Fürths, and read copiously. His tastes were catholic—a book on great mathematicians, detective stories, a survey of prehistoric archaeology. He did little research. His energy and enthusiasm of the previous fall had evaporated. Not participating in war research, he knew that he had been sidelined, out of the scientific loop. He once justified bothering Lindemann with a military suggestion by saying, "But you might consider my intelligence a little above average." It was a plea to help, not a boast.

One initiative for Born that summer was another attempt to help the Jews in Palestine. He wrote to Chaim Weizmann, who was president of the World Zionist Organization and a friend since they had met at Karlsbad five years earlier.

Born again wished to approach Lindemann (who had just become Lord Cherwell) and wanted Weizmann's opinion. Born felt it the duty of those Jews who had been saved "to help and ponder about the future."

With Weizmann's support, he urged Lindemann to aid the Jewish victims of torture by providing them a country in which to settle. That country, Born told him, could only be Palestine. Some Jews would return to their own countries after the fall of the Nazis, but, Born noted, "One must not be too optimistic, considering the fact that there is some anti-Semitism propaganda made, for example, by the Poles even here in this country." Cherwell responded that he would relay the concerns and apologized for a hasty note.

The war had entered a new phase. On June 22, 1941, Germany had torn up its nonaggression pact with Russia and attacked its former ally in force. Moscow, Leningrad, and Stalingrad came under a blistering strike. The Russians pulled back. Everyone worried whether Russia could hold out. Born was discouraged, particularly in light of these events, by Courant and Ladenburg's prognosis of no American intercession. In frustration, Born predicted to Gustav, "The Americans will get a thrashing from the Japanese, and then they will perhaps awaken."

Born invested his intellectual energy in a philosophical debate similar to one he had had in Berlin during the previous war. Physicists as well as British society in general were deeply divided over the issue of socialism. (Socialism was largely undifferentiated from communism, the only distinction being Marxist versus non-Marxist.) Britain had begun to tilt toward socialism during the Depression, when unemployment reached 22 percent in 1932. The Labour Party, which added left-wing intellectuals to its largely working-class base during this time, looked at this high rate, denounced "poverty in the midst of plenty," and pointed a shaming finger at capitalism. Its solution: introduce socialism and government programs; restrain capitalist greed and benefit the worker. All would work toward decreasing class divisions.

The war did not quiet these ideas; if anything, it promoted them. As one soldier said to another in the 1941 movie *Dawn Patrol*, "We found out in this war how we're all neighbors, and we aren't going to forget it when it's all over."

A founder of the Society for Freedom in Science, the Hungarian-born chemist Michael Polanyi, believed that communism subordinated freedom of thought—in particular, discovering scientific truth—to social welfare. For him, only a democratic, capitalist system could guarantee this freedom, although he did not object to economic planning as long as freedom of thought was the first priority. Born had received one of their pamphlets, which implied that socialism restricted freethinking. He sent a letter questioning this premise, and his old friend Polanyi responded by cajoling, "Socialists like yourself who wish to renew society on the economic side, while keeping mental freedom intact, should join us."

The debate was joined. Born retorted that he was "not a socialist, as you seem to think, if this expression means blind belief in Marxist theories." Dialectical materialism was to him "rubbish." Nevertheless, with the "western system of profit and vested interests," squalor and poverty existed for the masses and luxury for the few. The capitalist system—the unethical drive for profit—had supported the military buildups in Germany and Japan. Born wanted to temper "the ethical inferiority of the profit system" by merging the efficiency of free-market production with a regard for workers' rights. Based on reports he had read, he lauded the Russian system for "founding a new ethical background of economy, where there is no enmity between worker and management." To disprove the position of Polanyi's group, he cited Sweden's successful regulation of business with no effect on freedom of scientific thought.

Another of the society's pamphlets that Born received compared economic forces with those in physics and mentioned the three discoverers of quantum mechanics, "an Austrian, a Prussian, and an Englishman." In a two-page PS to a two-page letter, Born clarified his impression of this remark for Polanyi. "I suppose these are meant to be Schrödinger, Heisenberg, Dirac. I know that this is the official attitude as endorsed by the Nobel committee, apart from the inclusion of 'a Frenchman' (de Broglie); but Jordan and I are excluded."

Born then laid out the lineage of quantum theory: De Broglie and Heisenberg were the pioneers with the others, including himself, representing branches. Related to Heisenberg's ideas were Born's—which initiated the quantum mechanical formalism—and Jordan's—whom Born had asked to help—and some months later, Dirac's. From de Broglie's theory came Schrödinger's.

> There is no doubt that I was the first who ever wrote down a real quantum mechanical formula, not only $pq - qp = h/2\pi i$, but also $\dot{q} = \delta H/\delta p$, $\dot{p} = \delta H/\delta q$ as matrix equations. But the Americans always attribute them to Heisenberg (who, as a matter of fact, hardly knew what a matrix is, when I showed him these equations—he was my assistant at the time, and my paper with Jordan was written when he was absent from Göttingen during the long vacation). . . . I have never protested in public; for if ever a history of the development is written on the base of the original publications (not textbook) it will turn out all right. But I think you will not mind if I inform you privately. For we are both in about the same position, without the backing of a big nation.

Polanyi had never heard this version. He asked for copies of the articles to understand Born's claim of priority as well as for a reason why Heisenberg failed to correct the record. Born immediately replied that he did not claim priority; he just wanted his part recognized. "We were in close collaboration; he found a most important feature, not the final form of the theory." As for Heisenberg's actions, Born had a strong rationalization: If Heisenberg had made public remarks on behalf of a refugee, the Nazis would have punished him. "It would have cost him more than the prize, perhaps his existence." Born sent Polanyi a list of the original articles on the formulation of quantum mechanics and Heisenberg's letter of apology to him after the announcement of the Nobel Prizes.

After reviewing the articles, Polanyi took Born's side but disagreed with his defense of Heisenberg. To him, Heisenberg had had ample opportunity between 1926 and 1933 "to clarify the situation to which he alludes in his letter. He then could have told you 'I have stated the facts publicly. It is not my fault if they are not accepted.' I miss such a reference in his letter."

It was really the Royal Swedish Academy's award to Dirac by which Born felt most slighted. He equated his own contribution to the theoretical formulation as comparable to Dirac's, who had, like Born, used Heisenberg's breakthrough to develop a final theory. But he did not want Polanyi to cause trouble. He had enough problems on his hands, he said, with Raman, his former sponsor in India, who was both a friend and "a difficult fellow."

The reference to Raman denoted the nascent stage of the Born-Raman controversy, as it would come to be called. Born's earlier communications with Sir Lawrence Bragg and Kathleen Lonsdale had resulted in a thesis topic for one of Born's students. In a joint article, Born and his student explained that the spots recently found by Raman in the spectrum of diamond were due to thermal vibrations. It was a return to Born's early theory begun with von Kármán on specific heat and lattice dynamics—that atoms vibrate and exchange energy between them.

It was a straightforward explanation, but around this time, Raman published his own theory of the spectroscopic pattern—one that disputed Born's, thereby undermining a central principle of lattice dynamics. In 1912, when Born had modeled the structure of the crystal, he had ingeniously simplified his task by having his theoretical structure repeat itself so that the cells in the solid had no boundaries (edges). Otherwise, assumptions on energy distribution for the interior of the cell would not hold uniformly. Born's model leads to a frequency spectrum that is quasi-continuous. The Raman spectrum of diamond, however, is characterized by sharp vertical lines or peaks. To capture this spectrum, Raman developed a new theory that argued against Born's. (Ultimately, more sophisticated equipment showed that the sharp lines were a subset of a more continuous spectrum.)

When Born first began the exchange with Raman over lattice theory, he tried to be sensitive, writing to Polanyi, "It took me weeks to write a letter to 'Nature' which might not offend Raman; but he will be angry nevertheless." Kathleen Lonsdale commented that Born had treated Raman "mildly," leaving him room for retreat. Raman and his students, however, responded more confrontationally. European physicists, both theoreticians and experimentalists, sided with Born, which intensified the scholarly dispute. Certain that he was right, Raman saw the Europeans as colluding against him to support one of their own. He felt under attack.

The dispute had actually begun the previous year with Lonsdale taking on Raman over his statements about the effect of temperature on diamond. By early 1942, a friend of Born's observed, "The Bragg-Lonsdale and Raman armies are now confronting one another." Born was now equally involved in the battle, although the spirited Lonsdale was, in fact, the general. For the next few years of repetitive articles back and forth, she organized responses to *Nature* and the *Proceedings of the Physical Society* (major battlefields for the dispute), edited Born's letters and articles, and persuaded journal editors to publish Born's material in tandem with hers. She created an imposing European front, resulting in more separation with Raman. The battle over the same issues continued to wax and wane for years.

The intellectual fray ended Born's yearlong stretch of scientific hibernation. His paltry output of articles in 1941 mushroomed to eleven in 1942. It also cemented a long friendship between both Borns and Kathleen Lonsdale. Like Hedi, Lonsdale was a prolific letter writer and strong-minded Quaker. Within a year, she would earn a one-month stay in Holloway Prison for refusing both to register for fire-watch duty and pay the resulting fine (although she did her watch). She rebelled because the law did not exempt conscientious objectors.

Scientific work was a welcome but only partial escape for Born, as the tragedies in Germany continued to pile up. One Sunday evening in March 1942, the eighty-year-old Arnold Berliner, former editor of the *Naturwissenschaften*, hearing that his best friend had died, sat in his old armchair in his Berlin apartment and "went to sleep," as a friend described Berliner's suicide in a letter to von Laue. The Gestapo came a few days later and seized everything. For Born, it was the end of a relationship that had spanned more than forty years. It had been Arnold Berliner, who at a dinner party at the Neissers in Breslau, had bet the uncertain Born several bottles of wine that he would be a professor in ten years' time. Born wrote a tribute to him that recalled his heart, soul, and final tragedy.

Learning about Jews being deported from Germany to Poland had already traumatized Born. At the end of 1941, the Nazis had ordered 150,000 people onto freight cars. They allowed the deportees to take 100 marks with them then charged 90 marks for the ride. No one seemed to know where they had gone. Born listed his family still in Germany for Simon:

> a sister of my father, one of my mother, one of my stepmother, and three old aunts, cousins of my parents—6 old ladies all over 80; besides several children of these, amongst them a cousin Cläre whom I love very much. It is dreadful to think what they must suffer.

In spring 1942, Max heard that his cousin Cläre Kauffmann had died of a stroke, although he and Hedi, knowing nothing specifically, thought that the cause of death sounded strange. Then he received a letter from one of the "six old ladies," his father's sister, eighty-eight-year-old Aunt Selma Jacobi. She felt her death near and wrote to say good-bye. She knew but did not tell him that she would receive her notification of deportation soon. Max sensed that his entire family in Germany would perish. He felt an "intense hatred" and told Courant, "If Germany will suffer, I shall care only about the very few who are known to me to have been in opposition, like Laue."

Life went on. Hedi, a keen gardener, planted her victory garden: onions (no more waiting for the onion boat), potatoes, dill, red currant bushes, radishes, lettuce, purple kohlrabi, spinach, peas, celeriac, and French beans. Later, she added leeks. Gustav dug the plot, and Max helped spread manure around their apple trees. They waited for the Allies' "colossal production," as Max, mindful of the censors, referred to the anticipated Allied invasion to Maria Göppert-Mayer. So far, there was much talk but little evidence.

December 11, 1942, marked Born's sixtieth birthday. Hedi fixed him a lunch "like in peacetime" and baked him a chocolate layer cake. The math and physics departments held teas in his honor. Friends and former students sent telegrams and letters, many wistfully recalling his fiftieth birthday in Göttingen. Walter Heitler, who was at Schrödinger's institute in Dublin, wanted to compile a *Festschrift*—a volume of articles in honor of Born—but when he did not even hear back from Born's former assistant, Lothar Nordheim, he knew it was useless. Instead, he made a birthday wish that Born "be allowed for yet a long time to come to continue to work and to educate your pupils in this same spirit to which so many—including myself—owe so much." Born's best present came on the day after his birthday. He climbed to the top of Arthur's Seat with Gustav in less than an hour. He bragged to Franck that he still felt "quite young."

He did not mention the Japanese attack on Pearl Harbor, but the American entrance into the war must have heartened him as well.

1941

The new year started with Hedi recovering from a hysterectomy and Gustav preparing for the final round of medical exams that preceded his hospital residency. It was arduous for both of them. In the sitting room above Gustav's, his father sublimated his anxieties by practicing Schubert's exuberant Sonata in B-flat Major, interrupting the flowing melody by stumbling over the same measure every time. Expecting this, Gustav kept losing his concentration. Two weeks before the exams, he had to ask his father to stop playing.

Gustav became a resident at Western General Hospital just south of Edinburgh. Working twelve-hour days and having sleepless nights, he prepared to join the British army as a doctor. Some close relatives were already fighting. Gustav's cousin Franz, the only son of Hedi's brother Rudi, had just been wounded fighting for the Germans. With one Jewish grandparent, Franz was a "second-degree Mischling." (Rudi was a "first-degree Mischling"—one-half German and one-half Jewish.) The son of Max's cousin Hans Schäfer was also fighting for the Germans. Hans, who had married a non-Jewish Dutch woman, lived in Holland. When the Nazis captured the country, they forced his son into the army. At least one other cousin on the Kauffmann side also wore a German army uniform.

It was a strange fate indeed that Jewish cousins fought their kin for a country that wanted to exterminate them. A letter from French physicist Léon Brillouin was a reminder to Born of the serendipity of their fragile existence. Brillouin had managed to get himself and his family to America and secure a position at Brown University after twelve months of chasing visas to depart via Portugal. His French colleagues had suffered terribly: J. Solomon, shot by the Germans; Fernand Holweck, tortured and killed; Paul Langevin, Emile Borel, and Charles Mauguin arrested. Born understood the lucky timing of his appointment to Edinburgh that had thwarted the French offer to teach in Paris.

In spring 1943, the Durham Philosophical Society and the Pure Science Society invited Born to address their meeting. Born's friend, Frederick Donnan, had been trying to persuade him to write a book on "the philosophy of modern physical science" for a couple of years. Born had pleaded preoccupation with his research on reciprocity. Now the time was right for Born the philosopher. His topic was "Experiment and Theory in Physics."

Born wanted to discredit a new bias he saw arising in theoretical physics that proposed the "triumph of theory over experiment." Using a priori reasoning alone, physicists Arthur Eddington and E. A. Milne had tried independently to "solve the enigma of the world of atoms and of the cosmos." Born saw this as a danger to physics. He explained that the proper development of a theory was based on a synthetic process—that is, taking a set of principles, identifying an unexplained component, and making a guess informed by experience to produce a synthetic prediction. This prediction would advance theory if confirmed by experiment.

His example was Einstein's prediction of the deflection of light by the sun, which had started from a single unexplained empirical fact—the proportionality between mass measured by inertia and mass measured by gravity. Born cautioned that "to learn the art of scientific prophecy is not to rely on abstract reason, but to decipher the secret language of Nature from Nature's documents, the facts of experiment." Born had emphasized much the same point in his article in the *Frankfurter Zeitung* when the solar eclipse had proved Einstein's general theory of relativity. Born had been one of those who had shepherded the development of theoretical physics since its solidification in the early part of the century. That Eddington and Milne could undo the union of theory, mathematics, and experimentation disturbed Born.

When Einstein read "Experiment and Theory," he teased Born with the German proverb, "Young whores—old bigots." Born did not recognize himself in either of these, but he did see something else—Einstein's continuing speculative belief that "God does not play dice." When it came to quantum mechanics and indeterminism, Born thought it was Einstein who resorted to pure thinking, leaving behind his own reliance on experiment. Born granted him that privilege.

> I believe that you have the right to speculate, but that other people do not, myself included. Did I sin so in days gone by (or rather, as you put it, whore?) . . . It is my belief that when average people try to get hold of the laws of nature by thinking alone, the result is pure rubbish.

Edinburgh was becoming continental: Czech, Polish, French, Canadian, Norwegian, Dutch, and Belgian soldiers milled around the shops on Princes Street, the most Scottish thing being a bagpipe band playing on summer Saturday evenings in the Gardens. Irene tried to visit often with Hugh and Rona, leaving behind her isolated English village to enjoy the atmosphere. Gritli arrived with Johnny and her newborn, Sylvia, too. The grandparents introduced the children to the family tradition of exploring hills and fields, feeding ducks at the pond, and picking wildflowers. Hedi thought the tots were somewhat spoiled, but Max believed that they were just spirited. Whenever he wrote about the grandchildren, he glowed. They were "splendid creatures (of course)."

Hedi and Max spent their time together. He still withdrew upstairs to his study. She still vacuumed at 6:30 in the morning regardless of who was asleep. The wife of an Edinburgh colleague remarked that Hedi's "charity begins outside the house." Nevertheless they were a team, and Göttingen with its friction a memory.

Erwin Schrödinger's invitation for Max to lecture at the Dublin Institute of Advanced Studies in summer 1943 provided an opportunity to enjoy two summer weeks in neutral Eire (the Irish Republic), which excited them both. In the end, though, Max went alone; Britain's ban on regular travel to Eire made obtaining an exit permit for Hedi problematic.

Max's trip to Dublin began in Glasgow when he saw Renate Born and her mother. From there, he endured a miserable twenty-four-hour trip, with endless queues, two long train rides, and a rough steamer crossing to get to Belfast, where he met his old friends Paul and Ella Ewald (Ella being the *El* in *El Bokarebo*) for the final drive to Dublin.

Max had a close friendship with both Schrödingers. He found Erwin "lovable, independent, amusing, temperamental, kind and generous" with "a most perfect and efficient brain" and extremely egocentric. He enjoyed being around Anny, who had kept house for him during one of Hedi's sojourns in Göttingen. (Hedi did not share his feelings.) The Schrödingers' own relationship, which frequently involved a ménage à trois because of Erwin's adventures, was sometimes awkward for Born, but he largely overlooked it. After listening to Anny's life with Erwin for the past eight years, he was glad he and Hedi had each other and shared everything, "good and bad."

The two weeks of lectures, sherry parties, receptions, dinners, and discussions with Prime Minister Eamon de Valera (who, as a mathematician, came to some of Born's lectures) were, as he told Hedi, a "dull affair" without her. Overall, he found Eire "a strange world, quite separate from the big waves which shake the ships of the other nations. But they are a lovable people altogether." He was happy to return home at the end of July, bearing gifts of lemons and candy for Gustav's twenty-second birthday.

The year 1942 ended with Gustav's military call-up. He became a lieutenant in the Royal Army Medical Corps in September, underwent basic training, and in November started two weeks of embarkation leave in Edinburgh. It was a close family time. He and his father played flute sonatas, discussed music, and went for long walks, many to Arthur's Seat as they had done during his medical studies. With his parents, Gustav looked through childhood photos of holidays in the Dolomites and Engadine, choosing which ones to take with him. At one point, Max felt dizzy when coming home from morning classes, but the problem cleared up quickly. In mid November, Max and Hedi took Gustav to Waverly Station to travel to his first duty station.

Gustav soon wrote home from his post near London. At his father's behest, he looked up Niels Bohr, who was staying at the St. James Court Hotel under the name of "Mr. Baker." Gustav thought Bohr looked "much older and more worn" than when he had seen him a few years earlier in Edinburgh. Bohr recently had been flown from Stockholm aboard a British Mosquito Bomber and nearly died because he forgot to put on his oxygen mask. His son Aage, a budding physicist acting as his father's secretary and assistant, accompanied him to London. When German officials in Denmark had started to round up Jews in September 1942, Bohr, who was half Jewish, fled with his family—everyone except Aage, who spent several days hiding in the woods before escaping. Bohr said little more about himself, but Gustav realized that he was engaged in secret war work. Gustav had brought him a confidential message from his father, indicating his unwillingness to work on an atomic bomb.

Gustav soon was off on his own secret journey on a transport ship headed east. From the start, Max wrote him weekly letters, now addressed to Lieut. George V. (Vernon) Buchanan. Like all German-born soldiers in the British services, Gustav had to choose a British family as "wartime" next of kin. Retaining the same initials, he asked family friend Mrs. Buchanan to become his wartime mother, but the inside of his parents' letters still began with "Dear Buzi." (His cousin Rolli Ehrenberg, also in the British army, became Ralph Elliott; unable to coordinate, Rolli's Ehrenberg cousins Geoffrey and Lewis took the name Elton.)

Max's only birthday wish that year was Gustav's safe return. His year-end letter to his son described a quiet life with little music since Gustav was not there to accompany him. He ended with the New Year's sentiment "that your life, after this uproar, will be peaceful and devoted to the things of the spirit, which we love. My thoughts are always with you."

The beginning of 1944 brought their first communication from Gustav: It read in total, "Send love to Otto and Brenda." Decoded it meant that he was on his way to India where cousin Otto Koenigsberger lived, and he was going via Egypt where his friend Brenda was.

Born started the January term with a heavy teaching load of all the classes for regular students; his lecturer, Robin Schlapp, had the hundreds of cadets from the air force, navy, and artillery. In the middle of a lecture, Born "swooned." It was a feeling of giddiness and weakness, similar to the minor episode before Gustav's departure. The doctor diagnosed low blood pressure and exhaustion, sent him to bed for three weeks, and called in a cardiologist. The examination revealed nothing, and the final diagnosis was nerves: Born was worn down from teaching and from "Gustav's sudden disappearance to the East." The doctor ordered him to take the term off and rest—no work other than discussions with his research fellow Peng, with whom he had begun working on the quantum mechanics of fields. Hedi nursed him, and when he was allowed out of bed, they spent their time "very quietly with much sleeping, eating, going for little walks, reading, and writing."

Not knowing Gustav's whereabouts made Max anxious. For sleepless nights, he invented a game to occupy his mind since it was impossible for him to "think about nothing." He made up rhythmic nonsense, a string of words, "crazy rhymes," in German. The first sleepless night he would start with two lines that he repeated until he had memorized them then add a couple more. The second night he began with the original ones, practiced them, and added a couple more. By the time he had a dozen lines, he would fall asleep before the third repetition. He felt he had the one asset needed to make his invention work—no gift for poetry.

Before leaving, Gustav had asked his father to write down something of his life. Max now began this task, always keeping in mind the question, "Will this story interest my boy?" His other writing efforts were weekly letters to Gustav—and occasionally a second one with special news—he and Hedi always anxiously awaiting replies. Since delivery was unpredictable, Max began to number his letters in case some went missing. From January 1 to June 1, he sent twenty-seven. Some gave family news and responded to Gustav's requests such as sending music for his flute. Others were more searching.

> It is true that we lived as children in great style, having very wealthy grandparents. But that was greatly offset by having no mother, and in this respect you were so much more fortunate. On the other hand my father gave me very much, in fact my entire outlook on life; and I have tried to continue this tradition with you. One thing puzzles me. My father's life was relatively safe and quiet, compared with my own, he did not have to survive two world wars, inflation, revolutions, expulsion and emigration. But he died at the age of 49, a completely exhausted and broken man; at least that is how it appeared. His arteriosclerosis may of course have had purely bodily causes, but I doubt it. The small worries of life wore him out; the death of my mother was a big blow, but it was followed by innumerable little difficulties and disappointments, in private life and in his profession. So far I have carried on through all the catastrophes of our life with toughness. I am of course not as strong and energetic as I used to be; but still I feel quite unbroken and optimistic.

Dredging up the past for his memoirs provoked such reflection.

The memoirs pulled in the whole family. Irene, Gritli, and Hedi smoothed out Max's English. Hedi also criticized the substance: She approved of his storytelling, judged his personality during his student years in Breslau "very disagreeable," and found his general considerations to be poorly done. Max confided his suspicions to Gustav. "My views on life, the world and God are in many points very different from hers and . . . she does not like to argue about her fundamental views. You know our good Mutschi." He wanted Gustav to tell him who was right and hoped that Gustav would "have more understanding for my character, as you have inherited a good deal of it."

Letters to Gustav often bore signs of good-natured disputes. In a later letter, Hedi inserted along the margin, "Vati made me read this letter & so I cannot let it pass when he says wrong things about my attitude." Writing about trouble in Greece, Max had referred to "the shades of difference" between the two of them.

> She is of course not only violently opposed to British interference in Greece, as I am too (and most people are, even Conservatives) but she sees it all bright on one side, all black on the other. While I according to my temperament am inclined to think that matters are not as simple as that.

Hedi squeezed her rebuttal between Max's lines. "(This is not all true, I am not as silly as that, but I have a more definite stand than Vati, just as you have.)" Max enjoyed playfully provoking Hedi, or, as he had said a few years earlier, engaging in their "perpetual little war."

At the end of May, still on leave from teaching, Max was too anxious to play the piano. By June 13, he was practicing a Schubert sonata and a lot of Bach: When he was writing to Gustav on June 6, a friend rushed in and told him the news. "The invasion has begun! And yesterday Rome was taken." Born now listened to news reports several times a day, especially to learn about the Indian-Burmese front. Gustav was stationed at a pathology lab in Poona, south of Bombay near the west coast.

For the next few months, Max's letters to his son intermingled war news—the fall of Cherbourg, the Russians' breakthrough, some minor bombing of Göttingen, the assassination attempt on Hitler—with his musical interests—Brahms, Schubert, Bach. In the Brahms' symphonies remembered from youth, he heard a "noble form of romanticism . . . the key to all that is bitter and sweet, great and gentle in human life." Music remained an important outlet for his emotions.

In his memoirs relating to childhood, Max confides to the reader that he looked forward to getting to the sections on science. He much preferred discussing his research than describing people in his life. It was easier to unscramble what he observed in the physical world.

Born had a setback to his health in July and became depressed. He had just learned definitively from family members that the Nazis had killed his stepmother's sister Paula and his father's sister, Selma Jacobi. To Gustav he wrote,

> For me the news is horrible—for although they were 84 and 88 it is dreadful to think what these harmless old ladies must have suffered. Aunt Selma's hus-

> band, Dr. Jacobi, was an old soldier who fought in 2 wars for the same nation, which has now murdered his widow.

Ordered by the doctor to rest, he and Hedi took a holiday in Aviemore, a small village in the Highlands directly north of Edinburgh. While they hiked and biked, Paris was liberated by the Allies. Max came back stronger.

On the evening of September 17, 1944, downtown Edinburgh lit up: the blackout was lifted and crowds danced in Princes Street. The Allies were pushing through the Siegfried line and the Russians advanced in the Balkans. "Doodle-bugs"—noisy, pilotless bomber drones that the Germans had launched on London during the summer—had largely ceased.

A return to normalcy for the family, however, was remote. Gritli's husband, Maurice, was ordered to Canada to do secret research; and Brin Newton-John, Irene's husband, to Washington, D.C., with British intelligence. (Brin's orders were eventually canceled.) During the war, Brin had interviewed German prisoners and sometimes masqueraded as a German soldier to get information in POW camps. For a while, he worked at a country estate located near Irene and the children called Bletchley Park, where a secreted team broke the Germans' so-called unbreakable Enigma code. Still far from coming home, Gustav was now a captain in charge of his own lab in India. The beginning of fall term signaled the resumption of daily life at the university, but even that was odd. The world had changed. Although Born was finally back teaching, his class consisted of only one student—although "gifted"—whom he taught twice a week.

Ever since the daughters of Heinrich and Maria Rausch von Traubenberg had left Germany without their parents five years earlier, Hedi had watched over them, helping with boyfriends, education, jobs, and refugee problems. Dorothee had married a German refugee mathematician, and Helen was in nurses' training. With a reunion of the family seemingly assured by the Allies' advance, Lise Meitner wrote Hedi with terrible news.

On September 19, the SS had ordered Maria, who was Jewish, to board a train for a camp in Czechoslovakia. On the way to the station, Heinrich had chest pains. Maria persuaded him to return home while she went on. At the station, she explained to the SS that her husband was an "Aryan," and they let her go back home. There she found that Heinrich had died of a heart attack. The SS wanted to send her to Berlin, but Max von Laue and Otto Hahn were trying to persuade the Nazis that only she could organize her husband's valuable scientific papers. Eventually, the SS had her transported to the ghetto at Theresienstadt.

Meitner's other news was that Max Planck's son Erwin was in prison, implicated in the plot to assassinate Hitler. His life was in jeopardy.

Born was shaken on both accounts and his letter to Meitner mourned his old friend. But he also noted some good news—Otto Stern was to receive the reserved 1943 Nobel Prize in physics. Twenty-five years had passed since Stern had begun working on the molecular beam method in Born's tiny institute in Frankfurt—work that led to the Stern-Gerlach effect, which proved that atoms moving in an inhomogeneous magnetic field take only certain orientations in space.

Born wrote a congratulatory letter to Stern, who was at the Carnegie Institute of Technology. "It was really good news in these sad times. You have been my first candidate every time I have been asked by the Nobel Committee. And I feel a particular and rather mean satisfaction that you have not shared the honour with the Nazi Gerlach." Unlike Born's former assistant, Pascual Jordan, Walther Gerlach had, in fact, not joined the Nazi Party, but he was working on fission research for the Nazi regime. Clearly, Born had heard rumors.

In a letter to Born about Stern's success, Ladenburg hoped that those theoretical physicists passed over would get the prize next time. In spite of this kind of support, Born more and more belittled his own successes and abilities. Counseling Gustav about entering politics, Max said that one became a politician by having "some mysterious personal gift of influencing people" or by being recognized in one's field, then using the resulting influence. His own hopes to do that were "a miscalculation as I have become not much more than an average professor whose weight is negligible. Yet I think for people like us the second way is the only one."

Now was certainly the time to have influence. To Born science was beautiful and "a benefactor of human society," but when he looked around, he saw that it "had been degraded to nothing but a means of destruction and death." He laid out a case to Einstein for scientists to "unite to assist the formation of a reasonable world order." He looked at the physicians' code of ethics and thought that scientists should have a similar international code "by which our scientific community could act as a regulating and stabilizing power in the world, not as at present being no more than tools of industry and governments." Born felt powerless sitting in Edinburgh, especially when the people he wanted to work with on this initiative were unavailable. One, Ralph Fowler, was very sick; the other, Niels Bohr, was nowhere to be found.

Einstein's response aimed to discourage Born. He expected little success from such an undertaking. "The feeling for what ought and ought not to be grows and dies like a tree, and no fertilizer of any kind will do any good." Born replied that he was too tired to persist on his own.

As scientists' complicity in the horrors of Nazi Germany came out in the news, however, Born renewed his efforts. He wrote up a short description of examples of biological experiments performed by German scientists on Jews and others, using this to argue for a code of ethics among scientists. He then sent the material to the biologist Julian Huxley who, he learned, was very ill. He next tried Kingsley Martin, editor of *The New Statesman and Nation*, who liked the proposal and recommended that Born send it to A. V. Hill, secretary of the Royal Society. Born asked Hill for his personal opinion about an ethics code and about the Royal Society formulating an explicit policy to exclude scientists who made discoveries using inhuman methods.

Hill replied that the society dealt with "improving natural knowledge" not "morals and politics" but he would ask the opinion of the three other officers. Hill's request to them added the note, "Personally I feel rather sorry that so eminent a person can be so silly: but perhaps I am prejudiced!" Hill thought Born's ideas "rather fantastic."

The favorable Allied position in 1945 prompted many thoughts of the future. At a meeting of the Royal Society, Dirac asked Born to cosponsor Heisenberg for foreign membership. Taken by surprise, Born agreed, but quickly reconsidered; he felt that he would be unfaithful to his murdered relatives and friends if he participated in honoring a German. He wanted to wait and see where guilt and merit fell before making such decisions. Dirac told him, "Heisenberg's discovery will be remembered when Hitler is forgotten." Dirac proposed Heisenberg for membership. (He was not elected.) For Dirac, who mainly lived in the world of science, it was not such a distasteful proposal.

For Simon, whose opinion Born sought, it was just that. He agreed with Born that it was inappropriate to consider any German at that time—not even von Laue—and certainly not Heisenberg who, he noted, had "behaved very badly as a person." In Born and Simon's correspondence, Heisenberg's name cropped up, Born passing along rumored bits of information. One concerned Heisenberg's comment from a few years earlier, as the Russians successfully defended themselves against the German siege at Stalingrad—at a staggering cost to human lives—that the Germans would certainly win the next time.

Dirac's request only heightened Born's desire to get support for a code of ethics. Not knowing Bohr's whereabouts, Born wrote to him via Ladenburg. Thus far, only the Communists endorsed a code, Born told Bohr, and he felt that this was less out of ethical convictions than political ones, since Germany was their enemy. He reported about Hill and Dirac and recounted a recent article by a war correspondent who, after witnessing the horrors of the Buchenwald extermination camp, had visited the Zeiss optical works in nearby Jena. There he met four physicists, one of whom, Georg Joos, had been Franck's successor in Göttingen. In his letter, Born quoted the reporter, who asked them if they knew about Buchenwald. "Oh yes, they knew; but they are not politicians, they had no interest in such things. They had only the job to solve scientific things."

"There you have the situation," Born wrote. "Everywhere this attitude that scientists are not interested and not involved in the ethical problems of humanity. I think that is a most dangerous and pernicious attitude, and I hope you share this view." If Bohr answered, the letter did not survive.

The war was coming to a close, and the news of the death and destruction in Germany was emerging in full horror. Born was overwhelmed by personal losses. After the firebombing of Dresden, he had written to Courant,

> Now the Germans will feel it also and worse is to come for them. It is mostly the wrong kind of people who suffer most. But I cannot feel much for any of them. If they have not behaved like savages themselves, they have allowed the gangsters to get into power. When Dresden was destroyed I felt the barbarism of our time particularly strongly; for I loved this old baroque and rococo architecture. But behind this lovely façade there has developed such a devilish mind that there seems to be no other way than to destroy it altogether.

By the end of March, the Americans were on the road to Kassel and Göttingen; Frankfurt was destroyed; Heidelberg remained intact; and the Germans were leaving Holland. The Russians were attacking Breslau, coming up from the south through Grandfather Kauffmann's formerly grand Kleinburg estate. Hedi's nephew, Franz Ehrenberg, was in the German lines there, a Mischling sent to prove his bravery. He was killed. The Borns found out only later.

Max knew that his dreams of seeing his boyhood city again were destroyed; but he asked himself, "What does Breslau matter when the whole of Europe is being destroyed? The old world is gone, and we have to go on living." On April 22, which would have been his father's ninety-fifth birthday, he wrote to Gustav that his relatives had committed suicide rather than be sent to a concentration camp; he was glad that his father had not witnessed this. Max and Hedi had seen the horrors of the concentration camps in newsreels. Hedi thought that the camps must be the work of a small group of criminals—the German people being innocent—but Max found the extent of horror too great for only a few to be involved. He wrote to Gustav that if he had "done no other thing than to bring you and your sisters out of that cursed country, that under its lovely exterior has covered this pest, I have done enough."

The Allies were advancing into Germany and arresting Nazis and anyone else who might be of interest, including scientists. In a letter to Simon, Born happened to pass on a rumor about Heisenberg. On an earlier visit to Zurich, Heisenberg had "explained why one has to accept certain inconveniences of the Nazi regime in favor of its merits." Simon responded that the information was of interest "in some other connection." He had showed it to Lindemann and also to an organization "which was very much interested in your first letter. Are you keeping Dirac informed about this?" Born had no idea that in May, the Americans had taken Heisenberg and nine other German scientists into custody and transferred them to the British. As Born and Simon corresponded, British intelligence held the Germans at a country house near Cambridge called Farm Hall, debriefing them on war research and secretly taping their conversations. The British released them after six months.

Hedi captured the unfolding events in her diary.

> April 27: Mussolini captured.
>
> April 28: Himmler offers unconditional surrender to the British and Americans only. Not accepted.
>
> April 29: Himmler says Hitler is dying. Cerebral hemorrhage?! Mussolini executed.
>
> May 1: Hamburg Radio says that Hitler is dead; movie pictures of German *horror* camps.
>
> May 2: The Germans in Italy (& South Tyrol) surrender unconditionally!!!
>
> May 4: Germans in Holland, Denmark, & North Germany surrender!!!

(Max wrote to Gustav that victory was in the air. About a dozen rather dirty street children, having seen a pile of wood and branches at the side of the house, had rung the doorbell. They were getting ready for a bonfire when VE Day came and were excited when the Borns let them carry the pile away.)

> May 7—Germans in Norway surrender!! But crisis with Russians over Poland. Oh God help. Germans complete surrender, signed in Rheims!!
>
> May 8—VE Day

Max and Hedi waited for the Japanese collapse so Gustav could come home.

As the war ended, the civilized world stirred again. Schrödinger invited Born to lecture in Dublin (though Born hesitated when Prime Minister de Valera paid a condolence visit to the German ambassador on the occasion of Hitler's death). The Soviets asked thirty British scientists—one of whom was Born—as well as Americans, French, and others to a two-week celebration of the 220th anniversary of the Russian Academy of Science.

That trip preceded Britain's national elections on July 5 by two weeks. The conservative Tories were pitted against the socialist Labour Party—Churchill's victory in war against promises of full employment. Some, including Born, thought that Churchill's esteem as war leader made his election fairly certain;

but war had not completely overshadowed other issues such as education, health care, and social insurance, nor had people forgotten the reconstruction period after the last war. Hedi organized the Labour vote in Niddrie, where she knew everyone from her nursing. Max, who was supposed to leave for Dublin on July 3, dithered about staying to cast his four votes: as a resident, a member of the university, a proxy for Klaus Fuchs (who became a citizen in 1943), and a proxy for Gustav. (He ultimately would cast all four early and leave for Dublin as scheduled.) Even with all of Max's votes they did not think their district—which always voted Conservative by a large margin—would go to Labour.

Born took the train to London on June 13 to meet the rest of the scientific delegation for the trip to the Soviet Union. The following day, twenty-two scientists instead of the original thirty flew off to Moscow on the DC–3 Dakota. The British government, fearing for the return of the physicists involved in war research, barred eight from going.

Born's diary recorded days in Moscow that were a blur of long speeches heard from hard seats, sumptuous lunches and dinners with too much wine, and events that incorporated all the cultural wealth of the country. Curious to see communism first-hand, he observed something different from what he expected. "Expertocracy"—rule by "the ablest men, if you are an optimist," he wrote to Gustav—described the situation well (except that the mixture of Greek and Latin strained his sensibility). The highest level—political and military leaders, artists, and a group of about 400 academicians—lived in nice houses with servants. The populace in the streets, whose low salaries and ration cards provided life's necessities, wore peasant clothing and lived in small, low-rent apartments. Occasionally, they bought a luxury—an expensive dress or tie—with meager, hard-earned savings. People were stratified by different standards of living as opposed to traditional class differences. There was a large gap between workers and intellectuals. Born's diary chronicled the scientific meetings and special cultural events, a grand hospitality. He glimpsed another side only when children came up to his train car and begged for bread at a railway station between Moscow and Leningrad. The conductor chased them away.

Max saved his true feelings for a later letter to Gustav in which he described "a state based on power, a terrible power, exercised without pity against every enemy of the regime." He had discovered that two of his friends—Professor Romanov, an acquaintance from his 1928 trip, and Georg Rumor, his former assistant in Göttingen—were victims of Stalinist purges: the government had seized them. Friends warned him not to inquire.

When Hedi saw Max on the morning of July 2, after his all-night travel from the Soviet Union, she thought he looked better than he had in years. Old friends and the sense of scientific community had sparked his spirit. That day he and his collaborator Peng accepted the MakDougall-Brisbane Medal of the Royal Society of Edinburgh for their research on "Quantum Mechanics of Fields." Somehow, a year earlier, when he was resting and not working, he and Peng had written several articles on this topic. The two departed the next day for Dublin, a journey more arduous than the one to Moscow. The government had lifted the travel ban to Eire but had not licensed more ships for the crossing. Even with first-class tickets, Born and Peng had to sit "in a corner of the dirty dining room of the dirty steamer the whole night, with soldiers drinking and playing cards all the time."

Born did not sense much of a change in Dublin with the war's end except that everyone was seeing war movies that the censor had earlier forbidden. Comparisons with Russia were inevitable, and Born reported to Simon that theoretical physics prospered in both places. "Russia because they know it to be the foundation of scientific research and material prosperity, in Eire because it is one of the great achievements of the mind in praise of God's creation." Born did not know which he preferred. He considered the two attitudes "complementary aspects of the same phenomenon (like particles-waves)."

The day after Max's return home, election results were announced: Winston Churchill was out, and Clement Atlee was the new prime minister. It was a big win for the Labour Party—and the Borns had a Labour representative from their district.

On August 6, the United States dropped an atomic bomb on Hiroshima. The Borns were midway through their two weeks at Inverness. Born wrote to Simon, "Shall I congratulate you on this success? I am sure it had to be done in order to forestall the Germans. But I have little confidence that this terrible concentration of power will be used for human progress. The future of our race seems to me to be dark." With Gustav, he hashed through ideas about restricting scientific invention, coming down firmly on the side that the quest for knowledge could not and should not be stopped. "How could Lise Meitner who discovered U–235 and its fission by neutrons foresee that this minute effect could be developed into the most destructive explosive?" He did not wrestle with the morality of dropping the bomb on Japan. With the Nazis beaten, he thought it heinous, "nothing more than the purest military brutality."

Born wrote to his son, "If man is so constructed that his curiosity leads him to self-destruction there is no hope for him. But I am not convinced that he is so constructed, for besides his brain he has his heart. Love is a power just as strong as the atom."

THIRTEEN

A CURSE OF THE AGE

SOON AFTER THE WAR'S END, BORN SAT DOWN AND MADE A LIST. IT WAS HEADED "VICTIMS of the Nazis amongst my Relatives and Friends:"

> "Selma Jacobi: sister of my father. Over eighty years [old]; deported to Theresienstadt; died there or committed suicide.
>
> Charlotte Mugdan: sister of my mother; committed suicide when threatened with deportation; almost eighty.
>
> Paula Hirschfeld: sister of my stepmother; over eighty, and her son Bruno Hirschfeld: committed suicide to evade deportation. . . . "

He listed thirty-four names of relatives and friends, two-thirds of whom had committed suicide rather than face imprisonment in a concentration camp. Three of them—his nephew, Franz, and two cousins—were young men who had died fighting for the Germans. Born also included his grandmother Kauffmann's family, the Joachimsthals, all of whom "seem to have completely perished" though he did not know how.

Slowly, amid the postwar chaos in summer 1945, more letters—filled with both the bitter and the sweet—made their way to Edinburgh. The Borns learned that the Nazis had thrown Hedi's brother Rudi into a work camp and that the Americans had rescued him when he was on the brink of starvation. They also learned that Maria Rausch von Traubenberg had survived the Theresienstadt ghetto, imprisoned in the old fortress while organizing her husband's scientific papers. They heard from Max's cousin Hans Schäfer—his childhood pal whose idea of war, so long ago, was playing Bismarck to Max's General von Moltke—who had amazingly survived the Nazis.

Hans had a chilling tale of life under Nazi occupation, which for him started in 1940, when the Germans occupied Holland. At this time he was living in Delft and overseeing a tapestry plant for the Meyer-Kauffmann textile company. Within a very short while, the Nazis prohibited him from working there (turning the factory into a munitions plant), pinned a yellow star on him, and conscripted his son Klaus into the army. Somehow, Hans managed to buy a chemical firm with contributions from manufacturers who wanted him to continue his research. Then Klaus was killed on the Russian front. Owing to his son's death, Hans escaped the order for Jewish men married to non-Jewish women to be sterilized. One night late in the war, the Gestapo seized him for immediate transport to the Westerbrook camp—from which no one returned. Miraculously, his wife, Dora, and an industrialist friend persuaded the Gestapo to release him. Throughout all of this he and Dora hid other Jews in their apartment, sometimes for months. The Dutch police warned them when inspections by the security police were imminent, and others made sure they had enough food.

Fate, however, would not allow the two cousins an opportunity for reunion. With the loss of Klaus, reaffirming the community of family was important to Hans and he anticipated their meeting; but at the end of September, Max heard from Dora that Hans had died suddenly after an operation. She thought the true reason was his unremitting depression over Klaus's death. Max wrote to Gustav, "He was my oldest friend with whom I spent all my youth, and I feel the loss very much. Now I am the last of the group of (male) cousins surviving. All the Jacobis are dead." In fact, they were not all dead. Max learned that Selma's son Fritz had survived in southern Germany, sheltered by his non-Jewish wife.

With the slaughter at an end, Born began to mourn his losses. He had a mountain of sorrow and anger to resolve. Years later, he shied from places that revived these tortured memories. But gradually he came to terms with what must have been not only unimaginable but incomprehensible. His outlook in 1945 was hopeful, perhaps because of seeing his children and grandchildren flourish and knowing that his decision to leave Germany had given them a future. When he thought about his good fortune at having saved the family, he realized that he had nothing to complain about. His health improved, and he was able to walk long distances again. He saw many good prospects.

Yet one question haunted Born: "What have they done to my beautiful science?" It was science that had always given him both a sense of optimism and order in the world. He had believed all of his life that science was a "noble pursuit, like philosophy and art and true religion," but he acknowledged to Gustav that this belief was not easy to sustain when science, like religion, was "perverted to destruction and hatred." The Christian churches, he told Gustav, always claimed that killing in the struggle of conversion, whether the burning of heretics or the conquest of heathen nations, was a product of the frail human mind, not of the gospel of love. Now Born thought,

> Science is in a similar position—but not quite so bad. For people do not fight because of differences of scientific views, as they used to fight because of differences of religious dogma. Yet scientists are willing to assist the slaughter with their terrible weapons. That seems to me almost as wicked as to kill your neighbor for having another way of serving God or another idea about the mystical parts of the Bible. The real culprit seems to me the notion of Nationality, and the problem for us is to fight Nationalism in whatever form it appears.

He had watched a "morbid, insane" nationalism grow in Germany after World War I. With Eastern European countries under Russian control expelling longtime German residents, he feared its reemergence after this war. The idea of unity clearly did not direct men's aspirations. In his personal theory of social change, the turn from "the idea of nation" would occur through a "slow process of transforming [people's] convictions," as was the case with the disappearance of religious wars, when people became disgusted with the consequences of dogmatism. Then science would not be needed for defense.

In the fall, Born went to Oxford for a meeting and stayed with Francis Simon. Simon and others explained to him their part in the development of the bomb—that they had volunteered, certain that the Nazis were working nonstop on their own version. Once in the army, they became mere foot soldiers, following orders with no ability to control the direction. They had worked to counteract the Nazis, not to bomb the Japanese. Their advice was not heeded.

Not specifically mentioned but certainly known to some was that in June 1945, James Franck had headed a small group of scientists in Chicago who weighed the option of detonating the first atomic bomb on Japan. The report concluded that such action was "inadvisable." The group used possible negative reactions in America, moral reasons, and the future likelihood of launching an armaments race as justification. Instead, it urged the U.S. government to demonstrate the bomb's power on a barren island, thereby persuading the Japanese to surrender.

Simon and the others did not speak of guilt; rather, they argued that even without the war, nuclear power would have been developed in several countries, only more slowly and in a more competitive climate. American and British physicists who had worked on the bomb project strongly favored international control, while Secretary of State James Byrnes believed the United States would have a monopoly for many years. The Franck Report had recommended an international agreement backed with effective enforcement to preclude proliferation. As civilians, the scientists wanted to organize, rally their forces, and establish guidelines for the responsible use of atomic energy.

Born believed in their resolve and was cheered by the budding movement, and he saw a potential for bringing people together to create a safer world. Although with some ambivalence about accepting the invitation, he even debated atomic energy in the Edinburgh Student Union. The motion was "That this house considers atomic energy a curse of the age." Born's three-person team argued the opposite side, and they won by a large margin. At least for one night in Edinburgh, Born thought that atomic energy was a boon.

Born knew the position of many in his generation and felt comfortable with it; he was less certain about the younger generation. Some of the theoretical power behind the U.S. Manhattan Project to develop the bomb had been his students and assistants: Robert Oppenheimer, Victor Weisskopf, Enrico Fermi, Edward Teller, Eugene Wigner, Klaus Fuchs, and Maria Göppert-Mayer. He wondered to Maria "whether Oppenheimer or any of the others have ever spent a moment of consideration before starting that uranium work." He also asked himself whether so many of his students would have worked on the bomb had he more pointedly stressed a physicist's social responsibility. But science was too beautiful to him to have thought of that at the time.

Maria Göppert-Mayer was one of the younger scientists arguing for international control and world organization. Oppenheimer was with her, she told Born, but he could "never take an unambiguous stand." Just five months after Hiroshima, Oppenheimer supported the idea of international control, but he sometimes waffled on specifics. He had initially supported the May-Johnson bill, which gave control of the bomb to the military.

Göppert-Mayer's more immediate concern was Germany. She had a unique perspective—a non-Jewish German émigré who lived in the United States, was married to an American, and had worked on the Manhattan Project. She believed there were still decent Germans. "I am a German who refuses to believe that the people I have known and liked could have changed even under the Nazis," she said to Born. He was far less certain and did not know how to deal with former German colleagues and friends who thought that "with a few friendly words all the dreadful things which have happened would be forgotten."

One he was certain about was Pascual Jordan. A letter from Jordan had arrived at the Borns on the same day that the bomb fell on Hiroshima. Enclosed with this letter was a longer one that Jordan had sent to Niels Bohr. For Bohr's benefit, Jordan had accounted for his actions during his "black 12 years" under the Nazis. He told how he joined the Nazi Party to influence the course of events—moderating the strident attitudes of Lenard and Stark and trying to undermine their authority. He explained how he was constantly under suspicion because of his former teachers and his unwillingness to throw out relativity and quantum theory, the "Jewish science" loathed by Lenard and Stark. He rationalized how winning the war would have let the military "liquidate" the Nazis.

Jordan's letter to Born described personal details, which indicated, Born thought, a close relationship to the Nazis. Frau Jordan—who, he felt sure, was an ardent Nazi—and their children had been in Salzburg (Hitler's retreat) since February. "Who except the very strictest Nazis would have been allowed to live in the Redoubt?" Born asked Simon. To Born, Jordan was "an extremely clever fellow, and I quite believe that he has tried to double-cross the Nazis, but he has certainly been a Nazi and behaved like a Nazi." Born had no intention of answering his letter.

More difficult for Born were the motives of Werner Heisenberg. Born struggled with what to think. Fundamentally, he did not believe that Heisenberg was "mean" but felt that he "must have succumbed to some degree to the general madness of the Germans. There are too many reports to doubt it." Born had not forgotten Heisenberg's comments sympathetic to the Nazis that he had passed on to Simon.

Born heard from a friend in Switzerland, however, that toward the end of the war Carl Friedrich von Weizsäcker (who had studied with Born for a semester in 1931 and was one of Heisenberg's best friends) had been in Zurich. Von Weizsäcker told a trusted friend that he and Heisenberg always acted as 120 percent Nazis there. They had to behave this way because the Swiss federal police collaborated with the Gestapo. Likewise, the two had remained in their positions on the atomic bomb team in Berlin in order to delay studies. They considered their work "as clear and plain sabotage 'in de Fuehrers face.'" After reporting that, Born's friend asked him, "Do you believe in fairy-tales? I must leave it to you. It is obvious that everybody who has carried on all these years with the Nazis produces a more or less reasonable explanation of his attitude."

Clearly, Born's friend did not believe von Weizsäcker's account of his and Heisenberg's behavior, instead seeing the story as von Weizsäcker's excuse for working with the Nazis.

Born's answer to the question does not exist, but a little later he told Dirac that he would reconsider proposing Heisenberg as a foreign member of the Royal Society. Born decided that those like Heisenberg, whose reputations were tarnished but who were not committed Nazis, had perhaps been punished enough with starvation and national humiliation. His view happened to echo Hedi's. Born's change of heart was likely a product of discussions with her. From the beginning, she made a sharp distinction between the horrific actions of the Nazis and culpability of other Germans.

Rudi Peierls heard via Dirac what Born was considering and wrote immediately to dissuade him. He wanted to maintain a clear distinction between those such as Max von Laue or Otto Hahn, who had not ingratiated themselves with the Nazi government, and those such as Heisenberg, whose intentions were more ambiguous. Peierls, for one, did not even want social contact with Heisenberg "until we are able to take a more detached view of the past than we can do just yet." Born listened and agreed to do nothing further about the Royal Society membership.

Myriad questions and confusions accompanied the relief of the war's end in 1945. The refugees' road to emotional reconciliation was long. An issue that evoked reflection was the response to conditions in Germany. People there were starving. The Borns decided to send packages containing clothing and food to family and friends. But conditions were not easy in Britain either. Rationing was still in place, food being more scarce and queues longer for nonrationed goods than during the war. This meant that they had to put aside some of their own limited, rationed supplies and sometimes ate out in order to have extra to send.

Demobilization was slow and Born's teaching load therefore was light, giving him an opportunity for research. His institute quickly built up to ten doctoral students, all from different countries—the "League of Nations," as they called themselves. He worked with his research fellow H. W. Peng on quantum field theory and continued investigating crystals, writing an article or two to counter one of Raman's in their simmering controversy. His new Ph.D. student Herbert Green worked on Born's old idea to develop a rigorous statistical theory of liquids that included transformations into gaseous and solid states—the motion of evaporation and freezing rather than equilibrium. Born felt quite contented with his program when he heard about research at other places. Even in a place as isolated as Edinburgh, he judged his department was "pretty near the top and [has] not missed important points." He thought that his and Green's theory of liquids was probably more advanced than others.

Born had some new ideas on reciprocity, but he told Peierls that he felt "too old, nervous, and shaky to do the intricate calculations (which are necessary)," adding, as an afterthought, "I am still quite good in general reasoning." He constantly queried Peierls, "Where is Fuchs?" Fuchs, replied Peierls, was still at Los Alamos finishing up his research and would probably need a rest when he returned from five years of stressful work on the atomic bomb.

The Born family had not yet completely reassembled by the end of 1945. For the traditional Christmas dinner of goose and red cabbage in Edinburgh, the Newton-Johns were there with children Hugh and Rona, as were the Pryces with Johnny and Sylvia. Gustav, as a specialist in pathology, was still in India, about to join the British Occupation Force destined for Japan. The family did not expect to see him for at least another year.

A few weeks earlier, at the time of the announcement of the 1945 Nobel awards, a couple of newspapers asked Born to write an article on the new laureate Wolfgang Pauli, who was being honored for his discovery of the exclusion principle. This award was the first in the area of quantum theory since those for 1932 and 1933. Born was a bit "dashed." Not that Pauli was undeserving, but Born could not see why Pauli's work was more important than his own. Awaiting Gustav's return, though, cast these kinds of events in a different perspective. Born wrote to Gustav, "There is so much hunger, famine, hatred, illness in the world, and I have suffered nothing. I am grateful that my fate is so mild, and if I have a wish it is not for a prize but to see you return well and happy."

In spite of the ills Born recounted and the daily disruptions in life, the confined and meager existence of wartime had been transformed into a world of reunion and opportunity permeated by an aura of excitement. Soldiers returned home; colleagues and friends, who had been unsure if they would ever see one another again, met at conferences; and universities in far-off places sent invitations.

The University of Ankara, Turkey, proposed a one-year appointment or longer, if Born so wished. China and Israel threw in bids for his time. A little later, Arnold Sommerfeld asked him to take over his chair at the University of Munich—an offer Born declined because he could not bring himself to return to Germany. King Fouad University in Alexandria, Egypt, asked Born to take a professorship. He declined that as well but was pleased by it, telling Gustav, "You see! Your old daddy is not without chances in the world." He converted that invitation into a two-month stay at the university in Cairo, where, during spring 1946, he lectured and Hedi basked in the warmth and sunshine. In the off times the two dabbled in archeology.

That summer, eighty-eight-year-old Max Planck and von Laue traveled to London for the Royal Society's tercentenary celebration for Sir Isaac Newton. Born found it overwhelming to meet his old friends, especially von Laue, whose name throughout the war had represented courage and honor to his expatriated colleagues. Whether going to the cemetery for Arnold Berliner's burial or writing an obituary for Fritz Haber, von Laue had refused to buckle under the Nazis. Born had long talks with him but was surprised that "even this honest man and enemy of the Nazis lives in a different world from ours. In some points," he told Gustav, "it was difficult to understand his attitude." Born was referring to that same attitude he had seen in letters from "good" German colleagues—namely, that the blame for the war and the punishment should fall only on the Nazis.

Many old friends such as Lise Meitner, Niels Bohr, and Theodore von Kármán came to the Newton celebration. Much of the group, including Max and Hedi, traveled to meetings in Oxford and Cambridge that followed. Max and Hedi visited both daughters. Maurice Pryce was a professor at Oxford, and Brin Newton-John the headmaster (the youngest one in England, Max told a friend) at the Cambridge Grammar School. Gritli's dinner party in Oxford brought together more old friends such as Simon and Cherwell, and Irene's invitations in Cambridge brought Pauli and the Schrödingers. One guest at both affairs was Klaus Fuchs, who had just arrived from America to head up the theoretical section at the British government's new Atomic Energy Research Establishment in Harwell. He was the same quiet, unassuming person whom Mrs. Peierls had described as "'Penny in the Slot' Fuchs. You put your penny in and get a word back." With his new responsibility, he did not have time to help Born with his research on reciprocity.

Life in the colleges at Oxford and Cambridge in summer 1946 seemed like a fairy tale existence to Born, "as if no war had ever been and no people were starving!" The professors had garden parties in the afternoons and glasses of port in the evening. In Edinburgh, as in the rest of Britain, the government had just announced bread rationing.

In Oxford, Born was a guest at a meeting of the newly formed Association of Atomic Scientists, the main purpose of which was to keep nuclear power free from military influence and put into place restrictions on weapons. Scientists in the United States had founded a parallel group. A few months earlier, it had stopped congressional passage of the May-Johnson bill, which would have put atomic power under control of the military.

Born thought the scientists' discussions were "decent and sober, full of good will. But I could not suppress the feeling that they were trying to make good what they themselves had initiated." Bohr could not be there, but Born had talked to him earlier and found him "quite optimistic" that "war is an absolute impossibility and that even the most stupid politicians can be brought to understand this." Born hoped Bohr would be proved right but foresaw the possibility of another scenario—"a situation where the outbreak of a major war means certain mutual destruction." Born also worried about weapons even more disastrous than a bomb—"biological things, more in your line," he told Gustav. That this research was secret and not in the public's mind distressed him further.

What Born did not find at the meeting was:

> a discussion of the ethical principles involved in scientific weapons: is mass murder to be justified under any circumstances? Is the preparation for it not a crime? Can it be justified by the danger to one's own nation or to civilisation at large, as in the case of the Nazis? Most of my colleagues seem to have accepted the latter standpoint.

Born answered his own questions by reciting to Gustav a conversation with the Darwins. Charles Galton Darwin and his wife had said that if Born had lived under daily bombing rather than in peaceful Edinburgh, he would think differently about means of defense. Born saw this response as "the reaction of decent people to evil things"—evil being contagious and spreading over the world.

During the meeting, Born stayed in Magdalen College at Oxford on the invitation of its president, Sir Henry Tizard, who asked him to deliver the prestigious Waynflete Lectures in February 1948. Although the lectures on the relationship between physics and philosophy were eighteen months away, Born was already looking for a special inspiration so that they would not be dull and dry. Reflecting on his observations from the meeting, he thought he would emphasize

> the meaning and purpose of science, for the individual and for the community as well. There the question of values appears in its most elementary form. In my youth, and even 20 years ago, I would have enthusiastically emphasized the value of science for human life—but who can do that today without a careful scrutiny? We have now a terrible responsibility, we should do nothing without thinking where it may lead to, and we cannot retire to an "isolationism" or ivory tower. Yet I am quite convinced that the eternal value of science lies in things remote from any applications, good or bad, in finding the truth about reality, and a man is justified to spend his life on this search just like an artist or musician—provided he keeps himself in a sphere beyond any possible application. You may doubt whether such a sphere exists. Of course it does not exist in any absolute sense, just as even Mozart's art is not safe against being used as jazz and played on a cinema organ. Yet Mozart is still Mozart, as Einstein is still Einstein.

Shortly after Born returned home, he received an offer that rekindled bitter feelings. In Göttingen, which was in the British zone, the director of the Education Branch of the Military Administration wrote, asking Born to return to the university as a professor of physics. Born addressed his answer to Heisenberg, who had resettled in Göttingen because Leipzig, home of his former university chair, was now in the communist East.

Born was a scrupulous adherent of German letter-writing etiquette. Yet in this first contact between the two, he skipped the protocol of social greetings and began, "I had long intended to write you. Now a special occasion presents itself to do so." Each remark was direct, starting with the fact that the offer had arrived on the same day as the verdicts were issued in the Nuremberg trials. Injustices both big and small, said Born, were being atoned for. He had claim to both kinds and cited the loss of friends and family and the extermination of his ancestry at the hands of the Nazis for his rejection of the offer. He emphasized his satisfaction with Edinburgh and with being a British citizen. His familial connections to Britain underscored his emotional distance from Germany: His son was in the British army, his daughters had married Englishmen, and his grandchildren spoke only English. "Believe me," he told Heisenberg, "this transformation was not easy for me; even more difficult for my wife." Born never asked Heisenberg how he was or what he was doing, but he did end with a wish to hear from him again.

A peripheral reason Born gave for his rejection of the offer was that physics in Göttingen was brilliantly represented and had no place for an old man such as himself. But with lack of equipment and research personnel as well as serious food, clothing, and housing shortages, the physicists in Göttingen may not have felt very brilliant. Nonetheless, their scientific life was far better than in the rest of the country. By placing their headquarters for science there together with bringing in the former Kaiser Wilhelm Institutes, the British aimed to lift Göttingen science to its former level of excellence. The Kaiser Wilhelm Institutes brought with them Otto Hahn, Carl Friedrich von Weizsäcker, and Heisenberg. (British control of these men at Farm Hall made their relocation to the British zone a simple matter.) As for the physics institutes at the university, Richard Becker still had Born's old chair in theoretical physics, and Robert Pohl directed much of the experimental teaching and dominated the department.

On November 20, 1946, the Borns received a much anticipated phone call. Gustav's ship from Japan, the course of which they had followed in the newspapers, had docked. At 8 P.M. on November 21, they met him at the Princes Street station. His five-day visit could not compensate for a three-year absence—Gustav would not be discharged until the following August—but a couple of weeks later, he came back from his posting in London for his father's sixty-fourth birthday and stayed for a quiet Christmas, just the three of them.

The war was behind the Borns, and they could finally think to the future. The passage of the war years had brought a new dimension to their plans. Born would reach compulsory retirement age in six years. Any decisions that he and Hedi made had to recognize this looming transition.

Soon after the new year, Schrödinger asked Born if he would consider a professorship at the Institute for Advanced Studies in Dublin. Schrödinger outlined an offer for Max to work full time until age seventy-five and afterward receive a pension of about half his salary. The main drawback was the possibility of a lack of research fellows with whom Born could work, since Schrödinger's institute did not grant Ph.D.s.

Hedi took only one day to decide that the idea was quite desirable. Financially, it was certainly a major improvement over their British alternative. Max's

British pension—an annuity—would yield only about £400 a year because of his limited number of years in Edinburgh. This was a fifth of his present salary, and with no personal savings (having recently gotten out of debt) their retirement looked bleak. Furthermore, "Life in Britain is not easy and sometimes gloomy," Born told the much better off Courant. Housewives faced an average of an hour per day of queuing to buy nonrationed goods. Another benefit of Eire was a warmer climate. Although Max preferred the cool weather of Scotland, he told her that as long as they were together, he did not care where they lived.

The official offer from the Irish government differed from what Schrödinger had outlined. Although Born could teach past the age of seventy, his appointment thereafter had to be reconfirmed annually. If he became sick or the institute did not renew the contract, he would receive no pension, just a lump-sum payment of about £1,000. This pension policy was too risky for Born to accept, and he requested the Irish government to change it. They would not.

Hedi thought Max seemed disappointed. They resigned themselves to staying in Britain with Max planning to supplement their income through lectures and books. He admitted to Einstein a little later, "I will probably have to go on working right to my blissful end. Not a bad fate, really."

In midsummer, the Borns took their first trip back to the continent. A monthlong trip to Hohfluh, Switzerland, lush with food and physical luxuries, brilliant skies, four-leaf clovers, and long walks in woods that they had missed so much, dispelled their concerns about the future. They returned home, rested and relaxed, to a profusion of music. Summer 1947 was the inaugural of the Edinburgh Festival. The city's program of concerts and operas filled the void left by the destroyed continental music festivals. Born's old friend Artur Schnabel and the string trio of William Primrose, Pierre Fournier, and Joseph Szigeti gave one of the performances. Gustav came up from London to be Schnabel's official page-turner for Brahms's A Major Quartet. Except for a broken string on Szigeti's violin, it was a perfect performance and a perfect end to the summer.

At year's end, Werner Heisenberg spent two nights with the Borns. He was in England to give a lecture in Bristol, and Born invited him to come to Edinburgh and lecture to his students. It was the first time the two had seen each other since before the war. As they talked about his adventures during the war and current affairs in Germany, Hedi described the feeling of friendship as "unchanged."

Heisenberg gave a "splendid" lecture on superconductivity, spent several hours discussing physics with Born's students, and went shopping with Hedi to spend some of his lecture fee on necessities for his family. A few weeks later, he reported to Sommerfeld that he felt an undertone of hostility from some Jewish colleagues at other universities, but that the visit with the Borns had been "as good as in the old times."

In a letter to Gustav, Max gave his impression of Heisenberg.

> His philosophy of life is definitely somewhat infected by Nazi ideas. He has a kind of "biological" creed, "survival of the fittest" applied to human relations, and seems to regret more that the Germans have not turned out to be the fittest, than what we regard to be the sad and regrettable things. For instance, concerning India he said that the British withdrawal is quite wrong as it shows only weakness and lack of will. But in spite of that we liked him immensely. Mutti gave him a lot of things to take to Göttingen, for his family and for others.

With Heisenberg's departure, Max pulled out Gustav's old electric train set and village in anticipation of a Christmas visit from both his son and Irene's son, eight-year-old Hugh. To the grandfather's delight, the three generations of Borns spent many hours of the Christmas week on the floor, playing with the train.

A few weeks after the holidays, Max took off for a conference on thermodynamics in Brussels. The start of his 1948 schedule signaled a year of nearly constant travel. After Brussels, he traveled directly to Oxford to give the Waynflete Lectures at Magdalen College that he had agreed to almost two years earlier. He no longer planned to discuss the purpose of science as he had considered doing after the meeting of the Association of Atomic Scientists. Required to address himself to the lecture series' overall theme of the boundary between science and philosophy, he decided to define nature within a consistent system that is ruled by both cause and chance. As he said to Hedi, "If both are God's work, then is the point a crazy game of rule and chaos or of chaotic order?"

The hall for the first lecture at Magdalen was so crowded that Born and the president of the college had to push their way through the standing-room-only audience to get to the podium. After the introduction, Born began by observing that chance can follow rules and cause-effect relationships cannot predict the future with certainty. He told the audience that he would untangle these two ideas and show, through the course of the lectures, the common features of cause and chance through the example of physics.

Astronomy, the oldest science, was Born's first example. He explained that its foundations were deterministic, and according to his definitions, not causal since early theorists such as Ptolemy, Copernicus, and Johannes Kepler did not state a cause for the behavior of the planets, except God's will. Indeed, cause only became prominent when Galileo determined the rules for falling bodies and Newton, the laws of motion.

Born's lectures proceeded through electromagnetic fields, thermodynamics, kinetic theory of gases, and electrons to illustrate the interplay of causality, chance, contiguity, and antecedence in the development of physical theories.

Then he came to quantum theory and the statistical interpretation of the wave function, a theory, as he said, that is "frankly and shamelessly statistical and indeterministic." He asked, "Can we be content with accepting chance, not cause, as the supreme law of the physical world?" And he answered,

> To this last question I answer that not causality, properly understood, is eliminated, but only a traditional interpretation of it consisting in its identification with determinism. I have taken pains to show that these two concepts are not identical. Causality in my definition is the postulate that one physical situation depends on the other, and causal research means the discovery of such dependence. This is still true in quantum physics, though the objects of observation for which a dependence is claimed are different: they are the probabilities of elementary events, not those single events themselves. . . . We have the paradoxical situation that observable events obey laws of chance, but that the probability for these events itself spreads according to laws which are in all essential features causal laws.

Thus Born summed up his own contribution to the interplay of cause and chance. The rules of the game were causal, the events themselves probabilistic.

Born concluded his lectures with metaphysical beliefs—ones that could not be traced back to more fundamental ideas. To show the importance of the "act of faith," he quoted from two letters of Einstein's about the wish for a return to the deterministic world. A month earlier, Einstein had written that the statistical interpretation contained some truth, but, said Einstein,

> I cannot seriously believe it because the theory is inconsistent with the principle that physics has to represent a reality in space and time without phantom actions over distances. . . . I am absolutely convinced that one will eventually arrive at a theory in which the objects connected by laws are not probabilities, but conceived facts, as one took for granted only a short time ago.

Born had sent Einstein the manuscript to ask permission to use the quotes, and Einstein had sent it back annotated. He agreed with some of Born's points, for instance, that "of course statistical predictions can be trusted in the real world," but not with others, such as Born's criteria for "distinguishing subjective impressions from objective fact," which provoked Einstein to comment, "Blush, Born, Blush!"

The lectures took place over a six-week period. During that time, Born experienced pluses and minuses of life in Magdalen College—plenty of hot water, for instance, but no central heat. He sat in an overcoat in front of the gas fire in the evenings and asked Hedi to send him a hot water bottle to put in his bed at night so that he wouldn't freeze. Even so, he enjoyed his eclectic set of companions. One day, he listened to his former student and guest lecturer Linus Pauling. Another time, he heard war stories from his old Breslau friend Stanislaus Loria. He went walking with a new acquaintance, C. S. Lewis (of future *Narnia* fame), whom he described to Hedi as "a youngish, fat little man with humoristic eyes and some wit." Even Heisenberg was there. The British feared, whether reasonably or not, that the Soviets might try to kidnap him from Göttingen, which was not far from the border with the East. One evening, Born visited friends who were entertaining academics from the University of Heidelberg—a hotbed of fascism just a few years past. He found himself too uncomfortable around them to stay at the reception for long.

He spent time with Gritli, Maurice, and the children and occasionally got over to Cambridge to see Irene and the rest of the Newton-Johns. Irene was pregnant, giving her parents a slight concern since life had been a bit unsettled with Brin changing jobs. But as Max wrote to Hedi, "Irene is such a motherly creature, that she will get much happiness out of the new baby. . . . If this baby will be as fine as the other children I shall be most happy and grateful."

One night in Oxford, he had dinner with Klaus Fuchs and his father, Emil, who was visiting from Germany. Other than Fuchs finding the work at Harwell on nuclear technology rather dull, he was doing well. His father, though, worried about him because he was not married and his father thought him too shy to meet someone. Born asked Gritli and Maurice, who were friends with Klaus, to introduce him to some young women.

Fuchs occasionally dined with the Pryces, and Maurice always enjoyed chiding Fuchs about his communist philosophy. On one occasion, Fuchs related that he had nothing to do with the British Communist Party. The Germans had told him to keep away from the British because "they were a bunch of fools." Another night, Maurice, teasing, said that the government had not yet caught up with his communist tricks. Fuchs blanched.

World events kept crashing into the scholarly life at Oxford—first, Gandhi's assassination in India and then tensions between the Soviets and the Western Allies along the border of East and West Germany. Born thought the Soviets were reacting out of fear—of attack, as Marx's theory explained—and a need to set up defenses throughout Europe. As worrisome as the situation was, when he wrote to Hedi, he soothed her alarm by explaining that war on the part of the Russians would be too crazy. He had talked to experts and knew that Stalin had neither an atomic bomb nor defenses against it.

Born returned home in mid March and within two weeks was traveling again with Hedi, this time to Bordeaux to celebrate the twentieth anniversary of the Raman Effect. Raman was to receive an honorary degree from the university there—as was Born. Born saw a scientific message and some humor in this double presentation. Their controversy concerning the correct theoretical structure of a solid continued unabated. In spite of this, Born and Raman occasionally exchanged cordial letters on nonscientific topics, and their wives wrote each other a few times a year.

On the first morning, after Born and Raman received their honorary degrees, the assemblage went to the Cité Universitaire for a luncheon. Raman passed Born and commented negatively to him about a theoretical physicist who had tried an experiment that Raman regarded as poor, and Born replied (according to his account) with words to the effect, "But my dear Raman, what about the other way round, when experimentalists venture to make theories?" Raman exploded and was so angry that when he sat down to eat, he missed his chair and fell on the ground.

At lunch, Hedi sat next to Raman. She tried to start a conversation but could barely get a word out of him. Suddenly, with a furious look he burst forth, "Madame, your husband has treated me incredibly rudely and conceitedly. In fact, I consider leaving the congress at once. Excuse me if I do not want to speak at all." Then he told her what Max had said, "'You stick to experiment and leave theory alone.'" When Hedi tried to appease him, he accused Max of racial prejudice. He also referred to the ugly Senate meeting in Bangalore twelve years earlier, saying, as Hedi recorded in her diary, "He [Raman] had stuck to Max when Max had few friends left." Finally, he told her how badly the British had treated him since Lord Rutherford had died, how other Indian scientists had received honorary degrees in England but he had not.

For the next few days, all was quiet. Hedi noted that Raman gave a number of long, tactless speeches, the conceit of which made everyone uncomfortable. On the last evening, the mayor of Bordeaux presented medals of the city to both Raman and Born then took them to a gala at the opera.

The Borns left Bordeaux for Paris the next day, where Max was to lecture at the Institut Poincaré and attend a conference on chemical bonding. Raman also traveled there.

Alfred Kastler, an organizer of the Bordeaux meeting, had a dinner party in Paris one night for a number of physicists, including Raman and the Borns. After dinner, Kastler asked for everyone's attention and spoke of the debate between "our two friends Sir Chandrasekharan Raman and Professor Max Born" on the structure of diamond. Kastler pulled from his pocket two identical and difficult cubic puzzles, quickly disassembled them, and said, "I think we can now settle the dispute. Would Sir Chandrasekharan and Professor Born be kind enough to rebuild the diamond in its cubic form? We shall see who is right."

The two physicists stood on either side of the piano, Raman sneaking glances at Born's progress.

Born finally stopped in defeat, excusing his failure by stating, "I drank too much this evening." Raman, who had taken the challenge more seriously, feverishly kept trying but to no avail. The French physicist Cabannes took Born's puzzle to release Raman from his misery and, knowing the special trick, reassembled the cube. Putting a lighthearted touch on the competition, Kastler announced, "I am so glad that a French scientist has been able to solve the problem before the great physicists we have here tonight."

Once back in Edinburgh, Born heard that Raman had supposedly convinced him that he, Raman, was correct. Others heard it as well. Finding it surprising, Peierls wrote to ask Born for his version. It took Born only one day to answer that Raman was mistakenly convinced that the entire audience, Born included, agreed with his theory. Irked and frustrated, Born questioned Raman's sanity, but softened his comment, saying, "at least there is not a trace of common sense left."

The Born-Raman controversy quickly receded into the background as Born focused on the crisis in the Middle East—the fate of the Jews in Palestine. On May 14, 1948, the British, who had controlled Palestine for twenty-five years under a Mandate from the League of Nations, withdrew its troops. The next day, Israel declared itself an independent state. The British policy toward Israel under the direction of the foreign secretary Ernest Bevin alarmed Born. He characterized it as "a nasty game," in which the British had trained and armed the Arabs then pulled out to leave Arabs to do the "dirty work" of liquidating the Jews. Neither did Britain recognize Israel. Born felt "completely powerless and without influence in this country," but he asked Peierls, "Can we who have escaped Hitler, sit still and watch quietly this second mass murder, arranged by our government?" He wanted friends such as Peierls and Michael Polanyi to publish a protest with him.

Peierls found the situation much more complex than Born's portrayal. He looked at the claims of both sides, seeing neither Jews nor Arabs at fault, but both as victims, especially the women and children. Due to all the previous contradictory promises made by the British, both groups were convinced of their rights, he told Born, and no one could prove either side incorrect. He blamed the United States for the failure of the United Nations to act, not the British. As sad and helpless as he felt, he had no sense that the actions of a few refugee scientists would benefit the situation. Born agreed with Peierls "if I use my reason. But in such things feelings are stronger than reason. You have persuaded me to suppress my feelings for the time being."

In July, the Borns went to Hohfluh for what was becoming a regular vacation. This year they had the added pleasure of a visit from Richard Courant—

all the enjoyable elements of a family reunion, except for a haunting story Courant told about the death of Planck's son Erwin at the hands of the Nazis. The Borns knew he had been hanged for an alleged connection to the July 20, 1944, assassination attempt on Hitler, but nothing more.

Hedi recorded that Planck had written a petition to save his son's life. Heinrich Himmler, Hitler's chief of the SS, had told Planck,

> "We all, Göring, Goebbels and myself have signed a petition—your son is as good as saved. Now you sign this, too [referring to a certificate of loyalty to the Third Reich]." But Planck hesitated to sign and asked for time. When Hitler heard of this, he was infuriated—beyond measure and shouted with a foaming mouth that the son of Planck was to be hanged at once. Which was carried out.

Hedi was tormented by "this horrible story." She asked herself, "Can there be a more cruel choice than old Planck's? And what is the 'right' thing to do!" Max later repeated Courant's story to Gustav and said, "I would sign anything to save my children."

Erwin Planck's widow, Nelly, later challenged the accuracy of this story. When Born wrote a tribute to Planck, he included the story in order to attest to "the super human character of Planck" and also to counter "loud foreign voices who reproached him for obedience to the Nazi regime." Nelly told him that there was only old Planck's petition and not a loyalty oath. But Courant's source, Karl-Friedrich Still, with whom Born had conferred with, was sure that there was a petition and a loyalty oath.

From Switzerland, the Borns returned home to enjoy the Edinburgh Festival then crossed the English Channel again for a trip to Germany, their first visit since 1937, eleven years earlier. Born was to receive the Max Planck Medal from the German Physical Society at a meeting in Clausthal-Zellerfeld. Heisenberg wrote to congratulate him on the award, saying that Born's work on quantum mechanics had not been acknowledged as had Heisenberg's own. Born agreed, replying that although he had never doubted Heisenberg's "little paper contained the essential idea which brought the whole thing into motion," he resented British physicists crediting Dirac with the mathematical theory and American scientists crediting Heisenberg with matrices. Born asked Heisenberg, "If you have an opportunity to correct this, I should be grateful."

Born had mixed feelings about returning to Germany, especially since he had heard that former Nazis were resuming high places. Additionally, Richard Courant had surveyed science faculties there for the U.S. Navy and discovered that German scientists, like the majority of Germans, did not "really understand their position in the world." As Courant put it, "They have no clear conception of the misery inflicted by Nazi Germany on her victims. Self-centered, they indulge in criticism of the Allies and are unwilling to see the present plight of Germany as a consequence of Hitlerism rather than as of Allied mistakes." Scientists in both esteemed and mediocre faculties made statements such as, "We disliked Hitler, but the British and Americans are worse. They have replaced Hitler with Stalin." They accused the Allies of trying to starve the people. Born did not like what he was returning to.

The day before Max and Hedi were to leave for Germany, Max came down with a severe intestinal upset, and his doctor ordered that he limit his trip to one week. On September 6, the Borns flew to Frankfurt, where the tarmac was crowded with planes for the Berlin airlift. The effort to feed those in the Western section and break the Soviet blockade was two months old. Then they rode a bus to the destroyed rail station in the center of the city and took a train to Göttingen. To Hedi, it felt completely unreal. Both had the same reaction—as Max wrote to Courant, "We had no feeling of coming home, but [of] just visiting a strange foreign place." At one point, Hedi laughed at herself when she saw the Harz Mountains and, thinking of her Scottish hills, instinctively turned to Max to say, "Doesn't that look like the Cairngorms near Aviemore."

"Fat, jovial" Robert Pohl, the head of one of the physics institutes, and Dr. Ronald Fraser, the scientific liaison officer in the British zone, met them at the train station and drove them to see Herglotz for a few minutes. Hedi had heard that he had had a stroke nearly two years earlier, but that knowledge did not prepare her for the change in the man who had been such a central part of her life and emotions. She found an invalid, his "fine features washed out, eyes staring, face fat—only a faint shadow of his old self" and summed up with the final comment, "sad and moving." The following day, the Borns and Dr. Fraser were off to Clausthal for the award ceremony.

Clemens Schäfer, one of Born's first physics professors at the University of Breslau, presented him with the Planck Medal—a medal no longer made of precious metal, Schäfer explained, because they were too poor, but one that still embodied the tradition of the masters. Schäfer's touching tribute positioned Born as standing at the front with his work on relativity theory, on dynamics of crystal lattices, on quantum mechanics, and on the theory of the electron. When the old mentor turned to the past—not the "unspeakably horrible" previous fifteen years but those almost a half century earlier—he could still see Born sitting in the lecture hall together with his friends Toeplitz and Hellinger.

Born was deeply moved and sincerely thankful, but in his remarks he chose to go back to those fifteen years that Schäfer had avoided, giving a "stirring overview of . . . the abundance and complexity of the problems tackled," together with the names of his many excellent coworkers who hailed from all parts of the world. Born had flourished during those years, and he let his audience know it.

After a few more days in Göttingen, visiting old friends such as Otto Hahn, Max went back to Edinburgh while Hedi stayed a little longer. He wrote to her that he was "filled with impressions of Germany," pondering the unexpected and the strange, especially "that the new house [in Edinburgh] is so much more real than the old one." To Simon and Peierls, he wrote that the only reasonable people they encountered "who were not subject to nationalistic propaganda and draw correct conclusions of what is happening" were their former maid and housekeeper and his former secretary. The physicists were kindly, but the chemists like Eucken were "wildly nationalistic." Neither he nor Hedi ever wanted to live among Germans again.

Born had come back early on his doctor's orders. Miraculously, the day after returning to Edinburgh, his health returned to normal. He had a good lunch at the Horseshoe Restaurant and walked to the department to answer his mail. As far as he was concerned, Germany was a memory.

FOURTEEN

A TRIP TO STOCKHOLM

BORN'S THANK YOU NOTE TO OTTO HAHN FOR THE HOSPITALITY IN GÖTTINGEN PROUDLY announced, "On our return here, we discovered a new granddaughter." Baby Olivia had joined the Newton-John family in September 1948. Then Max and Hedi heard that Gritli was expecting twins in the spring. The British roots of the Born family were growing ever deeper.

Once in Edinburgh, atomic structure beckoned Born again. He decided it was time to develop a general theory for elementary particles—not one of photons, electrons, mesons, and such separately. For his underlying principle, he looked to the work he had done with Fuchs ten years earlier on reciprocity, which at that time mathematics doomed. Since then, Born's student Herbert Green had made some progress with the mathematics, and that fall they formulated a new theory of elementary particles. When Born presented his and Green's ideas at a colloquium in November, a colleague proclaimed, "We may possibly have attended an event that will make history in physics."

Born and Green's new quantum electrodynamics showed that there are an infinite number of elementary particles, with or without spin, and with definite, calculable restmasses. Max sent Gustav an enthusiastic explanation.

> You know that according to quantum mechanics there corresponds to each kind of particle a type of wave, or field, which determines the probability of the particle's occurrence. For instance, to the photon belongs Maxwell's electromagnetic field, to the electron the de Broglie wave etc. As long as the number of types of particles was small (electrons-protons) this was all right. But meanwhile this number has increased and is still increasing—we have, besides electrons and protons: neutrons, neutrinos, positrons, and mesons of many types, every week a new one is found. It seems to be almost preposterous that the space should be filled with so many different fields. Well, I think that we

> have overcome the difficulty by showing that these fields are special forms of a more general structure, which we have constructed, and we obtain all the main types of fields, together with the masses, spins etc. of the particles.

This new theory fit into Born's old notion that electrons must have purely electromagnetic mass. The harmony of the fit added to his convictions that he and Green had discovered something fundamental. He wrote to Courant, "This theory may be just as decisive a step as ordinary quantum mechanics was in 1925." When he wrote about the theory to Ladenburg, Bohr, and Patrick Blackett, his tone was jaunty; a renewed sense of confidence carried him forward. Hedi described him as being in a "blissful state all this year."

The years left to Born to make a scientific breakthrough were few, as were the opportunities to create moments of bliss. Perhaps the inevitable decline had been hastened by the deadly turn in his "beautiful" science. Several years had passed since he had the intellectual energy to do calculations, although, as he had said to Peierls a couple of years earlier, he could still reason. Born was the ideas man, and Green, a worthy collaborator, handled the mathematics.

Max began to look to Hedi for the sense of security and grounding that science had once given him, and he became more dependent on her and her judgment. She had always held strong opinions; and he had not usually paid too much attention to them, characterizing her thinking as black and white, while his own operated in the gray zone, as he weighed each side. Now he brought in her name to add weight to his side, something he had never before done.

Such was the case when Simon called the British scientific representative in Göttingen, Dr. Fraser, a "slavish admirer" of Heisenberg. (Some issues have lives of their own, and among Born's group, Heisenberg's politics was one of them.) Fraser, according to Simon, believed Heisenberg's story about the role of German scientists in the atomic energy project—a story that Simon said "we know does not correspond to the facts."

Born rejoined by accusing Simon of applying Fraser's favorable assessment of Heisenberg's scientific ability to Heisenberg's political views as well. At some point, Born decided to keep scientific ability and political views separate—as did Hedi. "Heisenberg is without doubt one of the most gifted men in theoretical physics, and nobody can resist his cleverness and charm. In regard to politics, Fraser is quite clear about Heisenberg's ideas, as my wife and I are, too." Born knew that Simon did not observe such a dichotomy.

Scientific propaganda in the USSR gave Hedi another opportunity to press her opinions. At the end of the year, the British pharmacologist Sir Henry Dale publicly resigned from the USSR Academy of Science to protest political attacks on genetics. Labeling genetics a "bourgeois pseudo-science," the academy, under the guide of agronomist Trofim Lysenko, had thrown out Mendelian theory and replaced it with something called Michurinian theory, a true pseudo-biology. When a reporter asked Born if he would follow Dale's example, Born said no, as long as the Russians did not attack physics. Hedi wrote in her diary that Max's view was "narrow and superficial." She wanted him to make a public statement encouraging reconciliation, not separation. She referred to the example of the members of the Academy of Science in Göttingen who did nothing to oppose the Nazis' expulsion of Max and then begged him to return after the war, behavior that showed the endurance of science.

On December 23, the London *Times* printed a letter from Born stating that he would not resign even if physics were attacked. He did not agree with the Marxist philosophy of science, but he would neither desert his friends and former students nor increase tensions. If the Russians chose to remove his name, as the Nazis had done, so be it. Now the same Germans wanted to include his name back on the list. "Governments and economic systems come and go while science remains and grows." Although Max may have agreed, it was, in fact, Hedi who wrote the letter.

At the start of the New Year, Hedi checked into a nursing home in close by Peebles because of insomnia and depression. Gustav diagnosed her health problem as a dissatisfaction with her life and work. The doctor's prescription was to "let her do as she fancies"; Max's was to offer love and tenderness. He knew from his own experience that reason did not work and that he would have to wait patiently for Hedi's depression to lift.

Four months into her stay, Hedi was reading Goethe's *Letters from Switzerland* when an old wisdom struck her: Humility is the key of life. In her diary, she scrawled thoughts of the turmoil resulting from her return from Germany to England, the loss of her "calm and balance" from a lack of humility with her friends in Germany, the realization of how insecure she was, of how she needed to begin again. Just as she finished, Max arrived for one of his regular visits. They chatted awhile, and as soon as he left, she took up her pen to continue her diary entry. "As God guides everything: I had hardly written all of this when Max came and said that he had a report from Pohl in Göttingen that Max will get his German pension (and I eventually a widow's pension) if we move back."

Max had contacted Robert Pohl about a German pension as a backup for her in case anything happened to him. (Hedi's reliance on sanatoriums and vacations had always kept the family pocketbook tight.) Pohl had written that Max was likely to be reinstated to emeritus status with all the benefits. Courant already had his status back. (Born's name would be officially entered on the emeritus list a few months later.)

The following day was the eve of the anniversary of their engagement, and to celebrate, Max sent her flowers, sweets, a cake, and a love note.

> Tomorrow is the day when 36 years ago we decided to embark together on the journey of life. Now we have weathered many a gale and had lovely sunshine. But the main result is that we love one another more than ever. I feel this so strongly every day.

At Max's visit to celebrate this anniversary, the tenderness turned to tension. Hedi told him of her yearning for a quiet country life, and for the smell of trees and flowers in spring. She had kept her spring fever at bay all those years with hectic activity. Now that lack of energy prevented her participation in social work, she sat lonely and listless, staring at the four walls at home. Her panacea: a return to Germany.

Max let out his fears to Gustav that evening.

> She had developed a plan: why should I go on working here in Edinburgh? I ought to resign in a year or two and live with her from the pension in a nice spot in Germany, near the Bodensee, for instance. Well, I was not enthusiastic about the idea, and that sufficed to make her rather excited and unhappy. I should perhaps have pretended to agree, but then she would have made me stick to my word and I would never get off. In fact, the idea is not bad in principle: if I had the means to retire in this country or in France or Italy, I would do it at once. But in Germany? There are too many bitter reminders for me to live there happily, quite apart from the difficulty of finding a decent home in the overcrowded country.

Hedi relied on Gustav for support as well, telling him of the emotional struggle that the previous fall's return to Göttingen had caused her. She knew that Max loved her deeply and needed her, but he did not understand "earthbound" things. He had been in a happy state all year and had never given a thought to her needs. "*Climate*" was the vital thing for her; she "*craved* for the seasons," as she knew them.

Four days later, she concluded that she could not face British suburbia. If Max stayed in Britain, she would try to get a little job with the Stills or German Quakers and move back to Germany by herself.

Max left for a conference on statistical mechanics in Florence. On the way, he went to Cambridge and Oxford to meet his three new granddaughters, Irene's Olivia and Gritli's twins, Lois and Susan. All sweet little babies, he wrote to Hedi, in one of his almost daily chatty letters from Florence that also recited names of old friends at the conference and places he and Hedi had seen together in Rome so many years ago.

In a serious letter, he brought up her need to return to Germany. He wanted her to do whatever gave her a fuller life. If that meant her moving to Germany and their occasionally being separated, he would forbear the lack of "her good, true presence." Financially, he thought they could manage. It appeared that she could use his pension in Germany even if he were not there, and that he could live on a regular consulting fee earned from a South African firm for his advice on diamonds.

He shared her nostalgia for landscape and spring aromas, but these could not fulfill his own longings for home, which was Breslau, now a pile of rubble and part of Poland. Göttingen was his second home, an intellectual one that held no childhood memories for him, as it did for her.

> That is why I am silent when you speak of it all so warmly and longingly. You can see for me what is lost. In my situation one must put armor around the heart. To me, your yearning is complete and living and conscious. On the other hand, you must not loosen my protective ring: my interest in philosophy, science, and music. Otherwise I become sad.

Usually their correspondence was in English. Max reverted to German only for subtle thoughts that were difficult for him to express in English. He wrote this letter in German.

On his return from Florence, Max stopped in London to see Käthe. She had just arrived from India, where she had lived with her son Otto for a few years. Back in Edinburgh, Max found Hedi well and cheerful. The rest and reassurances about Germany had restored her health.

In the midst of this tug between past and future, Max learned that his half brother Wolfgang had died of a heart attack after undergoing a hernia operation. He was fifty-six, seven years older than their father when he had died from heart disease. Wolfgang's death gave Max a violent shock, especially when he thought of life in Breslau when their father was alive; but he felt no deep grief, which made him sorry. Wolfgang and Max had a complicated relationship, one largely defined by Wolfgang asking for Max's advice and not heeding it. Max did what he could to help. They had little in common, but Born respected Wolfgang's talent as a painter and art critic. Gustav lent his father a shoulder for his pain. Hedi, who never liked Wolfgang, was unsympathetic.

Max took Hedi to Switzerland for most of the summer, and they came back strong and relaxed, ready to enjoy the Edinburgh Festival. Then Max began to prepare for a lecture series to be given by Niels Bohr—the prestigious Gifford Lectures—followed immediately by a conference on elementary particles.

Twenty-seven years had passed since the famous Bohr *Festspiele* in Göttingen that had entranced a generation of physicists. Edinburgh was a different stage in a different world and could not duplicate that unique electricity, but this academic community was enthusiastic. The university's 1,200-person auditorium was largely filled for Bohr's introductory lecture, "'Causality and Complementarity': Epistemological Lessons of Studies in Atomic Physics." Loudspeakers set up close to the dais were ready to boost Bohr's soft voice.

This gathering was not so charmed. The speakers did not amplify well, and most of the audience sat for one and a half hours trying in vain to grasp a word. The content, for those in front who could hear, was beyond most of them. Bohr took his abstractions from experiment rather than from everyday experiences to which audiences could relate. As planned, the remaining nine lectures were moved to a smaller hall, where about fifty attended faithfully. Born attributed what success these lectures had to Bohr's "personality of extraordinary charm and power." "To be frank," he told Gustav, "the lectures were not very good: no new ideas and delivered in an unimpressive, slow way with a low voice."

The lectures and the following conference on particles covered a span of four weeks. The two spent much time together, enjoying hashing through their scientific philosophies and other topics such as Heisenberg, who was to have spoken at the conference but canceled at the last minute.

In responding to Bohr's letter of thanks, Born revisited this subject.

> I was very happy that our outlook on the fundamental questions of science is so nearly the same, and (if I may mention a point of personal vanity) that you are aware that not all achievements usually connected with Heisenberg's name are really his. During the Nazi times I could not expect him to put this right, but when he did nothing after the end of the war, I felt a great disappointment. For I had loved and trusted him from the first day when he appeared in my study as a young boy of rare charm and genius. I am still more sorry about what you told me in regard to his attitude in the matter of the atomic bomb. I frankly confess that I was rather relieved when he informed me that he could not attend our conference.

Born's issue with Heisenberg was about attribution; Bohr's issue concerned a visit of Heisenberg's to Copenhagen in 1941. The only aspect of their meeting the two would later agree on was that Heisenberg broached the subject of an atomic bomb. Otherwise, Bohr always suspected that Heisenberg was trying to get information out of him. Heisenberg always felt that Bohr had misunderstood. The rift never healed, and the confusion was never resolved (the mystery surrounding this meeting later stimulated intense speculation by authors and playwrights).

The year 1950 gave Born little escape from issues of attribution. This was a double anniversary: Twenty-five years since the discovery of quantum mechanics and fifty since Planck's initial quantum formula. Scientific articles mushroomed in celebration. During the holidays, Born received two articles written by Germans that completely passed over his contributions and singularly glorified Heisenberg. Born blurted out to Simon, "I am not particularly ambitious, but if one's best idea in the whole scientific career is pinched in this way one cannot help [but] get annoyed." After reflecting on his annoyance for a few days, Born asked Simon to dismiss his letter as a "fit of temper."

Simon, however, decided to pursue the matter, obtaining a copy of the chapter from Born's "Recollections" that pertained to the quantum discovery. After reading it, he concluded that Born had been wronged; he also did not like the fact that "the Germans have become accustomed not to acknowledge the merits of the refugees." Simon proposed to Born that Bohr write an article for *Nature* to rectify the situation, but Born rejected the idea. He assumed that Bohr had been consulted on the original Nobel Prizes for quantum theory and that an article from him now acknowledging Born's contribution would be a contradiction. Simon then wrote to Léon Rosenfeld, Born's assistant in the late 1920s, who agreed to take up the task.

Rosenfeld wrote a short account of the development of quantum mechanics for *Nature*. To Born, he credited placing "at the disposal of the physicists the formidable arsenal of matrix theory." According to Rosenfeld, who was thoroughly schooled in the Copenhagen Interpretation, all inspiration on transitional amplitudes and the algebraic pairing of integers emanated from Bohr's correspondence principle through Kramers to Heisenberg while they had worked together in Copenhagen on dispersion theory. He did not recognize that Born had introduced difference formulas and that from Göttingen, Heisenberg had given the idea to Copenhagen through Pauli. Nor was he familiar with Born and Jordan's paper that identified transitional amplitudes as key factors and observability as a postulate. As for Born's interpretation of the wave function—a fundamental principle of quantum theory—Rosenfeld described it simply as "consonant with Bohr's view about the essentially statistical character of our description of atomic phenomena."

It took Born almost six months to thank Rosenfeld.

> This was actually the first time that my contribution to the birth of quantum mechanics was acknowledged. . . . For myself, the most interesting part is your report about the influence of Bohr in the development which was partly new to me. . . . My conviction that the ultimate form of quantum theory would be statistical was as old as the effect of Einstein's derivation of Planck's law with the help of transition probabilities and did not derive from Bohr's ideas, as the reader may infer from your article. As a matter of fact, when my first paper on collisions appeared I received a letter from Heisenberg . . . in

> which he really abused me as a "traitor" in having applied Schrödinger's wave function instead of sticking to the matrix formulation. Bohr himself would have seen the matter at once in the right perspective, as Heisenberg did soon enough.

Born could not really have been pleased. Twenty-five years earlier, he, along with Jordan, had taken Heisenberg's seminal idea and developed the full quantum mechanical formulation—hardly the impression left by Rosenfeld. Prior to that, Born had laid out, mainly in concert with Kramers' dispersion theory and Slater's virtual radiation fields, the basic foundations for Heisenberg to use.

Perhaps to the surprise of Born, an article by Heisenberg on quantum theory in honor of Planck's 1900 insight was more accurate and more generous. He described working in Göttingen with Born and Jordan to extend Kramers' dispersion theory and explained that these considerations led him to give up the concept of electron tracks and to consider amplitudes. He acknowledged that he did not know what matrices were, and he credited Born and Jordan with developing the final mathematical formulation of matrix mechanics. "It is incumbent on me," he wrote, "to stress how great the contribution of Born and Jordan, which has not always been adequately acknowledged in the public eye, has been to the foundation of quantum theory."

Heisenberg sent Born a copy of the article. Born responded, "I have read it with great pleasure and I wish particularly to thank you for having mentioned my own contribution to this development in such a nice way." His thanks were brief but warm.

The beginning of 1950 did not augur well for refugees. On February 3, Hedi heard over the wireless that Klaus Fuchs was in custody "for giving away atomic secrets in 1945 and 1947." Max came home for lunch and found her highly agitated. A reporter had just called to interview Max about the arrest.

At first, Max and Hedi could not believe that Fuchs was guilty, although Born had always thought it odd that Fuchs was assigned top-secret work, given his strong convictions about communism; Hedi, reflected simply that "he was always such a closed book." A research fellow in Born's department wrote to Fuchs saying that he believed his innocence, but the Borns saw Fuchs's reply to this and reported to Gustav that the only interpretation was, "He is guilty." "Upset about Klaus Fuchs," Hedi wrote. "For seven years he really has given all atomic information to the Russians. He confesses. His split nature." Max too was shaken.

Fuchs had gone to the United States with Rudolf Peierls to do atomic research during the war and in the Theoretical Division at Los Alamos his scientific contributions were an integral part of the work. Additionally, whenever the division needed a liaison with other branches, he was always the first to volunteer. Fuchs performed so well that the U.S. Navy requested his help with subsequent atomic tests on the Bikini atoll. When he returned to Britain in 1946, he worked at the nuclear research facility at Harwell. On September 23, 1949, the Soviets detonated an atomic bomb several years earlier than experts had expected. At the end of December, an officer from the British Secret Service asked Fuchs if he had given atomic secrets to the Russians. Fuchs answered "I don't think so."

Without a statement from Fuchs, the British government had no specific evidence on which to arrest him. There were indications of serious leaks when the British Research Mission was in the United States, but nothing to tie them to Fuchs. During the month of January, the Secret Service pressured him for a statement. According to Fuchs, they also gave him an easy way out: to leave Harwell a free man to teach at a university. He thought he even could have stayed at Harwell if he admitted "just one little thing and kept quiet about everything else."

During January, Fuchs's decisions of the previous eight years crashed down on him. "Controlled schizophrenia," he called it. He contemplated suicide. On January 27, he suddenly issued a full statement. Based on his statement, the British charged him with four counts of espionage.

The Russians had first contacted Fuchs in early 1942, while he was working with Peierls in Birmingham on atomic research. He forwarded to the agent only the product of his own research, which mainly involved the theory and mathematics of the gaseous-diffusion process for separating uranium isotopes. He was surprised when the agent asked him about the electromagnetic method of separating isotopes, something Fuchs had never thought about. In 1943, Fuchs went with Peierls to New York, got a new Russian agent, passed on material about the American program, and a year later went to Los Alamos. There he learned about the importance of plutonium and passed this information on to his handler. In June 1945, he gave Russian scientists a full description of the plutonium bomb that was to be tested in the desert of New Mexico a month later. The information included a sketch, its components, and all the important dimensions.

On March 1, 1950, Fuchs was convicted and sentenced to fourteen years in prison. (He would be released after nine for good behavior.) Since the Soviet Union was still a friendly nation, he was not subject to the death penalty. Lord Goddard, the Lord Chief Justice, said that he was sorry the law did not allow him to hang Fuchs.

A year later, the Borns saw Fuchs's father, who was now a Quaker, at the Annual Meeting. Hedi told him that they still had friendly feelings toward Klaus but stated clearly that his actions were wrong and that she could not understand how he could break a solemn oath. Emil Fuchs was astounded and replied, "But *that* was not wrong! He was convinced he was saving humanity by what he did and helping to avoid war. What *Klaus* considers wrong—what he is now suffering under—is that he deceived his friends."

The British public and most of the newspapers did not react emotionally to Fuchs's disloyalty. There were no repercussions for other refugees, although some feared that there might be. Colleagues of Born's referred any queries from journalists to him, but for whatever reason, they did not call.

The brouhaha did not come until a second incident. Bruno Pontecorvo, an Italian physicist who also had worked at Harwell, disappeared the following September, only to show up in Moscow a month later. The big bold headline on the front page of the *Sunday Express* on October 29, 1950, blared PERTURBED MEN. Underneath the subheading FOREIGN-BORN ATOM EXPERTS DISTURBED BY PONTECORVO CASE were pictures of six refugee scientists, including Born, Simon, and Peierls. The article gave the background of each scientist and told how they sought to assure the British of their loyalty. Simon wrote to Born, "Did you admire your picture in the 'Rogues Gallery'?"

This time the press swarmed over Edinburgh. Besides the Fuchs connection, Pontecorvo's brother Guido had been a professor of agriculture there. They pestered Born's secretary, Rachel Chester, his research students, and anyone else to pry out an imprudent quote, but they were unsuccessful. Hedi feared a "witch hunt" in Britain similar to the one in the United States. After a few weeks, quiet returned.

During the Christmas holidays, the Borns traveled to Oberstdorf, a village in the Bavarian Alps, for a vacation. Much to their surprise, when they disembarked at Harwich after the trip, they were directed by name to a customs official. He did nothing other than yell at them because they had £6 rather than the £5 allowed under the currency regulation; but he commented to Hedi, "It is people like you who would do that sort of thing." The remark did not go unnoticed by Hedi, and she later learned that customs had been instructed by the Secret Service to watch whether Prof. Born would return to Edinburgh when his German travel permit expired. It appears they half expected him to head farther east.

The midcentury year brought special events that closed two of life's circles. Born met James Franck for the first time in fifteen years, a treasured reunion, and Gustav got married. His bride, Ann Plowden-Wardlow, was from an old British family that had settled in the Scottish Highlands. They went to live in Oxford while Gustav finished his D. Phil. (Ph.D.).

Herbert Green, Born's main collaborator, was in Princeton for the year, and Born did not feel he had the energy to work on fundamental problems by himself. He was wistful about it, writing to Pascual Jordan (with whom he had reestablished contact), "It is rather a pity for I have some ideas which look to me quite hopeful." His attention turned instead to redoing the fifth edition of *Atomic Physics* and another of *Restless Universe* and to designing an unusual project.

The theorist Born had become caught up in something that, if not exactly practical, was concrete. Years earlier, Born had briefly flirted with designing a house—not just any house, but one based on a parabola that rotated while the occupants thought they were at rest. The purpose was to create "a bewitched world" where strange things happened. A friend mentioned the idea to the organizing committee for the Festival of Britain of 1951, which was to be a public showcase for Britain's progress in the last fifty years. Born sent in a proposal to explain what he called the "Newton-Einstein House."

> Einstein's theory of general relativity and gravitation is based on the principle of equivalence that states that an observer inside a box with walls prohibiting optical observation of the external world cannot decide whether the forces experienced are due to the motion of the box or to external gravitational fields.
>
> The present project is intended to illustrate this. As it is impossible to produce strong gravitational fields one can only use a moving box, which visitors are allowed to enter without being aware of its motion. It is expected that they will be under the impression of an enchanted world where the laws of gravitational mechanics are changed and surprising effects are experienced. While in fact everything that happens can be explained in terms of Newton's laws, I venture to predict that the psychological effect will correspond to Einstein's equivalence principle. The visitor will be convinced that he is in a room which is at rest and where the laws of mechanics are abnormal.

The effect would be produced with a parabolic floor rotating around a vertical axis such that the resultant of the gravitational and centrifugal forces was everywhere perpendicular to the surface. Born carefully worked out that there should be one rotation every sixth second. In his design, the surface would appear to the visitor to be a garden with many different kinds of ball games such as croquet, golf, and skittles, where the trajectories of balls near the rim would be influenced by Coriolis forces and fly in unexpected ways. In this world, people who stood near the axis would see those at the rim standing in "inclined" positions. Those at the rim would experience a feeling of increased weight. Born had figured out a way for people to enter this world without knowing they had undergone acceleration. He also envisioned other rooms, ones for experiments, for dancers, even a small restaurant where patrons who became seasick could sip brandy.

The committee's vote to build the room for the exhibit pleased Born immensely, but his pleasure was short-lived. Not only was there an "idiot" in charge, as Born complained to Gritli, but the Treasury was concerned about cost. The obstacles turned his excitement into frustration, which he dealt with by trying to learn *Faust* by heart. Like his father, he found comfort from having the verse available at will. In the end, the room was never built, and Born learned much *Faust*.

In fall 1950, a change in rules by the military government in Germany allowed Born to deposit his retirement income into a special German bank account. He could not take money out of the country, but he could give 300 DM to relatives and 35 DM per day to family members traveling in the three Western zones of Germany. Although he did not specify any plans to return to Germany, he told the Kurator that he did not need to worry about the future anymore.

His comment about the income perhaps was simply an offhanded remark, but within a week, he was writing to friends advocating a new tolerance toward Germany. Born chose Einstein for his first attempt at conversion, writing to counter an earlier essay of Einstein's that held the whole German people responsible for the Nazis' crimes. Born said,

> I think that in a higher sense responsibility *en masse* does not exist, but only that of individuals. I have met a sufficient number of decent Germans, only a few perhaps, but nevertheless genuinely decent. I assume that you, too, may have modified your wartime views to some extent.

Born's qualification—"only a few perhaps"—did not seem to denote strong conviction. Einstein promptly answered in the negative, citing the Germans' "dangerous tradition."

Using funds from their German bank account, Max and Hedi again spent the Christmas holidays in Oberstdorf. When they returned to Edinburgh, he reported to friends, including Einstein and Simon, how they had enjoyed good food and central heat. A side benefit—the good Bavarian beer—"worked like a fountain of youth." Hedi had welcomed the change from Britain, where she spent two to three hours a day tending to the individual coal fires in the house and where they still had food rationing. A rosy picture all around, except for

Born's comment, "We did not get in touch with many people of our own class, and one has to be careful if one does not want to spoil the enjoyment of the holiday." To avoid anti-Semitism, they limited their interactions to waitresses and drivers, who did not have such conversations with the guests.

Born's new enthusiasm for Germany preceded his telling Schrödinger that he would retire somewhere there. (Schrödinger was unlikely to be judgmental since he had similar thoughts with respect to Austria.) Maybe it would be in the south, maybe Innsbruck were he allowed to draw on his pension money. Göttingen he excluded from all possibilities; he didn't say as much, but apparently the prospect was still too much for him.

Francis Simon was another friend Born tried to convince about not condemning an entire nation for the acts of individuals. "I have changed my opinion," Born said to Simon, "or at least I try."

Simon agreed with him that the individual mattered in personal friendships, but he believed that the behavior of the collective group mattered in politics, especially when trying to predict the country's future actions. The majority of Germans, he stated, either condoned or supported the Nazis, and "*the Germans* have therefore to bear the collective responsibility for what has happened—with the exception of a very small minority." Simon said that Born's attitude allowed the Germans to excuse themselves from bearing any responsibility, an attitude that was the best way "of making sure that all is going to happen all over again."

Heisenberg served as Simon's example. In Göttingen, the science faculty had just appointed a person to a chair in chemistry who was "a bad Nazi and a 3rd rate scientist." The other contender was a first-rate scientist who had stood up to the Nazis. When Simon saw Heisenberg in Copenhagen, they had discussed the situation.

> He [Heisenberg] said quite shamelessly that of course in say 10 years time Germany will again be ruled by the Nazis, perhaps under another name; after all then the generation which was educated under the Nazis would be of the right age. And you can be quite sure that he would "mitmachen" [join in] with the greatest pleasure. I believe I have told you that a few years ago he told us here in our house: "If one would have left the Nazis for another 50 years, they would have become quite reasonable."

Neither Simon nor Einstein was going to change his opinion about the Germans or Heisenberg.

With the beginning of Born's seventieth year in 1952, he felt the weight of time. It was age and personal losses: Constantin Carathéodory, the charming Greek mathematician whom he had first met in Göttingen at dinner with the

Minkowskis, the Dutch physicist Hendrik Kramers, whose "simplicity and modesty" he would miss, Artur Schnabel, and Carl Still. The hardest was the sudden death of his oldest friend, Rudi Ladenburg. Born told Einstein, who spoke at the cremation ceremony, that it pained him not to be there. He explained to Gritli how it felt to get old, and in his case, tired.

> If one is younger one has no idea how it feels to be old. It comes quite suddenly. One day you are still quite vigorous, you make plans for the future and take the regular sequence of day after day as a matter of course. And suddenly you feel quite clearly that this sequence will end soon. It is not at all an unpleasant feeling, just because it is accompanied by that fatigue. Still I am full of interest in life, and in particular in the life of our descendants.

Descendants were much on his mind. In the previous year, the number of Born grandchildren had increased to eight. Ann and Gustav had their first child, a son whom they named Max. The grandfather took that as an indication that Gustav was pleased with being named after his own grandfather.

Reaching seventy would mark the certainty of retirement. The university was already making the office that he shared with Robin Schlapp more presentable for his new position in fall term as director of studies. Workwise, his excitement about the new theory on elementary particles with Green had gradually dissipated and ended when Green finally left to teach in Australia. Born concluded that the only thing he could do was that "philosophical stuff." This involved writing or rewriting six books: three were new editions of older works, two were German and English collections of popular lectures, and the sixth was a book on optics that was taking on a life of its own.

Optik, his original book, had been published in spring 1933 as the family was forced to leave Germany. Born assumed that very few copies had been sold. After the war, a number of scientists proposed that he update the old version or at least translate it into English. Born liked the idea and realized that with progress in the field, a revised edition was warranted. Paul Rosbaud, a former editor at Springer, the original German publisher, came onto the scene. Having remained in Germany during the war—spying for the Allies—he was now organizing a publishing firm in Britain. Then a coauthor appeared in the person of Dennis Gabor, a physicist at the Imperial College of Science and Technology in London. They planned a three-volume work with some chapters contributed by experts in particular fields, because, as Gabor explained, "There is nobody who could deal with the whole subject in such a masterly way as yourself." Born, who felt too tired to undertake the work, was to be the advisor.

Rosbaud's publishing firm dissolved; Gabor dropped out owing to lack of time. The only bright spot by 1951 was the young optics researcher Emil Wolf from the Cambridge University Observatory, who moved to Edinburgh to work as Born's private assistant on the book. Within two months, Born was telling Simon that he was extremely pleased with Wolf and the progress of the work.

A survey of his former assets in Germany already had alerted Born to the involvement of the U.S. government with the book. Before the war, U.S. scientists had spent $1.5 million on foreign (mostly German) scientific publications. Wanting to keep up with the German literature as much as possible during the war, the government's Office of Alien Property Custodian reproduced 116 foreign journals and licensed the reproduction of more than 700 books. When the war ended, the publishing firms, which held licenses, simply went on printing these books without paying royalties to the author or the original publisher. Each year, Born sent a letter to the custodian office to reclaim his property, pointing out that he had been a British citizen since before the war. The reply said that his request would be taken up in due course.

Inspired by President Harry Truman's intervention in a similar case concerning the composer Sibelius, Born wrote a letter to the *Manchester Guardian* explaining his plight as a naturalized British citizen—that the U.S. government had confiscated his book, had taken his copyright, paid him no royalties, and for six years offered no redress. Born then enclosed a copy of his letter in the *Guardian* to the British ambassador in Washington, the principal of the University of Edinburgh, the editor of the *New Statesman & Nation*, a French physicist for submission to a French journal, the Association of University Teachers, Lord Cherwell, and Sam Goudsmit, editor of the American journal *Physical Review*.

It did not take long for Born to hear from Mr. Thomas H. Creighton, Jr., of the Office of Alien Property, who responded with a six-page, single-spaced, none-too-friendly letter. Mr. Creighton had by this time received inquiries from the British embassy, the U.S. State Department, and various individuals. He was not pleased with Born's characterization of events and set about correcting "unwarranted implications." In particular, the U.S. government had not "confiscated" his copyright; rather, the book—which had sold 1,000 copies during and after the war—was "vested" by the government under the Trading with the Enemy Act, which provided for those who were eligible to regain their property. The government's qualm was whether Born was entitled to get back the copyright and royalties, which amounted to about $2,000. Creighton warned Born that he must apply to the U.S. government for a license if he wanted to use portions of *Optik* in a new book and must pay the U.S. government 2 percent of the retail price, since it owned the copyright.

After much irritation, Rosbaud and Born decided to continue with the new optics book "without taking any notice of the Department of Alien Property." A couple of years later, the United States resolved Born's claim, returning to him both his copyright and royalties of $2,332.25.

Born did not realize that problems were only starting. The new publishing house that Rosbaud joined, Pergamon Press, was owned by Robert Maxwell. It took another seven years for *Principles of Optics* (*Optik*) to be published and more than thirty years for coauthor Wolf to resolve all the financial and publishing tricks created by Maxwell—but the book became the major, enduring work in the field, with six revisions by Wolf.

Born wanted to have a big family party and avoid all public celebrations for his seventieth birthday. So children, grandchildren, nieces, nephews, and Käthe all gathered in Cambridge at noon in the parlor at Gonville and Caius College, where Born had stayed as a new Ph.D. almost forty-five years earlier. The gathering was cheer-filled and festive. The guests dined on grilled trout, fricassee of chicken and chestnut gateau elegantly served on the college china. In the late afternoon, after a short rest for Hedi and Max, all gathered again at Irene's for afternoon tea and presents. Two memorable ones were an electric heating pad from Gritli and an original recording of a flute concerto composed by a friend and played by Gustav. The following day, the Borns took off for two weeks of rest and good food in Oberstdorf.

On a clear Christmas Eve, after an invigorating walk through the alpine woods around the village, Born sat down to answer a long birthday letter from James Franck. Franck, who had turned seventy earlier that year, had written to thank Born for their fifty-year friendship—for having believed in him when others did not and for bringing him to Göttingen. Franck had realized when they saw each other in 1950 that the thousands of miles between them had changed nothing. It was as though they had just been riding bicycles to the physics institute in Göttingen.

At the end of his letter, Franck brought up the prize. In the back of their minds, this birthday was more than just the beginning of a new decade—it was the end of certain opportunities. Angry that Born had not received the Nobel Prize, Franck reassured his friend as best as he could. "You know who you are and what your work is worth. So you need not worry that the official stamp hasn't been received."

Born's reply echoed similar feelings of friendship, his thankfulness for Franck's steering him to physics so that he did not end up a second-rate mathematician, his satisfaction with the results, his leaning on Franck as much as Franck had leaned on him. But he too could not avoid speaking of the prize, saying that he felt annoyed at being passed over. He did not see a difference between his own work and that of de Broglie, Schrödinger, Pauli, and such.

On March 6, 1953, Born gave his final lecture, and Robin Schlapp brought him flowers. A few days later, Born wrote a long letter to Bohr underlining their basic agreement on philosophy and at the end spoke about his retirement. He had reserved a year to finish his books on crystals and optics and then would have time to help in "non-scientific activities." Revealing the loss of his sense of purpose, Born suggested, "I have of course no practice in such things and no authority as you have, but I might be useful to you to take some work from your shoulders as a kind of help-mate." Bohr accepted the offer.

Hedi was in Germany for this milestone organizing their retirement plans in Bad Pyrmont. This spa town where the Quakers held their annual meeting was the same place Hedi had rested before their marriage. During the last two summers they had gone to the meeting and enjoyed days filled with the culture of the festive vacation town and long daily walks. She was now there to buy property, hire an architect, and sell their house in Göttingen, which the German courts had finally restored to them. During this time, Herglotz died. She had seen him the week before, and he had not recognized her.

Just after Hedi returned home, Max experienced a tremendous loss—the death of Käthe. She had lost consciousness while standing in the open rear section of a London bus, fallen out, and cracked her skull. Max was grief stricken. He wrote to Gustav,

> Your sweet letter helped me a little in my grief, making me realize how much is left to me in the love of my children. Yes, it was a terrible shock and sorrow, and I cannot help crying like a child sometimes at night. For Käthe and I were inseparable as children, and although later each of us went his own way we never lost this feeling of belonging together.

Born had planned to visit Berlin later in the year. Only a few days after her death, he wrote von Laue that he could not come.

> Berlin is too full of sad memories. My mother's sister and my father's sister, old ladies over 80 and many other relatives had lived there, who perished all in the Nazi time. Göttingen has no such association. I do not really feel mentally robust enough to come to Berlin although I would like to see you and other friends again. You will understand that.

It was as if Käthe's loss weakened his strength to face all the old sorrows.

Göttingen had invited Born to the 1953 celebration of its 1,000-year anniversary. He, Franck, and Courant were to receive the "Ehrenbürger-Recht," a medieval rite bestowed on those who rendered the town heroic service, such as a general who saves the city. The town's honoring of the scientists with the "freedom of the city," was as much a symbolic gesture to all those exiled as recognition of their scientific achievements. One year earlier, to restore respect to this rite, Göttingen nullified the honors given to Hitler and other Nazi officials.

A brilliant, sunny Sunday morning in June greeted the Göttingen festivities. Born, Franck, and Courant received their awards in a ceremony at the medieval Rathaus, and in the afternoon they had front-row seats in the market square for a procession celebrating the town's history—elaborate horse-drawn floats, townspeople attired in period dress, and plumed cavaliers riding magnificent steeds.

After the celebration, Born began telling more friends that he and Hedi planned to retire to Bad Pyrmont, while retaining their British citizenship; the reason he gave was the lack of a pension in Britain and the pleasant living in Pyrmont—nothing about Hedi's need to return. Interestingly, he told Simon or Einstein last, perhaps hoping they would change their attitude toward Germany in the meantime.

Trying to convince Simon, Born wrote,

> I think that this is a very pleasant place to retire. The country is lovely, the Kurpark quite marvelous, there is a lot of music, theater, cinema, etc. and a group of cultured people—no scientists, but they are in easy reach in Göttingen. I can quite understand your aversion against the idea of living in Germany, but I have no choice, and Pyrmont is certainly a charming refuge for old people.

Simon held his tongue. But when Born gave much the same line to Einstein, Einstein did not. "If anyone can be held responsible for the fact that you are migrating back to the land of the mass-murderers of our kinsmen, it is certainly your adopted fatherland—universally notorious for its parsimony." The phrase "the land of mass-murderers" stung Born badly.

In the fall, Hedi felt strange returning to the house in Edinburgh. She was ready to move to Bad Pyrmont. She started giving things away to Gustav and Gritli and packing up the rest. Irene could not take much since the Newton-Johns were off to Australia right after Christmas. Brin was taking a job as the master of Ormond College in Melbourne.

One November evening, friends and colleagues gathered in the university's Common Room to present Born with a beautifully bound book of articles by colleagues and a list of all of his own publications—a total of 280. Hedi received a large bunch of chrysanthemums, and Max spoke about his life and work. When they got home, they felt, regretfully, the "closing of a long, important life's passage."

Max was not so ready to let go. One of the articles contained in the honorary volume was by Einstein, something Einstein called "a little nursery song about physics" that would amuse Born. Einstein thought that "it seems to be our lot to be answerable for the soap bubbles we blow. This may well have been contrived by that same 'non-dice-playing God' who has caused so much bitter resentment against me . . . "

At the base of the disagreement was Einstein's fundamental problem with the statistical interpretation of quantum mechanics. Max was not so much amused as eager to show Einstein why he was wrong. Within a couple of days, Born, as irrepressible as Einstein, was telling Simon that the article had led him to a new "very nice formulation of the problem of determinism against indeterminism."

The two old warriors still had a few more arrows to shoot. Born's initial response informed Einstein that he had chosen an incorrect solution for macroscopic solutions. Einstein replied that he was surprised at Born's point of view. Born wrote a formal paper critiquing Einstein's ideas and invited him to include any comments in the published article, and Einstein said he did not want to be "a fencing master" before the public. Meanwhile, Hedi complained that Max was preoccupied with Einstein's paper when he should be helping her.

Wolfgang Pauli's intervention finally ended what was becoming a testy scientific misunderstanding between old friends. Although Pauli did not agree with Einstein, he accused Born of not listening and thereby formulating a misdirected response, attributing ideas to Einstein that he did not hold. When Born rewrote his response, taking into account Pauli's criticisms, the dispute simmered down. Born and Einstein did not agree, but their disagreement became properly defined.

Running concurrently, the house got packed up, Hedi went ahead to Germany on her own, and Max stayed behind for another six weeks to take care of details.

At the time of Born's retirement, the BBC had begun airing the Reith Lectures by Robert Oppenheimer who spoke about atomic theory. Born was a faithful listener. The fourth in the series of six, "Atom and Void in the Third Millennium," was the story of quantum physics. Oppenheimer described Einstein's quanta, Schrödinger's wave equation, and Heisenberg's uncertainty principle, associating the name with the concept. For the interpretation of the wave function, he explained its statistical nature and how this concept led to a "statistical physics." He did not mention Born. It was a final reminder to Born that he was lost to history, although his theory was not. In the past he had considered the commutation formula in quantum mechanics as his most important scientific achievement, but of late, perhaps because of his correspondence with Einstein, he had begun to identify the statistical interpretation in this way.

Born wrote to Oppenheimer about his disappointment in being overlooked, speaking his mind on the subject for the last time. A few weeks later, burglars stole his academic medals. His comment to Gustav: "So vanishes vanity!"

The Borns found order in Bad Pyrmont. Max bought a small Beckstein piano that fit into their little house, and Hedi luxuriated in springtime blossoms. They began to live the quiet life they had planned "indoors with books and music, out-of-doors in the garden, the Spa's park and in the forests."

They were resting at 3 P.M. on a Sunday afternoon at the end of October when they heard a knock on the door. A young man was standing on the steps whom Max initially mistook for the son of a Swedish colleague. Explaining that he was a Swedish journalist, the young man said to Max, "I want only to tell you that you must make a trip to Stockholm soon." Speaking into Max's ear so that an accompanying photographer could not hear, he whispered, "Nobel Prize."

Hedi shed tears of joy. Max was stunned and, in his quiet way, overjoyed. The photographer left after taking lots of pictures of the Borns in the house and garden, and the journalist spent the rest of the afternoon sitting in the kitchen while the Borns showed slides of India to their neighbors, letting on nothing about their news. "All the time this secret knowledge," Hedi wrote in her diary that night. Afterward the journalist took the Borns into town for their first celebratory dinner.

For the next ten days, a steady influx of German and Swedish reporters from newspapers, radio, and television descended on the Borns and the peaceful little town. Still, not only was the news unofficial, but Max had no idea what research the prize honored. No announcement would arrive from the Swedish Academy until November 3. Not quite believing that he would really get the prize, Max waited anxiously. He awoke early one morning with difficulty breathing. The doctor diagnosed nerves.

On November 3, they waited throughout the day. Their postman wore a path to their door bringing congratulatory telegrams from around the world. Having apprised himself of the content of each telegram, he announced the name of the sender as he made his delivery. None was from Sweden. Finally, Max had to go into town for a radio interview. At 9:45 P.M., the postman knocked at the door, waved a piece of paper at Hedi, and said, "Now we have the telegram from Sweden." The delay occurred because the telegram had accidentally been routed to Göttingen.

In he came, accepting Hedi's offer of beer and cake and sitting with her in the kitchen to listen to Max's interview. Max ended the interview by expressing the moral and social responsibility of scientists toward mankind. He was just getting started.

The next day Born wrote to James Franck.

> After rumors and reports floating around for some time, the final news came yesterday that I, together with Bothe, have received the Nobel Prize. I have not expected it with my almost 72 years but now the joy is great. I know that essentially I have you and Fermi to thank. Hedi and I want to thank you very, very sincerely for your trouble.

In 1947 and 1948, Franck and Fermi, who were both at the University of Chicago, proposed Born for his research on the lattice of ionic crystals. Neither continued nominating Born but others did, and he received several nominations in each subsequent year. He became a favorite nominee of the Poles, for a couple years, of the Danes, and an assortment of others, including loyalists Heitler and Landé. Most of them nominated Born for the statistical interpretation of the wave function, but a few simply made lists of his research areas (a practice in contradiction of the Nobel nominating rules): specific heat, solid state, quantum mechanics, nonlinear field theory, reciprocity, and liquids. They could not decide what was most fundamental. Indeed, every field of research represented an important achievement.

Responding to the nominations, the Royal Swedish Academy's section on physics had reevaluated Born's research on crystal lattices and deferred consideration. In subsequent years, they passed the same sentence on Born's quantum contributions, favoring research that was more current. When the section on physics finally recommended Born in 1954 (a report that is sealed until January 2005), it was for his "fundamental research in Quantum Mechanics, especially for his statistical interpretation of the wave function." He was pleased that it was in an area that he had developed alone.

The activity surrounding the announcement was so intense that Max and Hedi escaped to the Harz Mountains to rest up before leaving for Stockholm on December 6. Only the two of them made the trip. For a reason never specified, Hedi would not allow Gustav and Gritli to participate in their father receiving the prize. The others honorees all brought their families.

The award ceremony was the day before Born's seventy-second birthday. In the late afternoon, with the pomp and fanfare of Old World Europe, amid music, applause, and royalty adorned with tiaras and military medals, Born was the first called to receive his honor. As he walked from the platform down a couple of steps to where the King of Sweden sat in the front row of the large audience, his foot caught under the edge of an Oriental carpet and he almost fell. With the poise of a person who had waited a long time for this moment, Born proceeded toward the king, bowed, and received the gold medal for the 1954 Nobel Prize in Physics. Next followed the award to his co-honoree in physics Walther Bothe (absent for health reasons), then Linus Pauling for chemistry, John Enders, Frederick Robbins, and Thomas Weller for medicine, and Ernest Hemingway for literature (also absent for health reasons).

At the banquet that evening, 750 guests dined in the glow of candlelight in the Golden Room of the Hotel de Ville. After dinner, Born was the last laureate to stand before the glittering audience to express his gratitude. His first words of appreciation recognized his debt to the other Nobel scientists now beside him and those who had come before. Prefacing his remarks on the transformation in thinking that occurred with the end of determinism, he said, in softly accented English,

> We are a great fellowship, men of all nations seeking after the truth. It is my greatest hope that the modern trend, to subjugate science to politics and to inhuman ends, and to erect barriers of fear and suspicion around national groups of scientists will not continue. For it is against the spirit of scientific research, as the mind can grow and bear fruit only in freedom.

Born finally had the soapbox that he had hoped for in order to condemn nationalism, certitude, and the use of science for destructive ends. On Christmas day, having settled back in Bad Pyrmont, he wrote to Gustav, "As I am too old to use the Nobel money for research, I think I will come nearest to Alfred Nobel's intentions by attacking the prostitution of science for war and destruction."

Born had the opportunity to expound the legacy left to him by his father fifty years earlier—of the promise of peace and of science. The horrors of the Nazis had made him waver about the uses of science, but the bomb on Hiroshima had shocked him into recognizing the need for an international community to fight against "isms"—especially by the early 1950s, nationalism. As he said after the bomb was dropped, "Love is a power just as strong as the atom."

EPILOGUE

BY THE MID 1950S, EUROPE WAS STRONGLY DIVIDED, THE IRON CURTAIN FIRMLY IN PLACE. Germany, part of the dividing line, was itself in pieces. Four foreign powers—the United States, Great Britain, France, and the Soviet Union—all had a presence, but that of the first three had lessened in time and that of the last had increased. France worried about West Germany increasing its strength while the United States and Great Britain encouraged its economic expansion and admission to the newly formed North Atlantic Treaty Organization. The Federal Republic of Germany, West Germany, which was poised to gain full sovereignty, was valuable as a buffer to Soviet aggression.

Born recognized the tensions in the two Germanys unique to its central position. To begin the new year of 1955, he wrote an article calling for German scientists to clarify their positions on nuclear weapons and research. The German science journal *Physikalische Blätter* published this and, in the years to come, many more. Only a few weeks after writing the article, he proposed to Otto Hahn (who had won the Nobel Prize for the discovery of fission) in Göttingen and Bertrand Russell in London a resolution to denounce war, signed by leading researchers such as Einstein and Hahn.

His overtures met with success. In Göttingen, Born, Hahn, Heisenberg, and Carl Friedrich von Weizsäcker drafted a statement, and in London, Russell wrote a similar one. As the time arrived to announce these statements, however, the effort had split. Of the ten signers to Russell's manifesto, two were known Communist sympathizers: Nobel physicists Frédéric Joliot-Curie and Cecil Powell. Hahn did not want to be associated with them.

On July 15, 1955, at the Nobel Laureate meeting in Lindau, Hahn, Born, and Heisenberg issued the Mainau Manifesto, which fifty-one laureates eventually signed. The declaration read in part, "All nations must bring themselves to the decision by which they voluntarily renounce force as the last recourse in foreign policy. They will cease to exist if they are not prepared to do this." A week earlier, the Russell-Einstein Manifesto offered a similar resolution. Einstein had signed it the day before he died, thereby turning Russell's news conference into "Einstein's Last Warning." Born was an original signer of both declarations.

Old age slowed Born down a bit, and he could not attend the outgrowth of Russell's manifesto, a meeting in the village of Pugwash, Canada, between scientists from the East and West to discuss the threat to mankind produced by weapons of mass destruction. But he still had one more campaign in him—to stop the Federal Republic of Germany from rearming with nuclear weapons.

In 1957, eighteen German physicists—the "Göttingen 18"—banded together to issue a declaration calling on the West German government to renounce the use of nuclear weapons. The scientists stated that they would not participate in any weapons-related nuclear research. Among the eighteen were Born, Otto Hahn, Max von Laue, Walther Gerlach, Werner Heisenberg, and Carl Friedrich von Weizsäcker, who was the main drafter and coordinator. Chancellor Adenauer prepared a statement in response saying, "the German Federal Republic has produced no atomic weapons in the past and will not produce any in the future." But the minister of defense, Franz Josef Strauss, tried to undermine the policy and tensions reflecting this split gradually pulled the Göttingen 18 apart. The Federal Republic, however, did not produce atomic weapons, even though at one point it proposed planting nuclear land mines along the border with East Germany.

As Born passed eighty, his health declined. He mostly stayed at home, translated the poetry of the German humorist Wilhelm Busch into English, and wrote philosophical articles. He never stopped trying to convince the world that absolutes will destroy mankind.

> I am convinced that ideas such as absolute certainty, absolute precision, final truth, and so on are phantoms which should be excluded from science. . . . The relaxation of the rules of thinking seems to me the greatest blessing which modern science has given us. For the belief that there is only one truth and that oneself is in possession of it, seems to me the deepest root of all that is evil in the world.

In winter 1969, Born was in a hospital in Göttingen. He had made it through his eighty-seventh birthday, and one evening Irene, who had temporarily returned to Britain from Australia, visited him there. At first, when she saw him lying in the hospital bed, his pale, taut face looking "calm and noble," she thought he had gone.

> But when I looked closely I saw a tiny pulse fluttering in his throat, his chest rising and falling almost imperceptibly. A strong emotion came over me, of

> love, the sense of loss, of all the things he had meant to me. The tears were streaming down my face. Suddenly he moved and opened his eyes, he must have felt me near him. "I was asleep," he said. "Is anything the matter?" "No," I said, "I only came to see how you are."

Max Born died on January 5, 1970, in Göttingen. Hedi Born died two years later. On their gravestone in the Göttingen Cemetery is carved "$pq - qp = h/2\pi i$," the fundamental commutation law of quantum mechanics first written down by Born in July 1925.

ACKNOWLEDGMENTS

Two families made this biography possible.

First, the Borns, many of whom helped out in ways large and small. My thanks to Olivia Newton-John for her friendship and for introducing me to her mother, who was such an inspiration for this book. My only regret is that Irene is not here to enjoy what meant so much to her. Olivia's sister, Rona, and brother, Hugh, have freely shared their memories and papers as well. Professor Gustav Born, entrusted by his father to oversee the family papers, has been as generous with these invaluable records and his time and energy as possible, relying too on help from his wife, Faith, and children, Carrie, Georgina, Matthew, and Sebastian. I am greatly indebted to Gustav for his friendship, enthusiasm, and support. His cousin Renate Koenigsberger sat with me for weeks translating years of Hedi Born's cramped, abbreviated handwriting in her diaries and patiently explaining the significance of many abstruse references. Anita Pollard, collector of wonderful Kauffmann memorabilia, was unselfish with her time and knowledge. Ralph Elliott (aka Rolli Ehrenberg) provided background on the Ehrenbergs and the Borns' early years in Edinburgh and kindly entertained Irene and me in Canberra. Without John Pryce's help I could not have interviewed his mother, Gritli, who was not well.

My own family lovingly offered opinions, advice, and reassurance whenever needed or asked for. Stanley, Sarah, Jake, and Elizabeth are the best.

Two other people helped on this project from beginning to end, from whatever country or continent in which they were living. Robert Metzke, physicist, linguist, and trusted teacher wise beyond his years, worked with me for days on end, patiently answering difficult, obscure questions. Another patient person, Richard Staley, foremost authority on Born's early scientific achievements, introduced me to the world of the history of science, its scholars and concepts. Both have been mainstays and more than generous with their time.

A number of archivists graciously helped. Many spent considerable time and effort in countless instances for which I am extremely grateful. A main reference source on the history of physics is the American Institute of Physics with the Niels Bohr Library; Joe Anderson and staff tirelessly extended themselves to accommodate me. More than fifty archives answered requests, the staffs of which made primary research as efficient to gather as possible, notably: David Farrell at the Bancroft Library, University of California, Berkeley; Len Bruno, Library of Congress, Washington, DC; Neal Guthrie, U.S. Holocaust Museum, Washington, DC; Diane Spielmann, Leo Baeck Institute, New York; Nancy Cricco, Bobst Library, New York University; Thomas Rosenbaum, Rockefeller Archives Center, North Tarrytown; Murray Simpson and Richard Ovenden, University of Edinburgh Library; Joanna Corden, Royal Society, London; Mandy Wise, University College, London; Karl Grandin and Tore Frängsmyr, Royal Swedish Academy of Science, Stockholm; Anders Bárány, The Nobel Museum, Stockholm; Finn Aasurud and Felicity Pors, Niels Bohr Archives, Copenhagen; Horst Schmidt-Böcking, Institute for Nuclear Physics, and Wolfgang Trageser, University Archives, Frankfurt; Ralf Hahn, German Physical Society Archives, Magnus Haus, Berlin; Ulrich Hunger, University Archives Göttingen; Helmut Rohlfing, University Library, Göttingen, Handschriften Abteilung; Wilhelm Füssl, Deutsches Museum, Munich; Michael Eckert and Karl Märker, the Sommerfeld Papers, Deutsches Museum, Munich; Helmut Rechenberg, Werner Heisenberg Institute, Max Planck Institute for Physics, Munich; Tilo Brandis and Eef Overgaauw, Staatsbibliothek zu Berlin, Berlin; and Andreas Walter, Max Planck Gesellschaft, Dahlem.

I am indebted to those who agreed to interviews: Hugh Begbie, Hans Bethe, Rachel Chester, William Cochran, Dorothee Fuchs, Brian Goldsmith, Rose Ibsen, Helly Koenigsberger, Kurt Lipstein, Philippa Ludlum, Maurice Pryce, Joseph Rotblat, Richard Silletto, Fritz Stern, Thomas Still, Edward Teller, Victor Weisskopf, Carl Friedrich von Weizsäcker, and Emil Wolf, who was one of the first to answer my call for help.

A number of people provided information, advice, and assistance: Arnold Beevers, Margie Billion, Cathryn Carson, David Cassidy, Marie Corrigan, Hans-Joachim Dahms, Angelika Ellmann, Robert Friedman, Paul Forman, Hubert Goenner, Horst Kant, Edward David Luft, Ed Mayberry, Dorothee Pfeiffer, Lolo Sarnoff, Michael Scheuring, Suman Seth, Skuli Sigurdsson, Fritz Stern, Klaus Summer, Roger Summerhays, Stephanie Taylor, Mark Walker, and last but not least the Washington biography group led by Marc Pachter.

Those who translated some of the thousands of German documents eased my struggle: Martin Berger, Ingrid Enzelberger, Katharina Jeremiah, Suse Martine, and Susan Richter. Margareta Storm kindly translated the Nobel records.

Tim Wells and Al Lefcowitz at the Writers Center in Bethesda, Maryland, began my education; and Laura Wexler, whom I met there, continued to support my efforts throughout the project.

Anne Hutchinson Wesp, my former college roommate, took care of me in New York; and Carol Verburg, my college friend, not only was willing to sort out difficult chapters but passed on her wisdom: You are still "Uncommon Women." Elisabeth and Reimer Eck gave me a home and friendship in Göttingen. Renate Hemleb did the same in Berlin. Dieter Hoffmann faithfully alerted me to sources of information about Born. Nancy Lewis enhanced my understanding of Jewish culture and history. Elisabeth Mait managed to translate even the most difficult Sütterlin. To these friends, a heartfelt thanks.

Joe Mait, Lyle Roelofs, and Ruth Sime graciously volunteered to read all or part of the manuscript and sacrifice leisure time; Roger Stuewer, too, who reads with the critical eye of a physicist as well as editor and historian. Their willingness to share scientific knowledge greatly improved this biography. I am indebted to them for their guidance and interest.

Robert Lescher, my agent, and David Shoemaker, my editor at Basic Books, patiently guided me through twists and turns, and I thank them for their steady hands.

NOTES

SPECIAL ENDNOTE

A story persists about an incident between Max Born and Werner Heisenberg. Sometime in the early 1950s when Born was visiting Göttingen, Heisenberg supposedly spit on the ground in response to Born's asking if he could return to teach at the university. Throughout my research I have looked for written evidence to verify this encounter, especially in Hedi's diaries, Max's letters to Gustav, and correspondence between the two men. I have found none. A close review of the letters between Born and Heisenberg shows no break in their tone. On a couple of occasions a few months elapsed between Born's visit to Göttingen and his thank-you note to Heisenberg, but the notes contained no tension.

I heard the story from the person who heard it from Hedi. I have no doubt that she and Hedi had a conversation about an unpleasant incident. A possible explanation is that in the telling the story became garbled as can happen with the ambiguous reference of a pronoun, especially if there is more than one story. The story certainly has overtones of the ugly institute meeting in Bangalore. Of course, it is possible that the incident occurred and nothing was written down. I cannot rule that out. I do know that by 1955, Born and Heisenberg were "on good terms again" because Born wrote that in a letter to Lord Cherwell (Frederick Lindemann). Heisenberg was planning to move his institute to Munich and Born regretted his departure.

ABBREVIATIONS

People

AE: Albert Einstein
FS: Francis Simon
GB: Gustav Born (son of Max Born)
HB: Hedi Born
INJ: Irene Newton-John

MB: Max Born
MGM: Maria Göppert-Mayer
PE: Paul Ehrenfest

Books

AUTO: Born, Max. 1978. *My Life: Recollections of a Nobel Laureate*. New York: Charles Scribner's Sons.

BOHR: *Niels Bohr: Collected Works*. 1972. Edited by Léon Rosenfeld et al. Amsterdam: New Holland.

DIRAC: 1995. *The Collected Works of P. A. M. Dirac, 1924–1948*. Edited by R. H. Dalitz. Cambridge: Cambridge University Press.

EINSTEIN: 1993. *The Collected Papers of Albert Einstein*. Vols 3, 4, and 5. Princeton: Princeton University Press.

LTRS: 1971. *The Born-Einstein Letters*. With commentary by Max Born. Translated by Irene Born. New York: Walker.

PAULI: 1979. *Wolfgang Pauli: Scientific Correspondence with Bohr, Einstein, Heisenberg, A. O.* Vol. 1, *1919–1929*. Edited by A. Hermann et al. New York: Springer-Verlag.

Archives

AHQP: Archives for the History of Quantum Physics, American Institute of Physics, the Niels Bohr Library.

AMPHIL-C: E. U. Condon Papers, American Philosophical Society Library, Philadelphia.

AMPHIL-D: Charles Galton Darwin Papers, American Philosophical Society Library, Philadelphia.

BERK-J: George Jaffé Papers, Bancroft Library, University of California, Berkeley.

BERK-L: Gilbert Lewis Papers, Bancroft Library, University of California, Berkeley.

BERK-S: Otto Stern Papers, University of California, Berkeley.

BOBST: Richard Courant Papers. Bobst Library, New York University.

BODLEIAN-N: Papers of the Notgemeinschaft deutscher Wissenschaftler im Ausland, Bodleian Library, Oxford University.

BODLEIAN-P: Rudolf Peierls Papers, Bodleian Library, Oxford University.

BODLEIAN-S: Papers of the Society for the Protection of Science and Learning, Bodleian Library, Oxford University.

BONN: Otto Toeplitz Papers, University of Bonn.

BRANDEN-L: Max von Laue Papers, Berlin-Brandenburgisch Akademie der Wissenschaften, Akademie Archiv.

CALTECH-D: Max Delbrück Papers, Archives of the California Institute of Technology.

CALTECH-E: Paul Epstein Papers, Archives of the California Institute of Technology.

CALTECH-K: Theodore von Kármán Papers, Archives of the California Institute of Technology.

CAMB: Lord Rutherford Papers, Cambridge University Archives.

CHICAGO-F: James Franck Papers, University of Chicago Library.

CHICAGO-P: Michael Polanyi Papers, University of Chicago Library.
CHURCH: Lise Meitner Papers, Churchill College, Cambridge University.
DM-G: Walther Gerlach Papers, Deutsches Museum, Munich.
DM-W: Wilhelm Wien Papers, Deutsches Museum, Munich.
DPG: Deutsche Physikalisch Gesellschaft, Magnus Haus, Berlin.
DUKE: Fritz London Papers, Duke University.
ETH-H: Adolf Hurwitz Papers, Eidgenössische Technische Hochschule, Zurich.
ETH-W: Hermann Weyl Papers, Eidgenössische Technische Hochschule, Zurich.
FAM: Private papers and letters of the Born family.
FRF: Archiv der Johann Wolfgang Goethe-Universität, Frankfurt, a.M.
GEHEIM: Geheimes Staatsarchiv, Preussischer Kulturbesitz, Berlin.
HARVARD-B: Percy Bridgman Papers, Harvard University Archives.
HARVARD-R: Theodore Richards Papers, Harvard University Archives.
HUMBOLDT: Humboldt University Archives, Berlin.
IEB: International Education Board Papers, Rockefeller Archive Center, North Tarrytown, NY.
KAP: Peter Kapitza Archive, Moscow.
LBI: Leo Baeck Institute, New York.
LOC-L: Irving Langmuir Papers, Library of Congress, Washington, DC.
LOC-N: John von Neumann Papers, Library of Congress, Washington, DC.
LOC-O: J. Robert Oppenheimer Papers, Library of Congress, Washington, DC.
LOC-V: Oswald Veblen Papers, Library of Congress, Washington, DC.
MICH: K. Fayans Papers, Michigan Historical Collection, Bender Historical Library, University of Michigan.
MIT: Norbert Wiener Papers, Institute Archives and Special Collections, Massachusetts Institute of Technology.
MPG: Archiv zur Geschichte der Max Planck-Gesellschaft, Dahlem.
MPG-H: Otto Hahn Papers, Archiv zur Geschichte der Max Planck-Gesellschaft, Dahlem.
MPG-Ha: Fritz Haber Papers, Archiv zur Geschichte der Max Planck-Gesellschaft, Dahlem.
MPG-L: Max von Laue Papers, Archiv zur Geschichte der Max Planck-Gesellschaft, Dahlem.
MUNICH: Werner Heisenberg Institute, Max Planck Institute for Physics, Munich.
NA: Public Records Office, the National Archives, London.
NBA: Niels Bohr Archives, Niels Bohr Institute, Copenhagen.
NBL: Niels Bohr Library, American Institute of Physics, College Park, MD.
NUFFIELD: Frederick Lindemann Papers, Nuffield College, Oxford University.
OSEEN: Oseen Family Archives, AXXIV:1, Regional Archives, Lund, Sweden.
ROYINST: William Lawrence Bragg Papers, Royal Institute, London.
ROYSOC: Francis Simon Papers, Royal Society of London.
ROYSOC-H: A. V. Hill Papers, Royal Society of London.
SAN DIEGO: Maria Göppert-Mayer Papers, Mandeville Special Collections Library, University of California, San Diego.

SOMMER: Arnold Sommerfeld Papers, Deutsches Museum, Munich.
STABI: Staatsbibliothek zu Berlin, Preussischer Kulturbesitz.
STILL: Carl Still Papers, Still family private collection.
SWEDACAD–1933: Reports and nomination letters for the physics Nobel Prize, 1930–1934, the Nobel Archive of the Royal Swedish Academy of Sciences, Stockholm.
SWEDACAD–1954: Reports and nomination letters for the physics Nobel Prize, 1948–1953, the Nobel Archive of the Royal Swedish Academy of Sciences, Stockholm.
SWEDACAD-O: Oseen Archives, Center for the History of Science, Royal Swedish Academy of Sciences, Stockholm.
UAG: University Archives, Göttingen.
UCLONDON-D: Frederick Donnan Papers, University College-London.
UGL-HA: University of Göttingen Library, Handschriftenabteilung.
USHMM: United States Holocaust Memorial Museum, Washington, DC.

NOTES

Prologue

1 "The spreading of the waves": Oppenheimer, 1954, 56.
2 "I was particularly happy": MB to Robert Oppenheimer, 11 December 1953, LOC-O.
2 "I am one of the last": Robert Oppenheimer to MB, 18 December 1953, LOC-O.

Chapter 1

5 "I got the couple of love-birds": AUTO, 8.
5 "It was the first experience": Ibid., 9.
5 "There was no mother": Ibid.
6 He had lovingly told her: Gustav Born to Gretchen Born, 1885, FAM.
6 "I have such a longing": Gustav Born to Gretchen Born, 26 May 1885, FAM.
6 "I am alone again": Ibid.
6 beginning a serious investigation: Born, Brandt, and Born, 1950, 469.
7 eyes that to his young nephew: Schäfer, 1933, LBI, 33.
7 Much sought after by: Helene Perlhöfter, interview with INJ, 1970, FAM.
7 "regally rustled out": Jacobi, 1917, FAM, 26.
7 "amiable, lovely personality": Ibid.
7 "a house in the sun": Ibid.
8 "There was never anybody": AUTO, 7.
8 he buried his frustration: MB to Gritli Born, 5 November 1950, FAM.
8 "full resounding voice": AUTO, 19.
9 "If one traveled the world": MB to Käthe Koenigsberger, 1948, FAM.
9 "dark, mysterious spruce forests": AUTO, 11.
9 "an intimate friend": Ibid., 13.

9 "From that moment": Ibid., 13.
9 From his father, Max learned: MB to GB, 12 April 1944, FAM.
9 I am certain that: AUTO, 14.
10 Nature did not follow: MB to GB, 20 August 1944, FAM.
10 "heard a scream from some living creature": Ibid.
10 He developed a method: Born, Brandt, and Born, 1950, 469–74.
10 Two frogs made love in a pond: "Der Froschlaich," 1884, FAM.
10 He treated Gustav with disdain: Erwin C. Froelich to Viktor Hamburger, 26 September 1947, FAM.
11 "The professor torments me": Gustav Born to Gretchen Born, 2 August 1882, FAM.
11 I do not love: Marcus Born to Fanny Ebstein, 10 November 1849, FAM.
12 "The loveliest feast": Jacobi, 1917, 19, 21.
12 "the beginning of a deep resentment": AUTO, 20.
12 "she was not charming": Helene Perlhöfter, interview with INJ, 1970, FAM.
12 very kind, shy, and lacking in social skills: Renate Koenigsberger, interview with author, 2002, London.
12 "How could Gustav": AUTO, 25.
13 Bertha's strong East Prussian accent: Helene Perlhöfter, interview with INJ, 1970, FAM.
13 "more German": Kurt Lipstein, interview with author. April 2000, Cambridge, England.
13 Barriers remained: Gay, 1992, 202.
13 "They believed they differed": AUTO, 26.
13 "[The families] obviously considered": Ibid.
13 Gustav's father, Marcus: Born Family, 1991; Heppner and Herzberg, 1929; Meyer, 1996–1998, 95.
14 A German census taker: Luft, 1987, ix–xiii.
14 last names were an important factor: Kaganoff, 1996, 21–23.
15 "How is it possible": Marcus Born to Dr. Freund, 1849, FAM.
15 "one syllable men": Ibid.; Marcus Born to Fannie Ebstein, 21 November 1849, FAM.
15 Gretchen Kauffmann Born's ancestors: Kauffmann, 1896; *Meyer Kauffmann Textilwerke*, Brilling, 1936.
16 The great violinist: Joseph Joachim to Max Born, 15 December 1897, FAM.
16 The Kauffmanns could not resist: Gustav Born to the Kauffmanns, 26 November 1898, FAM.
17 when the last wedge: AUTO, 34.
17 "Shall I tell you": Ibid., 35.
17 about trains crowded with soldiers: Ibid.
17 With "suppressed passion": Ibid.
18 "the meaning of the whole theory": Ibid., 32.
18 "Content to be just average": Ibid.
18 "Max, you are in love": Ibid., 43.

18 "deprived us of a natural outlet": Ibid.
19 "completely exhausted, broken man": MB to GB, 12 April 1944, FAM.
19 "Shadow of his suffering": AUTO, 46.
19 We profited from that: Ibid.
19 Gustav's own research: Born, Brandt, and Born, 1950, 475; Csapo, 1958, 40.
20 University officials asked Hasse: Helene Perlhöfter, interview with INJ 1970, FAM. Solomon Kauffmann, 18 July 1900, FAM.
20 "Professor Born is dead": Asch, 1900.
20 a "nightmare": AUTO, 47.

Chapter 2

21 "covered by a veil": AUTO, 47.
21 "He was then deeply interested": Ibid., 48.
22 "He was not afraid": Ibid.
22 "enemy of society": Ibid.
22 "the ancient metaphysical problems": Ibid., 49.
22 "what Hans doesn't learn": Gustav Born to Gretchen Born, 1885, FAM.
22 "All men are mortal": AUTO, 52.
22 In front of his fellow students: MB to GB, 27 September 1949, FAM.
23 concepts that "mean nothing": AUTO, 53.
23 "hen who sat": Renate Koenigsberger, interview with author, 1999, London.
23 "stupid little prig": MB to GB, 25 November 1944, FAM.
25 "happy-go-lucky": AUTO, 81.
25 "To my cousin Hans": Ibid., 68.
25 Testing their strength: Schäfer, 1933, 41–42, LBI.
25 Max as the classicist: Franck, 1962, CHICAGO-F.
25 a "peculiarly personal charm": AUTO, 68.
25 Max's "goodness and intelligence": Franck, 1962, CHICAGO-F.
25 "The clarity of [your husband's] thoughts": MB to Mrs. Hurwitz, 5 December 1919, ETH-H.
26 "a bad lecture": AUTO, 80.
26 At 2:45, Hermann Minkowski left: Debye, 3 May 1962, 2d sess., 8, AHQP; AUTO, 98. Planckstrasse commemorated the Göttingen law professor Gottlieb Planck, uncle of Max Planck.
27 "small and tame": AUTO, 81.
27 "There is one manuscript": Ibid., 82.
27 Properly specifying the problem: Reid, 1986, 103.
27 "a perfect formulation": Courant, 1964, 12, BOBST.
27 "the axioms of natural science": Born class notes, NBL.
28 Hilbert and Minkowski: Reid, 1986, 91; AUTO, 83.
28 The irrepressible and unorthodox Hilbert: Courant, 1964, 7–8, BOBST.
28 Most mornings, Born arrived: Reid, 1986, 104; Courant, 1964, 9, BOBST.
28 When Born had first arrived: MB to PE, 17 March 1909, AHQP.
28 as the valued young colleague: Reid, 1986, 103.

29 "guiding [the students]": AUTO, 85.
29 Hilbert's response: MB to GB, 27 September 1949, FAM. The proof assigned to Born was not solved until twenty-five years later, by former Hilbert student Carl Siegel. (Communication from Alan Baker to the author, 3 May 2003.)
30 "the mathematical treatment": Reid, 1986, 82.
30 "the pre-established harmony": Hermann Minkowski, "On Space and Time," 1908, Lecture at Society of German Scientists and Physicians, as quoted in Sigurdsson, 1991, 10.
30 "my science future": AUTO, 98.
30 "Dear Herr Born!": Felix Klein to MB, 14 March 1905, FAM.
31 "Klein's predilection for applied mathematics": AUTO, 100.
32 The small group in Göttingen: Pyenson, 1985, 101–36.
34 "If you got the prize": AUTO, 103.
34 What he did see: MB to GB, 27 September 1949, FAM.
34 "Why not?": AUTO, 100.
35 One end of a thin metal strip: Born, 1906.
35 "a moment of deepest satisfaction": AUTO, 104.
35 wearing the requisite formal dress suit: Irving Langmuir to Mrs. Langmuir, 16 December 1905, LOC-L.
35 After asking an hour's worth of questions: *Phil. Prom. Spec* B1, 13, UAG; AUTO, 105.
36 Klein had "broken": Courant, 5 September 1962, interview, 5, AHQP.
36 he encouraged Born: MB to James Franck, 24 December 1952, CHICAGO-F.
37 "sit in the lounge": AUTO, 116.
38 pushed the young German: Born, 1940a.
38 "Dr. Searle, something is wrong": AUTO, 118.
38 Ludwig Boltzmann had represented: Lindley, 2001, 73.
38 Gibbs's work presented Maxwell's and Boltzmann's statistical theories: Ibid., 195.
38 "wonderful, like a miracle": AUTO, 119.
38 In his first article: Born and Oettinger, 1907.
39 This approach would be: Staley, 1992, 87.
39 J. J. Thomson's lectures: Born, 1940a.
39 "real physicist": AUTO, 121.
39 "physicists are almost exclusively interested": Wilhelm Wien to Arnold Summerfeld, 11 June 1898, as cited in Seth, 2003, 5.
41 "You'll never become": Born, 17 October 1962, 3, AHQP.
41 When he mentioned some of these: Stanislaus Loria to MB, 12 June 1955, STABI.
41 Einstein sent them: Fölsing, 1997, 202.
41 Some physicists threw: Staley, 1992, 107–108.
42 "a lazy dog": Clark, 1971, 120; AUTO, 131.
42 Minkowski's 1908 article: Staley, 1992, 124–126.
42 Finding himself eclipsed: Stanislaus Loria to MB, 12 June 1955, STABI.

42 "Henceforth, space by itself": Minkowski, "On Time and Space," as translated in Reid, 1986, 112.
43 This topic was to be: Staley, 1992, 93–95.
43 Toeplitz helped Born: Born, 1940b.
43 "a happy time": Reid, 1986, 113.
43 a testament "to the modesty": Born, 1909a, 7.
43 After a day and night of writing: Reid, 1986, 116.
43 His words extolled Minkowski: Born, 1909b.
43 "friendly feelings in many hearts": AUTO, 132.
43 In a long obituary: Born, 1909a.
43 "With him also sank": Born, 1959, in Staley, 1992, 99.

Chapter 3

46 The aim of Born's lecture: Staley, 1992, 171.
46 His new definition allowed: Miller, 1998, 230.
46 "to study the mathematical literature": AUTO, 135.
46 "knowledge of physics": Ibid.
46 "I am interested": Ibid.
46 This consolation, in the most "wretched time": Ibid., 135–136.
47 "I, Max Born": Habilitation Lebenslauf, GEHEIM.
47 In 1909 Germany, only about twenty: Rowe, 1986, 428; Ringer, 1969, 54.
47 Göttingen had only broken the rule: Rowe, 1986, 428, 436.
47 Around this time another young physicist: PE to MB, 17 March 1909, AHQP.
47 for Einstein, Planck, and Minkowski: MB to PE, 5 July 1909, AHQP, in Staley, 1992, 155.
47 To define relativity as a generalized kinematics: Staley, 1992, 157.
47 "mathematical *tour de force*": Miller, 1998, 230.
47 One of his conclusions was that the charge: Staley, 1992, 184.
48 Before all, I need time: MB to PE, 5 July 1909, AHQP, in Staley, 1992, 187.
48 "I place such things": MB to David Hilbert, 4 January 1910, Cod. Ms. D. Hilbert 40 A, UGL-HA.
48 "I began to read it": PE to unnamed friend, undated draft, in Staley, 1992, 188.
48 Many physicists—including Einstein: Staley, 1992, 150–156.
49 "Einstein had already proceeded": Born, 1956, 107.
49 "You are one of the few": AE to PE, before 20 June 1912, EINSTEIN, vol. 5, 309.
49 "I have become imbued": AE to Arnold Summerfeld, 29 October 192, as quoted in Pais, 1982A, 216.
49 Born poetically introduced: Born, 1909c.
50 "mouthpiece of revolution": Gregor-Dellin, 1983, 167.
50 "rushed to me with open arms": Stephan Born, 1978, 118.
51 "He said a few words": Alfred Landé, 5 March 1962 interview, 2, AHQP.

52 not sure that he was at ease with the "pandemonium": AUTO, 149.
52 "Will you share a house": Ibid., 151.
52 there was "very little difference": Theodore von Kármán, 29 June 1962 interview, 2, AHQP.
53 Richard Courant, a mathematics student: Reid, 1986, 268.
53 They were a creative and eclectic bunch: Theodore von Kármán, 29 June 1962 interview, 9, AHQP.
54 "drowning in the formalisms": MB to Theodore von Kármán, 7 November 1922, CALTECH-K.
54 Einstein came up with an explanation: Lindley, 2001, 208–209.
55 Von Kármán disagreed, arguing that: Theodore von Kármán, 29 June 1962 interview, 3, 8, AHQP.
55 Born wrote that he could not "suppress certain doubts": Born and von Kármán, 1912, 309.

Born and von Kármán's article introduced all the basic concepts of lattice dynamics. These are: identifying "the independent degrees of freedom with the normal modes of vibration; the extensive use of three-dimensional Fourier analysis; the 'periodic' boundary conditions introduced to avoid the complication of surface effects, the analysis leading to the notion of acoustic and optical branches of the frequency spectrum, the limit of the long waves and the passage to a continuum description." Kemmer and Schlapp, 25.

55 Yet he also realized: Born, 1964, 2.
55 Shortly after publication, Arnold Sommerfeld: Arnold Sommerfeld to Karl Schwarzschild, June 1912, SOMMER.
55 Debye . . . had proposed a solid: Theodore von Kármán, 29 June 1962 interview, 8, AHQP.
55 Debye had explored: Debye, 1964, 11.
56 "the essentials of mathematical physics": AUTO, 141.
56 to "penetrate the mystery of nature": Born, 1968, 47–48.
56 Timid and awkward in their presence: MB to GB, 6 June 1944, FAM.
57 "it was so slow to turn": Otto Toeplitz to HB, 16 May 1913, FAM.
57 One of these, an old friend: Born and Born, 1969, 150.
57 "Today we have another large tea": MB to HB, 1912, FAM.
57 Explaining that his mother had used: MB to HB, 8 June 1913, FAM.
58 Hedi's parents had sent her: AUTO, 155.
58 While Hedi was there: notes written by GB in discussion with HB, 16 January 1947, FAM.
58 This perturbed Helene Ehrenberg: HB to GB, 1952, FAM.
58 Like his father: AUTO, 159.
59 At the turn of the nineteenth century: "Wolfenbüttel Speech," 1883, 4.
60 "a wedding in the": AUTO, 155.
60 They stayed long enough: Discussion of Doc. 16, EINSTEIN, vol. 4, 229.
61 Later that fall, Harald Bohr: Harald Bohr to Niels Bohr, fall 1913, BOHR, vol. 2, 567.

61 For some time, they had frequently seen: Livingston, 1973, 257.
61 there were . . . forces pulling: AUTO, 159.
62 Einstein's response to a colleague's request: AE to Alfred Kleiner, 3 April 1912, EINSTEIN, vol. 4, 285.
62 Sommerfeld . . . offered him a lecturer's position: Arnold Sommerfeld to MB, undated but after 4 September 1913, SOMMER.
62 A wreath of days: HB, 19 May 1914, FAM.
63 "I know, and my colleagues, too": Max Planck to MB, 12 August 1914, FAM.
63 "Joyous surprise": MB to Max Planck, undated draft from August 1914, FAM.
63 Planck answered that he would: Max Planck to MB, 18 August 1914, STABI.
63 For his part: MB to Max Planck, undated draft from August 1914, FAM.
63 "all internal life in Germany": MB to Jacob Laub, 25 March 1915, Autograph, I/318/1, STABI.
64 That same month, Rudi Ladenburg: MB to Frau Ladenburg, 27 October 1914, VcAbt., Rep1, MPG.
64 There was pain too: MB to HB, 21 September 1915, FAM.
64 "We were told every day": AUTO, 162.
64 "what the Dutch thought": ibid.
65 "It is not true": Fölsing, 1997, 345.
65 Such was the intensity: Stern, 1999, 113.
65 "overwhelmed with literature": "Eine Neutrale Stimme," newspaper clipping, FAM.
65 English and American scientists: Correspondence between Wilhelm Wien and W. H. Bragg, January 1916; Erklärung der Professoren Grossbrittanniens an die Deutschen Akademischen Kreise, DM-W.
65 Between summer and winter semesters: Barut, 1989, 39.
65 Born's courses listed: University course lists, 1914, UAG.
65 "If it's not nonsense": Alfred Landé, 3 May 1962 interview, 6, AHQP.
66 In the fall, Born decided: Born, 1914.
66 His goal was "to derive": Born, 1968, 28; AUTO, 162.
66 In the lighthearted spirit: HB to MB, Christmas 1914, FAM.
66 With deep regret: Max Planck to MB, 6 January 1915, FAM.
66 The Borns together labored over a reply: AUTO, 164.

Chapter 4

69 Born received an unexpected: MB to David Hilbert, 28 July 1915, Hilbert Nachlass, Cod. Ms. D. Hilbert 40A, UGL-HA.
69 Inspired by a demonstration: Van der Kloot, 149–150.
69 Five thousand soldiers: Fölsing, 1997, 354.
70 "Without a limitation": Born, 1968, 195.
70 His letter angered Haber: MB, "Erinnerungen an Fritz Haber," written for Jaenicke biography, MPG-HA.
70 "those who have died": MB to Wilhelm Wien, 2 February 1919, DM-W.

70 "Germany's strength is great": MB to J. Laub, 25 March 1915, Autogr. I/318/3, STABI.
70 With friends and colleagues dying: MB to HB, 15 September 1915, FAM.
70 His brother, Wolfgang: Renate Koenigsberger, interview with author, April 2000, London; statistics from Chickering, 1998, 129.
70 "We shall as German citizens": Meyer, 1996–1998, vol. 3, 361–363.
70 Forty-one-year-old Karl Schwarzschild: Rowe, 1986, 436.
70 On March 15: MB to J. Laub, 25 March 1915, Autogr. I/318/3, STABI.
71 Hedi—who blamed: HB to Frau Wachtel, 22 June 1916, FAM.
71 They slept in the same room: INJ, interview with author, January 1997, Melbourne, Australia.
71 For four weeks: MB to David Hilbert, 28 July 1915, Hilbert Nachlass, Cod. Ms. D. Hilbert 40A, UGL-HA.
72 Over mulled cider: MB to David Hilbert, 12 September 1915, Ibid.; MB to HB, 21 September 1915, FAM.
72 October found him: MB to Kurator, 25 May 1933, Kur PA Born, UAG.
72 The activities at the APK: MB to David Hilbert, 23 November 1915, Hilbert Nachlass, Cod. Ms. D. Hilbert 40A, UGL-HA.
73 "Then, I see the misery": Ibid.
73 "not a hair's breadth": Ibid.
74 Rather than sign: Fölsing, 1997, 345.
74 "for a victory": Clark, 1971, 184–186.
74 "born of the same source": Ibid., 106.
74 "He shook my hand": Born and Born, 1969, 117.
74 "abhorrence of the slaughter": MB to Mr. Davidson, draft letter, n.d. [c. 1950s], FAM.
74 Keeping faith: MB, notes for BBC Television on Einstein, FAM.
75 "a terrible wall": AUTO, 172.
75 "the greatest feat": Born, 1956, 109.
75 "convinced adherents": AE to Otto Stern, after 15 February 1916, EINSTEIN, vol. 8, 194.
75 "completely understood": AE to MB, 27 February 1916, LTRS.
75 "uncanny insight": AUTO, 167.
75 "the dark, depressing time": MB, notes for BBC Television on Einstein, FAM.
75 "His utter independence": HB, *Helle Zeit*, translated in Clark, 1971, 192.
75 "Hedi Regina": HB to AE, poem, 3 March 1917, FAM.
75 For his thirty-ninth birthday: HB to AE, 14 March 1917, FAM.
76 The fifty-eight-year-old Blaschko: Born and Banks, 1996, 42–43.
76 He made his most important public contribution: Roos, 2003, 87; Julia Roos, communication with author 14 December 2002.
76 "It was good you": MB to HB, 26 February 1918, FAM.
76 "skeptical with regard": AUTO, 178–79.
76 He shared my: Ibid., 179.
77 During the period when Max was going: MB to GB, 26 February 1953, FAM.

77 To feed the girls: INJ, interview with author, January 1997, Melbourne, Australia.

77 "a person with a thousand shaded sides": MB to Otto Toeplitz, 10 July 1922, BONN.

77 "any observation as a deception": AUTO, 173.

77 "the raw material": Ibid.

78 The first recruit: Ibid., 171; Barut, 1989, 39.

78 Born did not have a sense of urgency: AE to Otto Stern, 3 January 1917, BERK-S; Estermann, 1976, 41.

78 During Herkner's one-year stay: Born, 1918, 179.

78 "of this noble youth": Ibid.

79 "Physics does not have to work for the war": Walther Gerlach, interview with INJ, 1970.

79 "I am but a soldier": MB to Johannes Stark, 7 July 1917, STABI.

79 "beautiful paper on": MB to Niels Bohr, 12 May 1918, AHQP.

79 When responses were slow: MB to David Hilbert, 4 April 1916, 24 August 1917, and 14 September 1918, Hilbert Nachlass, Cod. Ms. D. Hilbert 40A, UGL-HA.

79 He asked Landé: Born, 1964, 4.

79 Sitting across from Born: AUTO, 181–182; Born, 1964, 4.

80 "pale and desperate": AUTO, 182.

80 "I thought you would never": MB, 18 October 1962 interview, 14, AHQP.

80 Born complied: Born and Landé, 1918a, 1067.

80 "The planar electron orbits": Born and Landé, 1918b, 216.

80 "second hint given": Born, 1964, 4.

80 "From that moment on": AUTO, 183.

81 "one could do": James Franck, 11 July 1962 interview, 12, AHQP.

81 The Borns depicted: AE to MB, 2 August 1918, LTRS; Hoffmann, 1972, 137.

81 the Borns saw a serious risk: MB to David Hilbert, 30 October 1918, Hilbert Nachlass, Cod. Ms. D. Hilbert 40A, UGL-HA.

81 the Borns knew the old government: MB to David Hilbert, 14 November 1918, Hilbert Nachlass, Ibid.

81 The streets in Berlin: Ibid.

82 "through streets full": AUTO, 184.

82 The three scholars: AUTO, 184–185; Goenner and Castagnetti, 1996, 37; AE to MB, 7 [November] 1944, LTRS (Born cites November in Born, 1949, rather than 7 March as in LTRS); MB to David Hilbert, 14 November 1918, Hilbert Nachlass, Cod. Ms. D. Hilbert 40A, UGL-HA. Born's recently discovered contemporaneous letter to Hilbert slightly contradicts the order of events as previously understood.

82 The next day, he: MB to David Hilbert, 14 November 1918, Nachlass Hilbert, UGL-HA.

83 "a free democratic": AUTO, 185.

83 One serious casualty: Gay, 1992, 243.

83 Eight days after: Max Planck to "Spectabilis," 19 November 1918, HUMBOLDT.
83 I have a request: Max von Laue to AE, 18 June 1917, EINSTEIN, vol. 8.
84 I have no need: AE to HB, 8 February 1918, LTRS.
84 Von Laue was extremely pleased: Max Von Laue to AE, 18 February 1918, EINSTEIN, vol. 8.
84 The Berlin faculty: Dekan, Frankfurt, 30 July 1918 to the Ministry; Dekan, Frankfurt, 25 November 1918 to the Ministry, Berlin, GEHEIM.
84 "I have received nothing": MB to Rudolf Ehrenberg, 28 December 1918, FAM.
85 "so charming and kind": MB, "Erinnerungen an Fritz Haber," written for Jaenicke biography, MPG-Ha.
86 Max held classes: MB to Alfred Landé, 12 February 1919, AHQP; Trageser, 1998, 4.
86 "In Frankfurt I hope": MB to Rudolf Ehrenberg, 28 December 1918, FAM.

Chapter 5

87 "It is ideal here": HB to Prof. Lichtenstein, spring 1919, NBA.
87 The Germans pointed out: Douglas Haig, Diary, 11 November 1918, National Library of Scotland, wall plaque.
88 Born's worry was for the hungry people: MB to Carl Oseen, 22 February 1920, OSEEN.
88 "the principle of justice to all": MB to Carl Oseen, 15 May 1919, SWEDACAD-O; Woodrow Wilson, address to Congress, 8 January 1918.
88 With resources for research: MB to Carl Oseen, 18 February 1919, OSEEN.
88 Plagiarism and piracy: MB to Carl Oseen, 24 April 1919, OSEEN.
88 Arnold Berliner, editor: Ibid.
88 Richards did not foresee: Theodore Richards to Carl Oseen, 5 April 1919, in Carl Oseen to MB, 8 May 1919, STABI.
89 We German pacifists: MB to Carl Oseen, 15 May 1919, SWEDACAD-O.
89 "greatly weaken if not destroy": MB to Carl Oseen, 22 February 1920, OSEEN.
89 "with tears in his eyes": AE to MB, 4 June 1919, LTRS.
90 Five hundred ninety-nine: Arnsberg, 1983, 31.
90 The number of postwar millionaires: Bethge and Klein, 1989, 30; Meyer, 1998, Vol.4, 331.
90 "lived in a princely style": AUTO, 192.
90 "extraordinarily beautiful": Walther Gerlach interview with INJ, 1970.
90 an institute that wartime: MB to Wilhelm Wien, 8 May 1919, DM; Bethge and Klein, 1989, 31.
90 Born organized the institute's lecture series: Trageser, 1998, 4.
90 He hired a young assistant: Trageser, 1998, 4–5; MB to Landé, 21 January 1919, STABI.
90 The staff encouraged: Walther Gerlach interview with INJ, January 1970.

91 “had thrown away”: Landé, 7 March 1962 interview, 25, AHQP; Landé lecture, 13 December 1918, Protokoll no. 10009, 1910–1921, DPG.
91 Landé then took a job: Barut, 1989, 41.
91 “Herr Professor”: MB to Alfred Landé, 21 December 1918, STABI.
91 Gritli and Irene adored: Meissner, 1990, 44.
91 “a ‘tame mathematician’”: AUTO, 190.
91 “this tiny restful department”: MB to Alfred Landé, memories, 1968, STABI.
91 one of nine papers: Forman, 1970, 172, 174; AUTO, 191.
92 Born wrote to Einstein: MB to AE, 1 July 1919, no. 1–34–1, Correspondence A-D, ltr. 12, MPG.
92 The university had kept: MB to Carl Oseen, 22 February 1920, OSEEN.
92 “My institute is”: MB to K. Fayans, 4 November 1919, MICH.
92 On May 29th of this year: Born, 1919.
92 “the human spirit”: Ibid.
92 “unfortunate and damaging”: Wilhelm Wien to Arnold Sommerfeld, 9 December 1919, SOMMER.
92 “to get the institute”: MB to Sommerfeld, 5 March 1920, SOMMER; MB to Carl Oseen, 22 February 1920, OSEEN. The sum cited in the Oseen letter is slightly different than that stated in the Sommerfeld letter.
93 And you Max: AE to MB, 27 January 1920, LTRS.
93 You have wrestled: MB to HB, 14 December 1919, FAM.
93 Max signed a contract: Springer contract, 3 February 1920, FAM.
93 “a black suit”: MB, “Erinnerungen an Fritz Haber,” written for Jaenicke biography, MPG-HA.
94 The unfortunate rupture: MB to Gilbert Lewis, 2 March 20, BERK-L.
94 Born explained that: Ibid.
94 Should you not take: Ibid.
94 “Several days ago”: MB to Irving Langmuir, 11 May 1920, LOC-L.
95 I hurry to answer: MB to Gilbert Lewis, n.d., BERK-L.
95 “shoot-outs”: MB to Carl Oseen, 25 March 1920, OSEEN.
95 “the further the hour”: HB to AE, 31 July 1920, LTRS.
95 “a real mother”: AUTO, 194.
96 “the most important”: AE to David Hilbert, 21 February 1920, Hilbert Nachlass, UGL-HA.
96 the faculty’s initial list: Faculty minutes, 21 February 1920, Philosophical Faculty, II, PH, NR. 36e, UAG.
96 “It is difficult”: AE to MB, 3 March 1920, LTRS.
96 Did he really want: MB to Gilbert Lewis, n.d., BERK-L.
96 “a real honor”: Ibid.
96 Wende agreed to create: Born, 26–27, commentary, LTRS; Jungnickel and McCormmach, 1986, 357, say that a mistake was not made in the account books but rather in the timing of the reassignment: however, they give no reference.
97 before departing for negotiations: MB to Geheimrat Wende, May 1920, STABI.

97 Franck liked the prospects: James Franck, 12 July 1962 interview, 10, AHQP.
97 After visiting Göttingen: MB to Geheimrat Wende, May 1920, STABI.
97 "The question 'Göttingen, yes or no?'": MB to Elsa Einstein, 21 June 1920, LTRS.
97 Born's anxiety about: MB to Theodore von Kármán, 7 November 1922, CALTECH-K.
97 "a collaborator to whom I feel": MB to Prof. Stille, 4 July 1920, UAG.
97 "Franck + Born are": Reid, 1986, 306.
98 He had asked the university: Kurator Voigt to MB, 3 July 1920, STABI; MB to Kurator Voigt, 10 July 1920, FRF.
98 Now the question of my successor: MB to AE, 16 July 1920, LTRS.
98 "Wachsmuth is agitating": HB to AE, 31 July 1920, LTRS.
98 At 6 A.M. on August 6: HB to the Blaschkos, 17 August 1920, STABI.
98 "a trace of the spiteful": MB to Carl Oseen, 14 October 1920, OSEEN.
99 "Poor, corrupt Deutschland": HB to the Blaschkos, 17 August 1920, STABI.
99 "2000 meters above the sea": Ibid., description on Suldenhotel letterhead.
99 "*Schlaraffen* existence": Ibid.
99 "fairy tale things": MB to Carl Oseen, 14 October 1920, OSEEN.
99 "heavenly restful": HB to the Blaschkos, 17 August 1920, STABI.
99 On August 24, 1920: Clark, 1971, 225–228; Holl, 1996, 91.
99 You must have suffered: HB to AE, 8 September 1920, LTRS.
100 Using nothing but elementary mathematics: Born, 1920.
100 Born had created: Hacohen, 2000, 18–19.
100 When Born returned home: Holl, 1996, 88.
100 "burning incense": MB to Elsa Einstein, 21 June 1920, LTRS.
100 After receiving von Laue's plea: Holl, 1996, 92.
100 The first printing: Ibid., 87.
100 Scientific questions: Heilbron, 1986, 117.
101 the government prepared: Clark, 1971, 263.
101 In Bathhouse 8: Forman, 1986, 19.
101 First came hours: Heilbron, 1986, 117–119; Clark, 1971, 263–264.
101 "I will . . . not allow myself": AE to MB, n.d., October 1920, LTRS.
101 Born knew that Einstein suffered: MB to Carl Oseen, 14 October 1920, OSEEN.
101 The Borns and many other friends: MB to AE, 13 October 1920, LTRS; HB to AE, 7 October 1920, LTRS; Ibid.
101 "Today I write principally": AE to MB, 30 January 1921, LTRS.
101 Hedi's six-page letter to Elsa: HB to Elsa Einstein, 18 November 1920, Albert Einstein Archives, Jewish National and University Library, Jerusalem.
102 "very much to heart": MB to AE, 12 February 1921, LTRS.
102 "correct choice of members": MB to Felix Klein, 21 November 1920, UGL-HA.
102 In another two years: Stark, 1922 and 1941. I thank Roger Stuewer for these references.

102 "The physics is so beautiful": MB to Gilbert Lewis, 27 November 1920, BERK-L.
103 Goldman, a first-generation American: "Henry Goldman, 79, Banker, Dies Here," *New York Times*, 3 April 1937.
103 "one sat the entire day": Walther Gerlach, interview with INJ, 1970.
103 "God be thanked": Ibid.
103 "One thing shines out": MB to Gilbert Lewis, 27 November 1920, BERK-L.
104 Hedi even mentioned: HB to AE, 2 October 1920, LTRS.
104 We are not going to pay: MB to AE, 12 February 1921, LTRS.
104 "a shattering view": MB to Gilbert Lewis, 27 November 1920, BERK-L; Serge Boguslavski to MB, 18 August 1920, LTRS, 46–49.
104 Born reported to Einstein: MB to AE, 12 February 1921, LTRS.
104 Wouldn't that be grand: MB to James Franck, 22 February 1921, AHQP.
105 They lost their silver: MB to AE, 12 February 1921, LTRS.

Chapter 6

107 "to bring Göttingen": MB to Carl Still, 8 December 1920, STILL.
107 excited to be with Franck: MB to Walther Gerlach, 11 May 1921, 13 May 1921, 16 May 1921, 23 May 1921, DM-G.
107 "A small boy, Gustav Born": MB to AE, 4 August 1921, LTRS.
108 Almost two years earlier: MB to Wolfgang Pauli, 23 December 1919, PAULI, 10.
108 "the greatest talent": MB to Kurator, 4 July 1921, Kur Alt 4 v.h.35, Bd1, UAG.
108 "infant prodigy": AUTO, 211.
108 Someone once rightly said: Laurikainen, 1985, 3.
108 "very heavy physics": MB interview, 18 October 1962, AHQP, 17.
108 "the quanta really are a hopeless mess": MB to AE, 21 October 1921, LTRS.
109 "Since the discovery": Sommerfeld, 1919.
109 "the Bible of": MB to Arnold Sommerfeld, 13 May 1922, 1977–28/A, 34(2), SOMMER.
109 You regard the application: MB to Wolfgang Pauli, 23 December 1919, PAULI, 10.
109 "superfluous elements and describe": Alfred Landé interview, 3 July 1962, AHQP, 25.
110 the article became for Born: MB to Arnold Sommerfeld, 5 March 1920, 1977–28/A, 34(1), SOMMER.
110 Born's guilt about: MB to Arnold Sommerfeld, 13 May 1922, 1977–28/A, 34(2), SOMMER.
110 "sighing" under the burden: Hund, 1983, 33.
110 Having failed to recover: MB to GB, 2 July 1942, FAM.
110 "an affinity for millionaires": James Franck interview, 13 July 1962, AHQP, 9.
110 A wealthy manufacturer: Reid, 1986, 306; MB to Carl Still, 8 December 1920, STILL; LTRS, 55.
110 "Poor hare!": MB to Walther Gerlach, 12 December 1921, DM-G.
110 The professors, who: LTRS, 55.

110 On this particular occasion: Reid, 1986, 313.
110 Hedi stayed home: HB to Bertha Born, December 1921, FAM.
111 The Vienna-born Pauli: MB to Arnold Sommerfeld, 5 January 1923, 1977–28/A, 34(3), SOMMER.
111 Born thought enough: MB to AE, 29 November 1921, LTRS.
111 "rocking slowly like a praying Buddha": AUTO, 212.
111 "Pauli was reached": W. E. Tisdale report, 13 December 1932, IEB.
111 "You probably have no idea": MB to Paul Epstein, 24 January 192[2], CALTECH-E. Born accidentally put the wrong year on the letter.
111 Emerich Brody, Born's assistant: MB to AE, 21 October 1921; MB to AE, 30 April 1922, LTRS.
112 "in death as in life": AUTO, 194.
112 "escapade into the Eldorado": AE to MB, 18 January 1922, LTRS.
112 Born enthused to Einstein: MB to AE, 30 April 1922, LTRS.
112 But a month later: MB to Alfred Landé, 29 May 1922, AHQP.
112 three of Bohr's young colleagues: George de Hevesy to James Franck, 29 March 1922, NBA.
113 For the first session: Oscar Klein interview, 20 February 1963, AHQP, 22; Friedrich Hund interview, 25 June 1963, AHQP, 1–3; Rudolf Minkowski, April 1962, AHQP, 1.
113 Bohr stood on the platform: Heisenberg, 1971, 38.
113 "an excellent overview": Friedrich Hund, 25 June 1963, AHQP, 1.
113 By the end of the lecture: Hund, 1985, 71.
113 At the end of another lecture: Cassidy, 1992, 127.
113 Bohr invited Heisenberg: Ibid., 131.
113 "The time is perhaps past": Born, 1922, 677–678, and translated in Cassidy, 1992, 146.
114 Born felt sure of himself: MB, 17 October 1962 interview, AHQP, 10.
114 Born wrote to Einstein: MB to AE, 6 August 1922, LTRS.
114 He wanted to start: Cassidy, 1992, 143.
115 Even the physicists: Werner Heisenberg to his father, 5 November 1922, ibid., 142.
115 "For once": Werner Heisenberg to his father, 16 November 1922, ibid.
115 "the Encyclopedia article": MB to Theodore von Kármán, 7 November 1922, CALTECH-K.
115 none of which: MB to Arnold Sommerfeld, 5 January 1923, SOMMER.
115 "a Mecca": Hanle, 1989, 107.
115 At the time, it had to be handled: MB to Theodore von Kármán, 7 November 1922, CALTECH-K.
116 [Discrimination against Jews]: AE to MB, 9 November 1919, LTRS.
116 [Einstein] often argued: LTRS, 17.
116 Born was politicking: MB to Arnold Sommerfeld, 5 January 1923, SOMMER.
117 Born and Heisenberg: Cassidy, 1992, 147–149.
117 "deep-going difficulties": Niels Bohr to MB, 9 April 1923, BOHR, vol. 4, 38–39.

117 "closer to the great mystery": MB to AE, 7 April 1923, LTRS.
117 "entire system of concepts": Born, 1923, 542, quoted in Cassidy, 1992, 149.
117 "The great majority": AUTO, 203.
117 For faculty: Protokoll Buch II, 20 December 1923, UAG.
117 A serious topic: Faculty meetings of mathematics-natural science, 1923, UAG.
118 Richard Courant's ingenuity: "Reminiscences on the Göttingen Mathematical Institute on the Occasion of R. Courant's 75th Birthday, (anon.) BOBST, 5.
118 Thirty-four hundred workers: "Meyer Kauffmann," STABI, 15.
118 With ribbons in their hair: INJ, interview with author, January 1997, Melbourne Australia.
119 From tomorrow: MB to AE, 25 August 1923, LTRS.
119 "One cannot solve": Reminiscences on the Göttingen Mathematical Institute on the Occasion of R. Courant's 75th Birthday, (anon.) BOBST, 6.
119 An emergency committee: Cassidy, 1992, 159.
119 The Rockefeller Foundation: James Franck, Max Born, Robert Pohl to Wickliffe Rose, 18 June 1924, box 34, folder 484, IEB.
120 The follies of the French: MB to AE, 7 April 1923, LTRS.
120 His lecture notes later: Born, 1925, foreword; Cassidy, 1992, 168–169.
120 Not only was Courant: Hendry, 1984, ch. 4, n.60.
120 Heisenberg reported it: Cassidy, 1992, 168–169.
120 "I realize ever more": Werner Heisenberg, 29 November 1923, ibid., 171.
121 "mathematics was cleverer": MB interview, 18 October 1962, AHQP, 19.
121 "'a Via regia, which led'": Frenkel, 1972, 293.
121 Schoolboy Niels Bohr: James Franck interview, 14 July 1962, AHQP, 14.
121 "Although I have only Heisenberg's": MB to Niels Bohr, 16 April 1924, BOHR, 24.
121 The BKS theory originated: Stuewer, 1975a.
121 John C. Slater, a young: Stuewer, 1975b, 11.
122 he declared to Born: 29 April 1924, LTRS.
122 Kramers's new theory: Dresden, 1987, 44.
123 "I have a gentle hope": MB to Alfred Landé, 14 July 1924, AHQP.
123 "It is odd": MB to AE, 25 August 1923, LTRS.
123 On July 28, 1924: Cassidy, 1992, 186, 188.
123 Kramers was imposing: Ibid., 184–185.
123 As Kramers finished: Ibid., 188–189.
123 Unable to resolve: Ibid.
123 Bohr later told: Dresden, 1987, 273–274.
123 the methods of the final paper: Van der Waerden, 1967, 16.
124 Their new objective: MB to AE, 15 July 1925, LTRS.
124 In between skiing: MB to David Hilbert, 30 March 1925, UGL-HA.
124 "A postulate of great reach": Born and Jordan, 1925a, 493.
125 Heisenberg had wavered: Cassidy, 1992, 197.
125 Heisenberg took these ideas: Ibid., 198–201.

125 Always thinking about physics: Ibid., 201–203.
125 On or about July 10: See Ian J.R. Aitchison, David A. MacManus and Thomas Snyder, 2004, "Understanding Heisenberg's 'magical' paper of July 1925: A new look at the calculational details," *American Journal of Physics*, vol. 72, 1370, for a full discussion.
125 he really did not know: Ibid., 203.
125 "I know you are fond": AUTO, 218.
126 "Heisenberg's Quantum Mechanics": MB Daybook, 23 July 1925, FAM.
126 "the first person": AUTO, 190.
126 "the climax of my research": Born's application for membership to the Royal Society, London, MGM.
126 "of the Heisenberg type": Tisdale Log, 17 January 1927, box 64, vol. 1, 14, RF Collection, IEB.
126 Born and Jordan's expanded: Van der Waerden, 1967, 38.
127 "Here, I got": Léon Rosenfeld, 19 May 1949, STABI.
127 "First of all": Wolfgang Pauli to Ralph Kronig, 9 October 1925, Van der Waerden, 1967, 37.
127 "bring Göttingen physics": MB to Carl Still, 8 December 1920, STILL.
127 The center for physics: Augustus Trowbridge, 8 June 1926, summary report, GEB Collection, box 9, folder log 2, 167–168, IEB.
127 Göttingen was the only: Chart, "Mathematical and Experimental Physics," box 10, folder 146, IEB.
127 Trowbridge boosted: Augustus Trowbridge, 8 June 1926, summary report, GEB Collection, box 9, folder log 2, 187, IEB.
128 I am so glad: MB to Niels Bohr, 10 October 1925, BOHR, vol. 5, 311–313.
128 Bohr saw it: Niels Bohr to Carl Oseen, 29 January 1926, BOHR, vol. 5, 238.
128 "it was just": MGM interview, 20 February 1962, AHQP, 1.

Chapter 7

129 "everything was perfect": AUTO, 226.
129 "What do you think": Richard Fowler to Paul Dirac, note on Heisenberg's article, DIRAC.
130 "gray ghosts out of haze": HB to her father, 13 November 1925, FAM.
130 The Borns spent: Ibid.
131 "the new theory is": MB to James Franck, 24 November 1925, FAM.
131 "only a further extension": Born, 1960, 128.
131 Born thought they could apply: MB to James Franck, 24 November 1925, FAM.
131 Their article: Pais, 1982b, 1194.
131 "grotesque, silly stuff": HB to father, 21 November 1925, FAM.
132 "one should not judge": MB to David Hilbert, 28 November 1925, UGL-HA.
132 at the end of December: HB to father, 30 December 1925, FAM.

132 As Max watched: MB to HB, 23 January 1926, ibid.
132 On January 22, 1926: Ibid.
132 It is a happy experience: Kemble, 1926, 424.
132 "it would perhaps be rash": Ibid., 423.
133 "the center of American physics": MB to HB, 25 January 1926, FAM.
133 When Born heard the music: Ibid.; movie filmed by Irving Langmuir, provided to author courtesy of Roger Summerhays.
133 Born was delighted: MB to HB, 25 January 1926, FAM.
133 "What could I do?": MB to HB, 3 March 1926, ibid.
134 "the fount of quantum": Karl Compton, *Nature* 139 (1937), 238–239, quoted in Cassidy, 1992, 263.
134 In the two-month tour: MB to HB, 1 February 1926, 3 February 1926, FAM.
134 "huge steel mills": MB to HB, 3 February 1926, ibid.
134 Woe to anyone: MB to HB, 20 March 1926, ibid.
135 The star attraction: GB, interview with author, April 2001, London.
135 Amid the serious problems: MB to Werner Heisenberg, 14 October 1947, STABI.
136 "always depressed conditions": MB to Kurator, 13 April 1926, Kur PA Max Born, UAG.
136 he described professorial life: Augustus Trowbridge, 9 October 1925 report, box 34, folder 482, IEB.
136 "When the gentlemen": MB to HB, 14 March 1926, FAM.
136 "In any case": MB to Niels Bohr, [December] 1925, AHQP.
136 "would be a great loss": Richard Courant to Kurator, 16 April 1926, Kur, UAG.
136 On July 4: MB to Kurator, 4 July 1926, Kur PA Born, UAG.
136 A ministerial official: Niessen to Kurator, 28 July 1926; MB to Kurator, 17 August 1926, Kur PA Born, UAG.
137 the purpose of which: Eck, 1989, 328.
137 For months, Gercke: Eck, 1989, 328–330; Dahms, 1987, 17–18; Pariser Tageblatt, 3 February 1934 (Wiener, reel 2), LBI.
137 "it could not be proved": MB to PE, 7 August 1927, AHQP.
138 Born had written: MB to AE, 15 July 1925, LTRS.
138 "was the most outstanding example": AUTO, 227.
138 "Through the investigation": Born, 1926a.
139 One gets no answer: Ibid.
139 "Statistical considerations": MB interview, 27 October 1962, AHQP, 27.
139 "ghost fields": MB to AE, 30 November 1926, quoted in Pais, 1982a, 443.
139 "the motion of particles": Born, 1926b, 208.
139 "The principle of causality": Forman, 1971, 65.
139 "assume that there are other parameters": Born, 1926b, 224.
140 Heisenberg called him a "traitor': MB to Léon Rosenfeld, 10 May 1951, NBA.

140 "The more I think": Werner Heisenberg to Wolfgang Pauli, 8 June 1926, Cassidy, 1992, 215.
140 "No genetic relationship": Erwin Schrödinger, *Annalen der Physik* 79 (1926), 734–756, ibid.
140 Schrödinger found Born's interpretation: Pais, 1982b, 1197.
140 Bohr so relentlessly: Heisenberg, 1985, 164.
140 "certainly imposing": AE to MB, 4 December 1926, LTRS.
140 Pauli wrote Heisenberg: Cassidy, 1992, 232–233.
140 "I am *very* enthusiastic": Werner Heisenberg to Wolfgang Pauli, 28 October 1926, ibid., 233.
140 Heisenberg wrote a fourteen-page letter: Werner Heisenberg to Wolfgang Pauli, 23 February 1927, ibid..
141 When Trowbridge arrived: Augustus Trowbridge, 2 July 1926, report, box 34, folder 482, IEB.
141 "people who were highly bizarre": Delbrück, "Erinnerung an Max Born," 2; Elsasser, 1978, 50.
142 One student spied Dirac: Meyenn and Schucking, 2001, 46.
142 the Berlin faculty: Faculty meeting, 18 June 1926, HUMBOLDT.
142 At the November 2 meeting: Faculty meeting, 2 November 1926, Ibid.
142 "deeper originality": Faculty report to the Minister, 4 December 1926, ibid.
142 Planck wrote a short: Max Planck to MB, 21 November 1926, STABI.
142 "I cannot break": MB to Max Planck, 23 November 1926, draft letter, FAM.
143 he complained he had no time: MB to Charles Galton Darwin, 7 April 1927, AMPHIL-D.
143 "I'm too old now": MGM interview, 20 February 1962, AHQP, 6.
143 "suddenly in the center": Delbrück, "Erinnerung an Max Born," 2.
143 "unapproachable and unsympathetic": WE Tisdale, 17 January 1927 report, 22, and 25 May 1927 report, 101, RF Collection, box 64, IEB.
143 H. P. Robertson: Infeld, 1941, 242.
143 "always seemed distracted": Elsasser, 1978, 68.
143 George Uhlenbeck simply thought: George Uhlenbeck interview, 30 March 1962, AHQP, 1, 3.
143 He had met Born: Kapitza Club records, 29 July 1926, AHQP; J. R. Oppenheimer to the Board of Research Studies, 18 August 1926, Smith and Weiner, 1995, 98.
144 E. U. Condon described him: Condon, Memoirs, 11–12, AMPHIL-C.
144 Some of his fellow students: George Uhlenbeck, interview, 30 March 1962, AHQP, 8.
144 "I felt as if he": Elsasser, 1978, 53.
144 Born, who later described: MB to Hans Thirring, 22 March 1957, STABI.
144 Oppenheimer later said: J. R. Oppenheimer interview, 30 November 1963, AHQP, 8.
144 "Unfortunately Born tells me": Smith and Weiner, 1995, 107.

144 "a man conscious": K. Hoffmann, 1995, 27.
144 Striding to the blackboard: MB, interview with INJ, 1967; AUTO, 229.
144 One morning Born found: Ibid.
145 On his dissertation: Math-Nat. Prom. Spec O, UAG; Michaelmore, 1969, 23.
145 "a bad habit": Léon Rosenfeld interview, 19 July 1963, AHQP, 1.
146 Oppenheimer, who was with me: MB to PE, 16 July 1927, AHQP.
146 Your information about Oppenheimer: MB to PE, 7 August 1927, ibid.
147 Niels Bohr delivered a speech: Bohr, 1928, 566.
147 Born, Kramers, Heisenberg: Ibid., 593–594.
147 Born sitting next to Madame Curie: MB to GB, 30 May 1944, FAM.
147 We regard *quantum mechanics*: Born and Heisenberg, Solvay Conference, Cassidy, 1992, 250.
148 During breaks, the physicist: Movie by Irving Langmuir provided to author by Roger Summerhays.
148 In his highly influential book: Heisenberg, 1930 (Born citation on p. 25 in German ed.); Cassidy, 1992, 265.
149 Pauli's 1933 *Handbuch*: Pais, 1982b, 1197–1198.
149 A few years later: MB to Kurator, 17 June 1929, 18 July 1929, Kur PA Born, UAG; Felix Block interview, 15 May 1964, AHQP, 29.
149 "We never dreamt": Pais, 1982b, 1198.

Chapter 8

151 Professor Born terribly: Augustus Trowbridge, 20 March 1928 report, GEB Collection, box 10, folder log 4 (1928), IEB, 118.
151 That spring Hedi: MB to PE, 14 May 1928, AHQP.
152 In a later letter: HB to PE, 27 September 1928, ibid.
152 Hedi is in [Bolzano]: MB to PE, 7 October 1928, ibid.
152 Why then a hang-dog mood: Ibid.
152 Yesterday evening I came back: HB to PE, 26 October 1928, ibid.
153 Years later, she confided: Rachel Chester, interview with author, November 1999, Penicuik, Scotland.
153 I shiver at the thought: MB to PE, 7 October 1928, AHQP.
153 Born anxiously announced: MGM interview, 20 February 1962, ibid., 1.
153 Around the same time: GB, interview with author, April 1999, London.
153 Born wrote to Einstein: MB to AE, [December] 1928, LTRS (LTRS date 20 February 1928 is incorrect); MB to Kurator, 20 December 1928, Kur PA Born, UAG.
153 I am here: MB to Norbert Wiener, 10 January 1929, MIT.
153 As he tried to rest: MB to Gritli Pryce, 5 September 1949, FAM.
154 "greasy businessmen": AUTO 240; LTRS, commentary, 112; MB to HB, 18 March 1957, FAM.
154 Born wrote to Bohr: MB to Niels Bohr, 11 February 1929, AHQP.
154 "a nervous wreck": W. E. Tisdale, 14 May 1929 report, RF Collection, Tisdale log 1929, IEB.

155 Hedi responded that: HB to PE, 10 May 1929, AHQP.
155 "But, gentlemen, that is not": Frenkel, 1996, 96.
155 "create clear external": HB to PE, 30 July 1929, AHQP.
155 Hedi took the advice to heart: Ibid.; MB to Gritli Pryce, 5 November 1950, FAM.
155 In a lighthearted letter: HB to PE, 6 August 1929, AHQP.
155 "always fighting between": INJ, interview with author, January 1997, Melbourne, Australia.
156 labeled others' reactions: HB diary, 28 February 1939, FAM.
156 "nest warmth": GB, interview with author, April 1999, London.
156 "We are firmly attached": HB to PE, 30 July 1929, AHQP.
156 "Mrs. Born does not like me": Hans Bethe, interview with author, July 1997, Ithaca, NY.
157 I am now in the Harz: HB to PE, 1 October 1929, AHQP.
157 By the end of her stay: HB to PE, 25 October 1929, ibid.
157 "to live a more": HB to PE, 24 January 1930, ibid.
157 She did not see: HB to PE, 5 October 1930, ibid.
157 "I hate that description": HB to PE, 4 March 1930, ibid.
157 She wanted to buy a dog: HB to PE, 5 June 1932, ibid.
158 Frustrated and impatient with: MB to Fritz London, 25 March 1929, DUKE.
158 Tisdale, who was in Göttingen: W. E. Tisdale, 21 January 1930 report, RF Collection, IEB.
158 To Victor Weisskopf: Victor Weisskopf, interview with author, June 1997, Newton, MA.
158 "What I wish": MB to MGM, 23 August 1935, SAN DIEGO.
159 "he was not a happy man": MGM in Franck interview, 13 July 1962, AHQP, 18.
159 "The dear God": HB to PE, 25 October 1929, AHQP.
159 "use of elementary": Born and Jordan, 1930.
159 many results of quantum theory: Pauli, 1930, 602, cited in Beller, 1999, 38.
159 "The layout of the book": Ibid.
159 "I make nothing of it": MB to Léon Rosenfeld, 14 March 1930, NBA.
159 "Naturally the book is": MB to Fritz London, 18 October 1930, DUKE.
160 "I've just gotten": Léon Rosenfeld interview, 1 July 1963, AHQP, 20.
160 "build a wall": HB to PE, 23 April 1930, ibid.
160 At the beginning of June: MB to Hermann Weyl, 1 June 1930, ETH-W.
161 The result was that: Gay, 1992, 253.
161 "In Göttingen": HB to PE, 5 October 1930, AHQP.
162 Compared to the national average: Rowe, 1986, 445.
162 "We are walking": HB notebook, FAM.
162 "the future looks dark": MB to PE, 29 September 1930, AHQP.
162 If Hitler were: MB to PE, 5 January 1931, ibid.
162 but in a letter: MB to Percy Bridgman, 13 September 1930, HARVARD-B.
162 Twelve o'clock: AUTO, 87.

162 She kept it: HB to PE, 13 March 1931, AHQP.
163 I have to send: MB to HB, 2 December 1930, FAM.
163 Born wrote: MB to AE, 22 February 1931, LTRS.
163 He pursued an offer: MB to Theodore von Kármán, 27 March 1931, 4 June 1931, 23 June 1931, CALTECH-K.
163 "Merely being a housewife": HB to PE, 22 January 1931, AHQP.
164 Her "quiet hope": Ibid.
164 Born's book on relativity: Carl Friedrich von Weizsäcker, interview with INJ; von Weizsäcker, interview with author, September 1997, Sternberg, Germany.
164 "The next time": MB to Joe Mayer, 11 August 1931, MGM.
164 When she returned: MB to Joe Mayer, 5 May 1932, MGM.
164 Hedi decided to tell: HB to PE, 29 October 1931, AHQP; INJ, interview with author, January 1997, Melbourne, Australia.
165 In 1928: Dahms, 1987, 16–19.
165 1 percent of the total: Bentwich, 1953, 2; Gay, 1992, 247.
165 "reject all attempts": Protokoll Buch III, 23 January 1930, UAG.
165 By 1931, Nazi students: Rowe, 1986, 446.
165 Communists placed leaflets: Jungk, 1958, 32.
165 The Ministry of Education's austerity: Becker, 1987, 411–414; Protokoll Buch III, 5 December 1931; 14 January 1932, UAG.
166 When he asked: Protokoll Buch III, 5 December 1931, AUTO, 248.
166 "Did I miss": HB to PE, 26 December 1931, AHQP.
166 During the holidays: Becker 1987, 411–14.
166 Born began the faculty meeting: Ibid.; Mathematical Natural Science faculty meeting, October 1931-October 1932, UAG.
167 "I wish sometimes": MB to Fritz London, 30 January 1932, DUKE.
167 Hedi, who had little sympathy: HB to PE, 16 January 1932, 26 February 1932, 5 April 1932, AHQP.
167 She walked: MB to Lou Andreas-Salomé, 3 February 1932, private collection.
167 In 1911, she: HB to GB, 25 January 1951, FAM.
167 "a wonderful, shameless": HB Daybook 100–101, ibid.
168 The day after: HB to Lou Andreas-Salomé, 15 February 1932, private collection.
168 Ehrenfest was the only person: HB to PE, 26 February 1932, AHQP.
168 In a sanatorium: HB Daybook, 100–101, FAM.
168 The first of them: MB to MGM, 2 March 1932, SAN DIEGO.
168 I praise the day: MB to HB, 8 May 1932, FAM.
169 "a joint life": Ibid.
169 You can imagine: HB to PE, 5 June 1932, AHQP.
169 "slowly climbing out": Ibid.
169 He took pleasure: MB to HB, 11 August 1932, 13 August 1932, FAM.
170 "You should live": MB to HB, 25 August 1932, ibid.
170 "The mysticism includes": HB to PE, 22 August 1932, AHQP.
170 Having discussed this arrangement: MB to HB, 13 August 1932, FAM.

170 Max had fantasized: MB to HB, 24 August 1932, ibid.
170 he acknowledged to Hedi: HB diary, 4 January 1933, ibid.
170 Half of the voters: Rowe, 1986, 445, Lemmerich, 1983, 26.
171 When Max heard: MB to HB, 25 August 1932, FAM.
171 She now thought: HB to PE, 25 March 1933, AHQP.
171 "A life without conflict": HB Daybook, 102, FAM.
171 Max watched Gritli: MB to HB, 24 September 1932, ibid.
171 "hinder all later experiences": HB to PE, 4 November 1932, AHQP.
172 "as could only": Kurt Tauss to MB, 12 June 1952, FAM.
172 "a personification of": INJ, notebook 1967, ibid.
172 And the matrix: Hans Bethe, interview with author, June 1997, Ithaca, NY.
172 One guest, who later reminisced: Kurt Tauss to MB, 12 June 1952, FAM.

Chapter 9

173 even down to stripping: Edward Teller interview with author, 30 December 1997, Palo Alto, CA.
173 the SA, and his personal: Dawidowicz, 1975, 50–51.
174 On a train: Beyerchen, 1977, 21.
174 When they were tried: Flavin, 1996, 1–2.
174 he spoke out publicly: HB to PE, 25 March 1933, AHQP.
174 Max finished his book: MB to Max von Laue, 12 March 1933, MPG-L; MB to Charles Galton Darwin, 7 April 1927, AMPHIL-D.
174 She dreamt about: HB diary, endnotes for 1932, FAM.
174 Hitler's storm troopers: Rose Ibsen, 4 April 2002, Rockville, MD.
174 A ringing phone: AUTO, 250.
174 When their friend, Maria Stein: Maria Stein interview, 28 May 1974, AHQP, 4.
175 an American student: Saunders MacLane, 1995, 3.
175 Huge placards posted in Berlin: "Five Years of German-Jew Boycott, April 1st, 1938" Reel 33A, Weiner Library, LBI.
175 By the next year: Hentschel, 1996, lv.
175 "an inner necessity": James Franck to the Minister of Education, 17 April 1933, GEHEIM.
175 "We Germans of Jewish descent": James Franck to the university rector, 17 April 1933, GEHEIM, trans. in Beyerchen, 1977, 17.
176 The Borns spent: HB diary, 18 April 1933, FAM.
176 The rumors were: James Franck to Herr Kühn, 24 April 1933, GEHEIM.
176 "We are unanimously agreed": Beyerchen, 1977, 19.
176 The Ministry of Education: Becker, 1987, 415; Protokolbuch II, 1927–1946, 19 January 1934, UAG.
176 The names on the letter: AUTO, 248.
176 he told his mother: Mrs. Jordan to her daughter, 20 April 1933, enclosed with letter from Pascual Jordan to MB dated 15 August 1948, STABI.
176 these students wrote: To the Prussian Minister, 7 December 1933, GEHEIM.
176 "Communist-infested": Beyerchen, 1977, 29.

176 Richard Courant later attributed: Richard Courant, 1964, 18–19.
177 "Max suspended": HB diary, 25 April 1933, FAM.
177 All I had built: AUTO, 251.
177 "Now Let Us": MB, Italian radio interview, 1955, FAM.
177 "supposed political agitation": Rudolf Ladenburg to Niels Bohr, 24 May 1933, AHQP.
177 "I shudder when": MB to PE, 19 May 1933, ibid.
177 Born wrote to: MB to Max Reich, 28 April 1933, Kur Alt 4.1.122, UAG.
177 Max made an overnight trip: HB diary, 6 May 1933, FAM.
178 "to bring Göttingen": MB to Carl Still, 8 December 1920, STILL.
178 With bands playing: "Fighting the Fires of Hate." 30 April–13 October 2003 Exhibit, USHMM.
178 As the Borns rode: GB, interview with author, April 1999, London.
178 In Göttingen, crowds: *Göttingen Tageblatt*, 11 May 1933, USHMM; INJ, interview with author, January 1997, Melbourne, Australia.
178 Afterward, as Frau: Diary of Ilse Neumann-Graul, USHMM.
179 Max, recovering from: MB to James Franck, 18 May 1933, AHQP.
179 This is a completely new: MB to PE, 19 May 1933, AHQP.
180 "well-formulated attitude": MB to PE, 11 June 1933, ibid.
180 "Surely," Heisenberg wrote: Werner Heisenberg to MB, enclosure in MB to PE, ibid.
180 According to them: MB to James Franck, 27 May 1933, ibid.
180 "But now I": MB to Werner Heisenberg, 11 June 1933, MUNICH.
180 Ladenburg followed up: Rudolph Ladenburg to Niels Bohr, 24 May 1933, AHQP.
181 Landé asked him: MB to Alfred Landé, draft, undated, FAM.
181 In a somewhat circumscribed: MB to AE, 2 June 1933, LTRS.
181 Among all the other possibilities: MB to PE, 7 June 1933, AHQP.
181 "to create a life": Ibid.
181 he had had the idea to return: MB to Lord Rutherford, 29 July 1933, MS.Add.7653, CAMB; MB to AE, 2 June 1933, LTRS.
181 Cambridge University answered: MB to MHA Newman, 16 June 1933, FAM.
182 To Ehrenfest, Born confided: MB to PE, 16 June 1933, AHQP.
182 "a very special one: Ralph Fowler to MB, 20 June 1933, FAM.
182 In the spring: MB to PE, 8 July 1933, AHQP.
182 "There is little one can do": Max von Laue to Rudolph Ladenburg, 25 April 1933, BRANDEN-L.
182 Conditions are very difficult: AE to MB, 22 March 1934, LTRS.
182 The solution of the Emergency Committee: Memo, "A Society for the Relief of Distressed German Intellectuals Outside Germany," BODLEIAN-N.
183 Prof. Stern wants to bring: W. E. Tisdale, 24 July 1933, series 12.1, box 64, folder log 7, IEB.
183 Fears have been expressed: memo, "The Problem of Refugee Scholars," IEB.
183 Heitler, Nordheim, London: MB to PE, 19 May 1933, AHQP.

184 "Teller is an unusually": MB to PE, ibid.
184 "to take Teller's name to heart": MB to Frederick Lindemann, 2 June 1933, NUFFIELD.
184 Just before Born departed: Frederick Lindemann to MB, 19 June 1933, ibid.
184 he had decided to accept: MB to PE, 8 July 1933, AHQP.
184 Even though Born had traveled: Theodor Valentiner to MB, 22 May 1933, FAM.
184 "I agree with Franck's opinion": MB to Valentiner, 24 May 1933, Kur PA, UAG.
185 To David Hilbert: MB to David Hilbert, 23 July 1933, Hilbert Nachlass, UGL-HA.
185 To Ehrenfest: MB to PE, 20 July 1933, AHQP.
185 He wanted to apply quantum theory: Born, 1934, 410–411.
185 "Nearly a mathematical": MB to Arnold Sommerfeld, 1 September 1933, SOMMER.
186 Wolfgang had become: Renate Königsberger, interview with author, April 2000, London.
186 "I am so horrified": MB to Carl Still, 29 August 1933, STILL.
186 Irene had already gone back: INJ, interview with author, January 1997, Melbourne, Australia.
186 Max stayed behind: MB to Lauder Jones, RF Collection, 12 September 1933, record group 1.1, series 401 D Univ. of Camb. Born, box 43, folder 553, IEB.
186 "You will understand": Ibid.
186 Ironically when staying: HB diary, 19 September 1933, FAM.
187 Valentiner thanked the Ministry: Kurator Valentiner to HB, 28 September 1933, 29 September 1933, MB to Kurator, 9 October 1933, Kur PA Born, UAG.
187 "As I came to the clinic": HB diary, 27 September 1933, FAM.
187 "This tragedy was a horrible shock": AUTO, 250.
187 On July 16: PE to MB and HB, 16 July 1933, AHQP.
187 "The future was not so dark": AUTO, 264.

Chapter 10

189 "On the Quantum Theory: Born, 1934.
190 Born was writing: MB to Frederick Lindemann, 22 December 1933, NUFFIELD; MB to James Franck, 13 December 1933, STABI.
190 "enthusiastic attitude": Infeld, 1941, 193.
190 an "intense desire to discover": AUTO, 255.
190 Articles in Swiss papers: *Journal de Genève*, 17 November 1933; *Wiener Journal*, 23 November 1933, FAM.
190 During the summer: MB to James Franck, 7 September 1933, STABI.
191 "I should be there": Feldman, 2000, 186.
191 "I was deeply hurt": MB to GB, 20 November 1945, FAM.
191 Dear Herr Born: AUTO, 220.
191 Born later said: MB to J. R. Oppenheimer, 11 December 1953, LOC-O.

191 At the beginning of quantum mechanics: MB to PE, 20 July 1933, AHQP.

191 the fundamental commutation law: MB, Royal Society application, SAN DIEGO.

192 "only a formal step": Born, 1940b, 617.

192 In the four years preceding: Crawford, Heilbron, and Ullrich, 1987. (The Royal Swedish Academy now allows access to the records fifty years after the prize year.)

192 From 1922 until 1944: Friedman, 2001, 169–170.

192 "from a logical point of view": Ibid., 171.

192 Oseen offered a number of reasons: "Utredning Rörande W. Heisenberg och andra Atomteoretiker" (1930), 33–34, SWEDACAD–1933.

192 Oseen maintained that: Friedman, 2001, 172.

193 "Dirac is in the forefront": "Utredning om P.A.M. Dirac" (1933), SWEDACAD–1933, 2727.

193 He described Born's contributions: Oseen, "Utredning om M. Born" (1934), ibid., 12.

193 As for Born's work: Ibid., 13–21.

193 In 1928, Einstein suggested: Pais, 1982a, 515.

193 "Don't make yourself sick": Hermann Weyl to MB, 23 December 1933, STABI.

194 The leading spirit is: MB to David Hilbert, 23 December 1933, Hilbert Nachlass, UGL-HA.

194 The girls were still in Germany: Ibid.

194 "somewhat fragile": Richard Courant to James Franck, 22 February 1934, BOBST.

194 A month earlier, Hedi: HB diary, 1934 endnotes, FAM.

194 Planck almost wished: Max Planck to MB, 12 February 1934, ibid.

195 Planck's letters to Born: Ibid.

195 Planck's new caution: MB to David Hilbert, 23 December 1933, Hilbert Nachlass, UGL-HA.

195 Teller's former mentor: Arnold Eucken to MB, 31 December 1933, STABI.

195 "nationality, blood, and race": G. H. Hardy, *Nature*, 18 August 34, 250, reel 5, F1, Wiener Library, LBI.

195 "In reality, as with everything": Hentschel, 1996, 100.

195 The acceptance of Mark: Friedrich Hund to MB, 16 May 1934 (enclosure of MB letter to Franck 18 May 1934), STABI.

195 The letter infuriated Born: MB to James Franck, 18 May 1934, ibid.

196 "in spite of that stupid": MB to James Franck, 25 October 1934, ibid.

196 "Almost every week": MB to AE, 8 March 1934, LTRS.

196 On Easter Sunday: HB to MB, 1 April 1934, FAM.

196 "the perfect and nice": MB to James Franck, 18 May 1934, STABI.

197 but he later told: MB to HB, 4 May 1934, FAM.

197 While they were in the garden: AUTO, 269.

197 May Week approaches: MB to HB, 30 March 1934, FAM.
197 "good friends; by the way": MB to James Franck, 13 June 1934, STABI.
197 "clinical and fanatical": MB to James Franck, 18 May 1934, ibid.
197 "not to visit anyone": MB to HB, 9 May 1934, FAM.
197 "speak too much": Ibid.
197 "there is a crack": MB to James Franck, 18 May 1934, STABI.
198 he hoped that it was: MB to James Franck, 13 June 1934, ibid.
198 But within months: MB to Richard Courant, 6 December 1934, BOBST.
198 "clearinghouse": Hoch, 1983, 13.
198 "If German scientists": Ibid.
198 He was wealthy: Moore, 1989, 267–268.
198 "energy, wit, sometimes": AUTO, 260.
198 Lindemann advised Born: Frederick Lindemann to MB, 10 December 1934, NUFFIELD; Richard Courant to MB, 25 January 1935, BOBST.
198 The League of Nations': High Commission for Refugees, "Progress of the Work," September 1934, Wiener Library, Roll 41, LBI.
199 "a first-rate Mathematical": C. V. Raman to MB, 27 January 1934, STABI.
199 Born's reply noted: MB to C. V. Raman, 18 February 1934, ibid.
199 the German government: MB to Frederick Lindemann, 14 November 1934, NUFFIELD.
199 He agreed but did not think: Max Planck to HB, 5 December 1934, STABI.
199 "Prof. von Ihering": Kurator to Ministry, 18 December 1934, Kur PA Born, UAG.
199 "hopeless disorientation": Hans Bolza to MB, 15 December 1934, FAM.
200 [Hedi and his daughters]: MB to Frederick Lindemann, 14 November 1934, NUFFIELD.
200 "not envisage all": Frederick Lindemann to MB, 10 December 1934, ibid.
200 "He has got a Foundation": MB to Frederick Lindemann, 13 July 1935, ibid.
201 He reported to Franck: MB to James Franck, 3 April 1935, STABI.
201 "that a deep insight": Born 1969b, 167.
201 "It is odd to think": Born, 1935, 1.
201 Along the side of: GB, interview with author, April 1999, London.
201 Born had stopped collaborating: MB to Richard Courant, 26 May 1935, BOBST.
201 Max received a friendly: Richard Courant to MB, 17 June 1935, ibid.
202 "You did not mention Hedi": MB to Richard Courant, 13 July 1935, ibid.
202 He pointed out: Richard Courant to MB, 26 July 1935, ibid.
202 Max had been sufficiently: Ralph Elliott, interview with author, January 1997, Canberra, Australia; INJ, interview with author, January 1997, Canberra, Australia.
202 He wrote to Maria: MB to MGM, 23 August 1935, SAN DIEGO.
202 According to your request: Adolf Hitler to MB, 23 July 1935, FAM.
203 "magical countryside": HB diary, endnotes 1935, ibid.

203 "cloudless skies": MB to MGM, 7 November 1935, SAN DIEGO.
203 "'1001 Nights'. . . ": HB, 25 December 1935, letters from India, FAM.
204 He wrote a couple: MB to R. W. Lawson, 8 March 1951, ibid.
204 "lively, interesting": MB to MGM, 7 November 1935, SAN DIEGO.
204 "tired, old, miserable": HB diary, 14 October 1935, FAM.
204 "clearing up" Raman's institute": MB to Lord Rutherford, 22 October 1936, CAMB.
204 Hedi kept reminding: HB diary, 10 October 1935, FAM.
204 "The state of the lives": MB to MGM, 7 October 1935, SAN DIEGO.
205 The National Tsing Hua University: P. Y. Chou to MB, 15 January 1936, FAM.
205 If a man comes here: MB to FS, 10 November 1935, ROYSOC.
205 Fowler counseled Born: Ralph Fowler to MB, 14 January 1936, STABI; MB to Lord Rutherford, 26 January 1936, STABI.
206 "who was rejected": Jayaraman, 1998, 112.
206 Mrs. Aston came: HB diary, 1–3 February 1936, FAM.
206 Aston had quietly: MB to Lord Rutherford, 22 October 1936, CAMB.
206 he indulged in such practices: MB to Edward Appleton, 5 October 1953, FAM.
206 "a terribly complicated system": MB to Lord Rutherford, 22 October 1936, CAMB.
206 Although Born no longer wanted: MB to Michael Polanyi, 19 February 1936, CHICAGO-P.
207 "A chrysanthemum cannot": HB diary, 6 February 1936, FAM.
207 "You are like a blossom": HB letters of India, 96, ibid.
207 He thought that during: HB diary, 27 February 1936, ibid.
208 "rare combination of": Scroll "To Professor Max Born," 14 March 1936, ibid.
208 "The dreams of the promised land": HB diary, 16 March 1936, ibid.
208 The council had not: MB to Peter Kapitza, 9 May 1936, KAP.
208 The faculty extended Max's position: MB to Peter Kapitza, 9 May 1936, ibid.
209 Hedi has an almost sick: MB to MGM, 1 July 1936, SAN DIEGO.
209 "then she will try": Ibid.
209 His only glimmer of hope: Ibid.
209 "of playing a nasty trick": Peter Kapitza to MB, 26 February 1936, KAP.
209 An independent position: MB to Peter Kapitza, 9 May 1936, ibid.
210 Likewise, he had concerns: Ibid.
210 But the attitude of governments: Ibid.
210 Kapitza responded psychologically: Peter Kapitza to MB, 22 May 1936, ibid.
210 "History shows that": MB to Peter Kapitza, 2 [June] 1936, (Born mistakenly wrote May), ibid.
210 Learning that having a Russian visa: Ibid.
211 A week later, Born: MB to Peter Kapitza, mid July 1936, ibid.
211 The Institut de Biologie: correspondence between Louis Rapkine and Walter Adams, 8 July 1936, 12 July 1936, 15 July 1936, 20 July 1936, BODLEIAN-S.

211 When Hedi asked why: HB diary, 25 July 1936, FAM; Born and Born, 1969, 149.
211 He was one of only three: Hoch, 1983, 13.

Chapter 11

213 Hearing this: Hugh Begbie, interview with author, November 1998, Edinburgh.
213 Dear James: MB to James Franck, 21 December 1936, STABI.
214 "What are you still doing": HB diary, 3 August 1936, FAM.
214 "Jews are not wanted": HB diary, 4 August 1936, ibid; GB, interview with author, April 1999, London.
214 "God gave us a peaceful": HB diary, 17 August 1936, FAM.
215 Upon returning to Britain: MB to Lord Rutherford, 17 October 1936, CAMB.
215 Now he was coming: Ralph Elliott interview with author, January 1997, Canberra, Australia.
215 Dark gray days: MB to Rudolf Peierls, 14 January 1937, BODLEIAN-P.
215 "built for eternity": Born, Hedi, 1963.
216 One sees exactly: MB to Alfred Landé, 19 January 1937, AHQP.
216 Born wanted Landé's: Ibid.
216 a scholar of quaternions: Birse, 1994, 81, 115.
216 physics was very respectable: MB to Lise Meitner, 17 February 1938, CHURCH.
217 "On an elastic bridge": AUTO, 282.
217 The Scottish students: Richard Sillitto, interview with author, May 1997, Edinburgh.
217 When Born had met Goldman: MB to George Jaffé, 12 December 1945, BERK-J.
217 He was unable to support: Henry Goldman to MB, 2 April 1937, FAM.
217 My best wishes: MB to GB, 28 July 1937, ibid.
218 Who dies for you: "Nazi March Marks Göttingen's Fete," *New York Times*, 27 June 1937, reel 6, Wiener Library, LBI.
218 "for occupying this": Richard Becker to MB, 12 July 1936, STABI.
219 Born described the twenty-five-year-old Fuchs: AUTO, 285.
219 On outings to the movies: Renate Koenigsberger, interview with author, April 2000, London, and Ralph Elliott, interview with author, January 1997, Canberra, Australia.
219 "hopped around like Trixi": MB to HB, 4 November 1937, FAM.
219 "true loving marriage:" MB to HB, 30 October 1937, ibid.
219 "as I do not criticize": MB to HB, 7 June 1944, ibid.
219 God, he said: MB to HB, 30 October 1937, ibid.
220 He knew from cosmic rays: MB to AE, 11 April 1938, and commentary, LTRS, 134.
220 he was unsure: MB to Richard Courant, 9 September 1938, BOBST.

220 Born confided his worry: MB to FS, 30 March 1938, ROYSOC.
220 The Lindemann-Churchill friendship: Fort, 2003, 94.
221 "a tiredness and lack": MB to HB, 25 May 1938, FAM.
221 "never think of the home country": MB to HB, 17 May 1938, ibid.
221 Erwin Schrödinger, who had returned: Moore, 1989, 337; FS to MB, 3 May 1938, ROYSOC; MB to HB, 25 May 1938, FAM.
222 The United States, which had a strict quota: "Evian Conference," www.yad-vashem.org.
222 Since in many foreign countries: *New York Times*, 13 July 1938, 16, in http:christianactionforisrael.org.
222 On September 2: MB to AE, 2 September 1938, LTRS.
222 The Jew Born: Gestapo to Kurator, 25 February 1939, PersAkt Born, UAG.
222 Since summer 1937: Finance office-Hannover to Kurator, 23 July 1937, Kur PA Born, UAG; Deutsche Bank to Finance Ministry, 8 September 1939, ibid.; HB diary, 3–4 September 1938, FAM.
223 "fanaticism versus fanaticism": MB to HB, 25 May 1938, ibid.
223 "a small, extremely fast": MB to FS, 6 October 1938, ROYSOC.
223 At the time of the Anschluss: MB to FS, 30 March 1938, ibid.
223 completely threadbare basis: FS to MB, 7 October 1938, ibid.
223 Born contemplated escape: MB to FS, 6 October 1938, ibid.
223 "How long will we be able": MB to Richard Courant, 26 November 1938, BOBST.
223 they received a standardized letter: Deutsches Konsulat to MB, 22 October 1938, FAM.
224 On November 15: *Reichssteuerblatt*, Nr. 101, 15 November 1938, ibid.
224 A few weeks later: Rector to Mathematics-Natural Science Faculty, 19 December 1938, UAG.
224 "Switzerland, which has as little use": "Origins of the 'J' Passport," newspaper article, 1954, reel 32, Wiener Library, LBI.
224 There was no discussion: MB to MGM, 13 December 1938, SAN DIEGO.
224 "In the camps": MB to Richard Courant, 11 November 1938, BOBST.
224 "In my family": MB to MGM, 13 December 1938, SAN DIEGO.
225 "British musicians": AUTO, 269.
225 she launched the Domestic Bureau: Minutes of the meetings, private collection, Edinburgh.
225 Born arranged a physics lecture: MB to MGM, 13 December 1938, SAN DIEGO; Dorothee Fuchs, interview with author, July 1997, Ithaca, NY.
225 "German-Aryan, German-Jewish": MB to Richard Courant, 30 April 1940, BOBST.
226 Every morning, he stopped: Reinhold Fürth, "Reminiscences of Max Born," FAM.
226 Much of his and Fuchs's: Herbert Green, 2 April 1970, notes to Nicholas Kemmer for Royal Society article on Born, ibid.
226 Born had no interest: Frederick Donnan to MB, 3 July 1939, UCLONDON-D.

226 the Scottish crown jewels: Jeffrey, 1992, 9, 11.
227 I actually do not know: James Franck to MB, 20 July 1941, STABI.
227 They were in the Borns' sitting room: Dorothee Fuchs, interview with author, July 1997, Ithaca, NY.
227 A census taken in Edinburgh: "1401 Aliens in Edinburgh," *Edinburgh Evening News*, 19 April 1940, Edinburgh Public Library; Bentwich, 1942, 41.
228 "The matter is": MB to FS, 14 December 1939, ROYSOC; FS to MB, 19 December 1939, ibid.
228 in the European tradition: GB, interview with author, April 2000, London.
229 Then the slavery question: MB to GB, 16 March 1940, FAM.
229 "Political ideologies are": MB to Richard Courant: 30 April 1940, BOBST.
229 "deadly danger": Richard Courant to MB, 13 May 1940, ibid.
229 He thought his reciprocity ideas: MB to AE, 10 April 1940, LTRS.
230 "All male Germans": HB diary, 11 May 1940, FAM.
230 Hedi's nephew Rolli: Ralph Elliott, interview with author, January 1997, Canberra, Australia.
230 They lived in makeshift: Bentwich, 1942, 44.
230 The policy resulted partly: Bentwich, 1953, 28–43.
230 Max's niece, Renate: Renate Koenigsberger, interview with author, April 1999, London.
230 The authorities gave: Domestic Bureau minutes, 13 June 1940, private collection; Philippa Ludlum, interview with author, May 1997, Edinburgh.
231 "to use them": MB to John von Neumann, May 1940, LOC-N.
231 The last weeks have not: MB to FS, 2 June 1940, ROYSOC.
231 "fiendish cleverness of": Ibid.
231 This concern led him to write: MB to Frederick Lindemann, 16 May 1940, NUFFIELD.
231 appropriate authorities: Frederick Lindemann to MB, 26 May 1940, ibid.
231 [It] could be directed: MB to FS, 2 June 1940, ROYSOC.
232 In September 1943: Mets, 1987, 9, 10, 19.
232 All totaled: Bentwich, 1942, 44, 46.
232 [I want] to express: Klaus Fuchs to MB, n. d., STABI.
233 The Nazis may succeed: MB and HB to Otto Oldenburg, 17 June 1940, LOC-V.
233 I stand with the British: MB to Richard Courant, 17 June 1940, BOBST.
233 "If we have to submit": MB to MGM, 17 June 1940, SAN DIEGO.
233 Göttingen friend and physician: Maria Stein, interview, AHQP, 8; Philippa Ludlum, interview with author, May 1997, Edinburgh.

Chapter 12

235 The police told the Borns: MB to Frederick Lindemann, 9 July 1940, NUFFIELD.
235 "nightmare of anyone": Ibid.
235 "to an extra class": Ibid.

235 Max asked Simon: MB to FS, 11 July 1940, ROYSOC.
235 Ladenburg wrote that: Rudolf Ladenburg to MB, 12 August 1940, STABI.
236 "In these days": Frederick Lindemann to MB, 17 July 1940, NUFFIELD.
236 Lindemann's personal fear: FS to MB, 15 July 1940, ROYSOC.
236 A number of brave: Bentwich, 1953, 29.
236 "The situation that seemed hopeless": MB to John von Neumann, 20 August 1940, LOC-N.
236 Otto Oldenburg wrote: Otto Oldenberg to MB, 1 August 1940, STABI.
236 Born turned his attention: MB to FS, 25 July 1940, ROYSOC.
236 "Fuchs is really": MB to John von Neumann, 20 August, 1940, LOC-N; Percy Bridgman, 30 August 1940, HARVARD-B.
236 Fowler could not do much: Ralph Fowler to MB, 9 September 1940, STABI; A. V. Hill to MB, 12 August 1940, ibid.
237 There was to be no physics: GB, interview with author, April 2000, London.
237 By the start of September: Taylor, 1965, 501–502.
237 Home Guard in: Jeffrey, 1992, 70.
237 It was the last time: Taylor, 1965, 499.
237 the government allocations: Briggs, 1983, 272; Taylor, 1965, 462.
237 "a brilliant Chinese and": MB to MGM, 4 September 1940, SAN DIEGO.
238 refugees Rudolph Peierls: Sime, 1996, 299.
238 "hush hush business": FS to MB, 4 September 1940, ROYSOC.
238 "not of the kind": MB to Patrick Blackett, 29 July 1940, ibid.
238 Trying to improve melting theory: ibid.; MB to Alfred Landé, 23 October 1940, AHQP; MB to MGM, 4 September 1940, SAN DIEGO.
238 He wrote after seeing Bragg's article: MB to W. L. Bragg, 31 October 1940, ROYINST.
238 "found to my bewilderment": MB to W. L. Bragg, 5 November 1940, ibid.
238 with Lonsdale correcting some: Kathleen Lonsdale to MB, 27 November 1940, STABI.
238 Born wanted his researchers: MB to FS, 24 October 1940, ROYSOC.
239 "it might be the only": MB to FS, 22 April 1941, ibid.
239 "If we have it": FS to MB, 26 April 1941, ibid.
239 When Fuchs went to: MB to GB, 2 September 1945, FAM.
239 he became a naturalized: Telegram, 1 March 1950, FO953/642, NA.
239 "But you might": MB to Frederick Lindemann, 16 May 1940, NUFFIELD.
240 Born again wished to approach: Chaim Weizmann to MB, 31 July 1941, FAM.
240 "One must not be too": MB to Frederick Lindemann, 7 August 1941, NUFFIELD.
240 Cherwell responded: Frederick Lindemann to MB, 19 August 1941, ibid.
240 "The Americans will get": MB to GB, 15 August 1941, FAM.
240 Britain had begun to tilt: Briggs, 1983, 276, Taylor, 1965, 346–350.
240 "We found out": Briggs, 1983, 273.
240 "Socialists like yourself": Michael Polanyi to MB, 29 July 1941, STABI.

241 "not a socialist": MB to Michael Polanyi, 31 July 1941, CHICAGO-P.
241 To disprove the position: MB to Michael Polanyi, 30 January 1942, ibid.
241 "an Austrian, a Prussian": Ibid.
241 There is no doubt: Ibid.
241 He asked for copies: Michael Polanyi to MB, 12 February 1942, STABI.
242 "We were in close collaboration": MB to Michael Polanyi, 14 February 1942, CHICAGO-P.
242 "to clarify the situation": Michael Polanyi, 20 February 1942, ibid.
242 "a difficult fellow": MB to Michael Polanyi, 14 February 1942, ibid.
243 "It took me weeks": ibid.
243 Kathleen Lonsdale commented: Kathleen Lonsdale to MB, 20 November 1941, and 27 February 1942, STABI.
243 "The Bragg-Lonsdale": Frederick Donnan to MB, February 1942, ibid.
243 One Sunday evening in March: Cäcilie Heidczek to Max von Laue, 6 April 1942, BRANDEN-L.
243 Born wrote a tribute: Born, 1942, 285.
244 a sister of my father: MB to FS, 7 December 1941, ROYSOC.
244 She felt her death: Selma Jacobi to MB, 14 June 1942, FAM.
244 "If Germany will suffer": MB to Richard Courant, 25 February 1943, BOBST.
244 "colossal production": MB to MGM, 14 August 1942, SAN DIEGO.
244 "be allowed for yet": Walter Heitler to MB, 28 November 1942, STABI.
244 "quite young": MB to James Franck, 13 December 1942, AHQP.
245 In the sitting room: GB, interview with author, April 2000, London.
245 Brillouin had managed: Léon Brillouin to MB, 22 March 1943, STABI; FS to MB, 29 July 1940, ROYSOC.
245 Born's friend, Frederick: Frederick Donnan to MB, 7 January 1941 and 26 January 1941, STABI.
245 "solve the enigma": LTRS, 151, commentary.
246 His example was Einstein's prediction: Born, 1956, 14.
246 "to learn the art": Ibid., 44.
246 Born had been one of those: Sigurdsson, 1996, 58.
246 "Young whores": AE to MB, 7 [November] 1944, LTRS. (Date given in LTRS is September but is incorrect.)
246 I believe that you have the right: MB to AE, 10 October 1944, LTRS.
246 "splendid creatures": MB to MGM, 20 January 1943, SAN DIEGO.
246 "charity begins outside": Maurice Pryce, interview with author, December 1999, Vancouver.
247 "a most perfect": AUTO, 270.
247 "good and bad": MB to HB, 13–27 July 1943, FAM.
247 "a strange world": MB to HB, 19 July 1945, ibid.
247 Max felt dizzy: HB diary, 11 November 1943, ibid.
247 At his father's behest: GB to MB, 21 November 1943, ibid.
247 Gustav had brought: GB, interview with author, April 2000, London.

248 "that your life, after this": MB to GB, 22 December 1943, FAM.
248 "Send love to Otto": HB diary, 13 January 1944, ibid.
248 "Gustav's sudden disappearance": MB to Gritli Pryce, 5 September 1949, ibid.
248 "very quietly with much sleeping": MB to GB, 20 March 1944, ibid.
249 "think about nothing": MB to Gritli Pryce, 5 September 1949, ibid.
249 "Will this story": MB to GB, 12 April 1944, ibid.
249 It is true that we lived: Ibid.
249 "My views on life": MB to GB, 15 October 1944, ibid.
250 "have more understanding": Ibid.
250 She is of course: MB to GB, 14 January 1945, ibid.
250 "(This is not all)": Ibid.
250 "The invasion has begun!": MB to GB, 6 June 1944, ibid.
250 "noble form of romanticism": MB to GB, 20 March 1944, ibid.
250 For me the news is horrible: MB to GB, 18 July 1944, ibid.
251 On September 19, the SS: Dorothee Fuchs, interview with author, July 1997, Ithaca, NY; Lise Meitner to MB, 22 October 1944, CHURCH.
251 Meitner's other news: Ibid.
252 "It was really good news": MB to Otto Stern, 11 November 1944, BERK-S.
252 In a letter to Born: Rudolf Ladenburg to MB, 2 December 1944, STABI.
252 "some mysterious personal gift": MB to GB, 4 February 1945, FAM.
252 "a benefactor of human society": MB to AE, 15 July 1944, LTRS.
252 "The feeling for what ought": AE to MB, 7 [November] 1944, LTRS (the date in LTRS is 7 September, but Born cites it contemporaneously as November).
253 Born asked Hill: MB to FS, 25 February 1945, ROYSOC-H.
253 Hill replied: A. V. Hill to MB, 8 December 1944, ibid.
253 "Personally I feel rather sorry": A. V. Hill to RS Officers, ibid.
253 "Heisenberg's discovery will be remembered": MB to FS, 1 March 1945, ROYSOC.
253 "behaved very badly": Ibid.
253 One concerned Heisenberg's comment: MB to FS, 24 April 1945, ROYSOC.
253 He reported about Hill and Dirac: MB to Niels Bohr, 17 April 1945, NBA.
253 "There you have the situation": Ibid.
254 Now the Germans will feel: MB to Richard Courant, 23 February 1945, BOBST.
254 "What does Breslau matter": AUTO, 106.
254 On April 22, Max: MB to GB, 22 April 1945, FAM.
254 "done no other thing": MB to GB, 13 May 1945, ibid.
254 on an earlier visit: MB to FS, 16 May 1945, ibid.
254 "in some other connection": FS to MB, 27 April 1945, ibid.
254 "which was very much interested": FS to MB, 21 May 1945, ibid.
255 Max wrote to Gustav: MB to GB, 5 May 1945, FAM.

255 Born hesitated when: MB to GB, 13 May 1945, ibid.
256 "Expertocracy"—rule by: MB to GB, 8 July 1945, ibid.
256 Born's diary chronicled: Born, "Journey to Russia," 13 June 1945–1 July 1945, ibid.
256 "a state based on power": MB to GB, 20 August 1945, ibid.
257 "in a corner of the dirty": MB to GB, 8 July 1945, ibid.
257 "Russia because they know": MB to FS, 22 July 1945, ibid.
257 "Shall I congratulate you": MB to FS, 7 August 1945, ROYSOC.
257 "How could Lise Meitner": MB to GB, 20 August 1945, FAM.
257 "If man is so constructed": Ibid.

Chapter 13

259 "Selma Jacobi: sister": Born, list of family relatives killed by the Nazis, FAM.
259 Hans had a chilling tale: Hans Schäfer to MB, 2 April 1945, ibid.
260 "He was my oldest friend": MB to GB, 30 September 1945, ibid.
260 Max learned that Selma's son: MB to GB, 28 January 1946, ibid.
260 When he thought about: MB to GB, 2 September 1945, ibid.
261 Science is in a similar: Ibid.
261 "morbid, insane" nationalism: Ibid.
261 Simon and others explained: MB to GB, 29 November 1945, FAM.
261 James Franck had headed: James Franck, "The Report of the Committee on Political and Social Problems," 11 June 1945.
262 Born believed in their resolve: MB to GB, 30 September 1945, and 29 November 1945, FAM.
262 The motion was: MB to GB, 20 November 1945, ibid.
262 "whether Oppenheimer or any of the others": MB to MGM, 16 December 1945, SAN DIEGO.
262 He also asked himself: Born, 1965, 195.
262 could "never take": MGM to MB, 8 January 1946, SAN DIEGO.
262 Oppenheimer supported the idea: I thank Kai Bird for this information.
262 "I am a German who": MGM to MB, 8 January 1946, SAN DIEGO.
262 "with a few friendly words": MB to Niels Bohr, 7 August 1945, NBA.
262 For Bohr's benefit, Jordan: Pascual Jordan to Niels Bohr, May 1945, NBA, in D. Hoffmann, 2003, 29–32.
263 "Who except the very": MB to FS, 7 August 1945, ROYSOC.
263 "must have succumbed to": MB to Niels Bohr, 7 August 1945, NBA.
263 Born heard from a friend: MB to FS, 9 January 1946, with copy of letter from Kurt Tauss, 23 December 1945, FAM.
263 Born decided that those: MB to Rudolf Peierls, 13 June 1946, BODLEIAN-P.
264 "until we are able to": Rudolf Peierls to MB, 12 June 1946, ibid.
264 Born listened and agreed: MB to Rudolf Peierls, 13 June 1946, ibid.
264 His institute quickly built up: MB to James Franck, 17 April 1946, AHQP.
264 His new Ph.D. student: MB to GB, 20 November 1945, FAM.

264 he judged his department: MB to GB, 3 August 1946, ibid.
264 "too old, nervous, and": MB to Rudolf Peierls, 20 June 1946, BODLEIAN-P.
264 "Where is Fuchs?": MB to Rudolf Peierls, 14 February 1946; Rudolf Peierls to MB, 20 February 1946; Rudolf Peierls to MB, 24 February 1946, ibid.
265 "There is so much hunger": MB to GB, 20 November 1945, FAM.
265 Arnold Sommerfeld asked him: Arnold Sommerfeld to MB, 24 April 1947, SOMMER.
265 "You see!": MB to GB, 15 October 1945, FAM.
265 Born had long talks with him: MB to GB, 3 August 1946, ibid.
265 that the blame for the war: Sime, 1996, 335.
266 He was the same quiet: MB to GB, 3 August 1946, FAM; Edward Teller, interview with author, December 1997, Palo Alto, CA.
266 "as if no war had ever been": MB to GB, 3 August 1946, FAM.
266 "decent and sober, full of": Ibid.
266 a discussion of the ethical: Ibid.
267 "the reaction of decent people": Ibid.
267 the meaning and purpose of science: MB to GB, 12 September 1946, ibid.
267 "I had long intended": MB to Werner Heisenberg, 2 October 1946, MUNICH, with thanks to Mark Walker for the copy.
268 Nonetheless, their scientific life: Artin and Courant, 1947a, 1947b.
269 "Life in Britain is not easy": MB to Richard Courant, 11 October 1947, BOBST.
269 Housewives faced an average: Taylor, 1965, 270.
269 The official offer from the Irish: MB to HB, 15 May 1947, FAM; MB to FS, 10 June 1947, ROYSOC; MB to Lise Meitner, 8 August 1947, CHURCH.
269 "I will probably have to": MB to AE, 4 March 1948, LTRS.
269 A monthlong trip: MB to Richard Courant, 11 October 1947, BOBST; GB, interview with author, April 2000, London.
269 As they talked about his adventures: HB diary, 27 November 1947, FAM.
269 he reported to Sommerfeld: Werner Heisenberg to Arnold Sommerfeld, 5 January 1948, SOMMER.
270 His philosophy of life: MB to GB, 7 December 1947, FAM.
270 "If both are God's work": MB to HB, 25 May 1947, ibid.
270 Born began by observing: Born, 1949, 1.
271 "frankly and shamelessly statistical": Ibid., 101.
271 To this last question: Ibid., 101–103.
271 "I cannot seriously believe": Ibid., 123.
271 He agreed with: LTRS, commentary on 164.
272 "a youngish, fat little man": MB to HB, 19 January 1948, FAM.
272 Even Heisenberg was there: MB to AE, 13 March 1948, LTRS.
272 "Irene is such a motherly": MB to HB, 18 February 1948, FAM.
272 Fuchs occasionally dined with: Maurice Pryce, interview with author, December 1999, Vancouver.
273 Born saw a scientific message: MB to AE, 13 March 1948, LTRS.

273 "But my dear Raman": AUTO, 277, 78.
273 Raman exploded and was so angry: Alfred Kastler to Jean Claude Pecker, 8 December 1967, private collection.
273 "Madame, your husband has treated me": HB diary, 5 April 1948, FAM.
273 "our two friends Sir": Personal correspondence with Jean Claude Pecker.
274 "I drank too much this evening": Alfred Kastler to Jean Claude Pecker, 8 December 1967, private collection.
274 Born heard that Raman: Rudolf Peierls to MB, 24 April 1948, BODLEIAN-P.
274 "at least there is not": MB to Rudolf Peierls, 26 April 1948, ibid.
274 He characterized it as: MB to AE, 22 May 1948, LTRS.
274 "Can we who have escaped Hitler": MB to Rudolf Peierls, 22 May 1948, BODLEIAN-P.
274 Peierls found the situation: Rudolf Peierls, 29 May 1948, ibid.
274 "If I use my reason": MB to Rudolf Peierls, 1 June 1948, ibid.
275 "We all, Göring, Goebbels": HB diary, 5 July 1948, FAM.
275 "Can there be a more cruel choice": Ibid.
275 "I would sign anything": GB, interview with author, April 2000, London.
275 "the super human character": MB to Nelly Planck, 2 June 1957, copy in Meitner files, CHURCH.
275 Nelly told him that: Nelly Planck to MB, 13 June 1957, copy in Meitner files, ibid.
275 "little paper contained": MB to Werner Heisenberg, 8 June 1948, STABI. Heisenberg's letter to Born does not exist, only Born's summary of it in his reply.
275 Born had mixed feelings: MB to Max von Laue, 19 June 1948, MPG-L.
275 "They have no clear conception": Artin and Courant, 1947b, 2–3.
276 "We disliked Hitler, but": Ibid.
276 To Hedi, it felt: HB diary, 7 September 1948, FAM.
276 "We had no feeling": MB to Richard Courant, 17 September 1948, BOBST.
276 "Doesn't that look like": HB to GB, 9 September 1948, FAM.
276 "fine features washed out": HB diary, 7 September 1948, 8 September 1948, ibid.
276 Clemens Schäfer, one of Born's: Moritz, 1948, 391.
276 Born was deeply moved: Ibid., 392.
277 "filled with impressions of Germany": MB to HB, 17 September 1948, FAM.
277 To Simon and Peierls, he: MB to FS, 18 September 1948, ROYSOC; MB to Rudolf Peierls, 23 September 1948, BODLEIAN-P.

Chapter 14

279 "On our return": MB to Otto Hahn, 14 October 1948, MPG-H.
279 "We may have possibly": HB diary, 26 November, 1948, FAM.
279 You know according to quantum mechanics: MB to GB, 12 November 1948, ibid.

280 This new theory fit: MB to Niels Bohr, 8 November 1948, 10 November 1948, AHQP.
280 "This theory may be": MB to Richard Courant, 13 November 1948, BOBST.
280 "blissful state all this year": HB to GB, 12 May 1949, FAM.
280 Dr. Fraser, a "slavish admirer": FS to MB, 22 October 1948, ROYSOC.
280 "Heisenberg is without doubt": MB to FS, 20 October 1948, ibid.
281 Max's view was: HB diary, 27 November 1948, FAM.
281 "Governments and economic systems": Born, Letter to the London *Times*, 23 December 1948, ibid., 5.
281 Although Max may have agreed: HB diary, 23 December 1948, FAM.
281 a dissatisfaction in her life: MB to GB, 16 January 1949, ibid.
281 "let her do as she fancies": MB to GB, 9 February 1949, ibid.
281 "As God guides everything": HB diary, 6 May 1949, and endnotes, ibid.
282 Tomorrow is the day: MB to HB, 7 May 1949, ibid.
282 Hedi told him of her yearning: HB diary, 8 May 1949, ibid.
282 She had developed a plan: MB to GB, 8 May 1949, ibid.
282 She knew that Max: HB to GB, 12 May 1949, ibid.
282 Four days later, she: HB to GB, 16 May 1949, ibid.
283 That is why I am silent: MB to HB, 21 May 1949, ibid.
283 Wolfgang's death gave Max a violent: MB to GB, 19 June 1949, ibid.
284 The speakers did not amplify: MB to GB, 23 October 1949, ibid.
284 "To be frank": MB to GB, 19 November 1949, ibid.
284 I was very happy: MB to Niels Bohr, 26 December 1949, AHQP.
285 "I am not particularly ambitious": MB to FS, 6 January 1950, ROYSOC.
285 After reflecting on his annoyance: MB to FS, 10 January 1950, ibid.
285 "the Germans have become accustomed": FS to MB, 12 January 1950, ibid.
285 He assumed that Bohr: MB to FS, 14 January 1950, ibid.
285 "at the disposal of the physicists": Rosenfeld, 1950, 883.
285 This was actually the first time: MB to Léon Rosenfeld, 10 May 1951, NBA.
286 "It is incumbent on me": Heisenberg, 1951, 52.
286 "I have read it": MB to Werner Heisenberg, 17 April 1951, STABI.
286 Hedi heard over the wireless: HB diary, 3 February 1950, FAM.
286 the Borns saw Fuch's reply: MB to GB, 10 February 1950, ibid.
286 "Upset about Klaus Fuchs": HB diary, 10 February 1950, ibid.
286 Max too was shaken: MB to GB, 10 February 1950, ibid.
287 Additionally, whenever the division: Hans Bethe, interview with author, July 1997, Ithaca, NY.
287 Fuchs performed so well: Klaus Fuchs to W. A. Akers, 8 February 1946, ES 1/493, and W. A. Akers to Klaus Fuchs, 18 February 1946, ES 1/493, NA.
287 "I don't think so": Klaus Fuchs interview with William J. Skardon, 26 January 1950, CRIM 1/2052, ibid.
287 Without a statement from Fuchs: Mandate, 1 March 1950, FO 953/642, ibid.
287 to leave Harwell: Klaus Fuchs to Genia Peierls, 6 February 1950, BODLEIAN-P.

287 "Controlled schizophrenia": Ibid.
287 The Russians had first contacted: M. W. Perrin, Record of Interview with Dr. K. Fuchs, 30 January 1950, AB 1/695, NA.
288 "But *that* was not wrong!": HB to GB, 15 August 1951, FAM.
288 The big bold headline: Sidney Rodin and Joseph Garrity, 29 October 1950, *Sunday Express*, 1, 7.
288 "Did you admire": FS to MB, 4 November 1950, ROYSOC.
288 Hedi feared a "witch hunt": HB to GB, 27 October 1950, FAM.
288 He did nothing other than yell: HB diary, endnotes for 9 January 1951, ibid.
289 "It is rather a pity": MB to Pascual Jordan, 29 June 1950, ibid.
289 "a bewitched world": Frederick Donnan, 3 July 1939, STABI.
289 Einstein's theory of general relativity: Born, draft proposal for Newton-Einstein House, FAM.
289 The effect would be produced: Ibid.
290 The committee's vote to build: MB to Gritli Pryce, June 1950, FAM; MB to Richard Courant, 23 January 1950, BOBST.
290 he told the Kurator: MB to Kurator, 17 August 1950; MB to Kurator, 26 August 1950, Kur PA Born, UAG.
290 I think that in a higher: MB to AE, 4 September 1950, LTRS.
290 "dangerous tradition": AE to MB, 15 September 1950, ibid.
290 "worked like a fountain": MB to AE, 4 May 1952, ibid.
291 "We did not get in touch": MB to FS, 11 January 1951, ROYSOC.
291 his telling Schrödinger that: MB to Erwin Schrödinger, 7 February 1951, AHQP.
291 "I have changed my opinion": MB to FS, 18 April 1951, ROYSOC.
291 "the *Germans* have therefore": FS to MB, 11 August 1951, ibid.
291 He [Heisenberg] said: FS to MB, 11 August 1951, ibid.
292 If one is younger: MB to Gritli Pryce, 22 January 1952, FAM.
292 Born concluded that all he could do: MB to FS, 28 March 1953, ROYSOC.
292 "There is nobody who could deal": Paul Rosbaud to MB, 28 September 1949, FAM.
293 Within two months, Born: MB to FS, 2 March 1951, ROYSOC.
293 Before the war, U.S. scientists: Report to the President, Office of Alien Property Custodian, 1 November 1945, FAM.
293 Mr. Creighton had by this time received: Thomas H. Creighton, Jr. to MB, 18 February 1952, ibid.
293 "without taking any notice": MB to Samuel Goudsmit, 11 March 1952, ibid.
293 the United States resolved: HB diary, 22 October 1954, ibid.
294 Franck, who had turned seventy: James Franck to MB, 3 December 1952, STABI.
294 "You know who you are": Ibid.
294 Born's reply echoed similar feelings: MB to James Franck, 24 December 1952, ibid.

295 "I have of course no practice": MB to Niels Bohr, 10 March 1953, AHQP.
295 Your sweet letter helped me: MB to GB, 15 April 1953, FAM.
295 Berlin is too full: MB to Max von Laue, 14 April 1953, MPG-L.
296 I think that this: MB to FS, 21 July 1953, ROYSOC.
296 "If anyone can be held": AE to MB, 12 October 1953, LTRS.
296 The phrase "the land of mass-murderers": Emil Wolf, interview with author, March 1997, Rochester, NY.
297 When they got home: HB diary, 24 November 1953, FAM.
297 "It seems to be our lot": AE to MB, 12 October 1953, LTRS.
297 "very nice formulation of the problem": MB to FS, 28 November 1953, ROYSOC.
297 Born's initial response: MB to AE, 26 November 1953, 22 December 1953, 2 January 1954, AE to MB, 3 December 1953, 1 January 1954, 12 January 1954, LTRS.
297 Wolfgang Pauli's intervention: Wolfgang Pauli to MB, 3 March 1954, 31 March 1954, 15 April 1954, ibid.
297 For the interpretation: Oppenheimer, 1954, 56–57.
298 "So vanishes vanity!": MB to GB, 8 January 1954, FAM.
298 "I want only to tell you": HB to Born children, 6 November 1954, ibid.
298 Not quite believing: MB to FS, 21 November 1954, ROYSOC.
298 He awoke early one morning: HB diary, 29 October 1954, FAM.
298 "Now we have the telegram": HB to Born children, 6 November 1954, ibid.
299 After rumors and reports: MB to James Franck, 4 November 1954, CHICAGO-F.
299 In 1947 and 1948, Franck and Fermi: SWEDACAD–1954.
299 Responding to the nominations: Ibid.
299 In the late afternoon: Video, "The Nobel Prize in Konserthuset", Stockholm, 10 December 1954; *Stockholms-Tidningen*, 11 December 1954.
300 We are a great fellowship: Born, 1955b, 57–58.
300 "As I am too old": MB to GB, 25 December 1954, FAM.
300 "Love is a power": MB to GB, 20 August 1945, ibid.

Epilogue

301 Born recognized the tensions: Born, 1955a, 1.
301 Only a few weeks after writing: MB to Otto Hahn, 1 February 1955, MPG-H.
302 "Einstein's Last Warning": Monk, 2000, 379.
302 I am convinced that: Born, 1969b, 182–183.
302 But when I looked closely: Personal papers of INJ.

BIBLIOGRAPHY

Arnsberg, Paul. 1983. *Die Geschichte der Frankfurter Juden, III.* Darmstadt: Hans-Otto-Schembs.

Artin, Natalie, and Richard Courant. 1947a. "Report on Impressions of Scientific Work in Göttingen and Hamburg, Germany—July 1947." BOBST.

———. 1947b. "A Summary Report on Conditions of Science in Germany—August 1947." BOBST.

Asch, 1900. "Professor Born." *Schlesische Ärzte-Korrespondenz* 22. FAM.

Barut, Azim. 1989. "Alfred Landé." In *Physiker und Astronomen in Frankfurt,* edited by Klaus Bethge and Horst Klein. Frankfurt: University of Frankfurt.

Becker, Heinrich. 1987. "Von der Nahrungssicherung zu Kolonialträumen: Die landwirtschaftlichen Institute im Dritten Reich." In *Die Universität Göttingen unter dem Nationalsozialismus,* edited by Heinrich Becker, Hans-Joachim Dahms, and Cornelia Wegeler. Munich: K. G. Saur.

Beller, Mara. 1999. *Quantum Dialogue: The Making of a Revolution.* Chicago: University of Chicago Press.

Bentwich, Norman. 1942. "Wartime Britain's Alien Policy." *Contemporary Jewish Record: Review of Events & Digest of Opinion* (February): 41.

———. 1953. *The Rescue and Achievement of Refugee Scholars: The Story of Displaced Scholars and Scientists, 1933–1952.* The Hague: Martinus Nijhoff.

Bethge, Klaus, and Horst Klein, eds. 1989. *Physiker und Astronomen in Frankfurt.* Frankfurt: University of Frankfurt.

Beyerchen, Alan D. 1977. *Scientists Under Hitler: Politics and the Physics Community in the Third Reich.* New Haven: Yale University Press.

Birse, Ronald M. 1994. *Science at the University of Edinburgh.* Edinburgh: University of Edinburgh, Faculty of Science and Engineering.

Bohr, Niels. 1928. "The Quantum Postulate and the Recent Development of Atomic Theory." With discussions by Born, Kramers, Heisenberg, Fermi, Pauli. *Atti del Congresso Internationale dei Fisici, 11–20 Settembre 1927.* Vol. 2. Bologna: Nicola Zanichelli.

Born Family. 1991. *Notes and Letters, 1790–1886.* Collected by Hedi Born. Translated by Marianne Sutcliffe. Unpublished. FAM.

Born, G. V. R., and P. Banks. 1996. "Hugh Blaschko." *Biographical Memoirs*. London: The Royal Society.

Born, Hedi. 1963. "Mein Hertz ist im Hochland, Ein Erlebnisbericht." *Ausburger Allgemeine*, 26 October. FAM.

Born, Max. 1906. *Untersuchungen über die Stabilität der elastischen Linie in Eben und Raum unter verschiedenen Grenzbedingungen*. Ph.D. diss., University of Göttingen.

———. 1909a. "Hermann Minkowski." Unpublished obituary. UGL-HA.

———. 1909b. "Beim Begräbnisse von Prof. Hermann Minkowski, 14 Januar 1909." FAM.

———. 1909c. "Über das Thomson'sche Atommodell." Habilitations-Vortrag. FAM.

———. 1914. "Über die Stabilität des Bohr'schen Atommodells." Unpublished seminar paper. Hilbert Nachlass no. 690. UGL-HA.

———. 1915. *Dynamik der Kristallgitter*. Leipzig: Teubner.

———. 1918. "Herbert Herkner." *Die Naturwissenschaften* 15: 179.

———. 1919. "Raum, Zeit, und Schwerkraft." *Frankfurter Zeitung*, 23 November.

———. 1920. *Die Relativitätstheorie Einsteins und ihre physikalischen Grundlagen*. Berlin: Springer.

———. 1922. "Über das Modell der Wasserstoffmolekel." *Die Naturwissenschaften* 10: 677.

———. 1923. "Quantentheorie und Störungsrechnung." *Die Naturwissenschaften* 11: 537.

———. 1925. *Vorlesungen über Atommechanik*. Berlin: Springer-Verlag.

———. 1926a. "Zur Quantenmechanik der Stossvorgänge." *Zeitschrift für Physik* 37: 863. Translated in *Quantum Theory and Measurement*, edited by John Archibald Wheeler and Wojciech Hubert Zurek, 51. Princeton: Princeton University Press. 1983.

———. 1926b. "Quantenmechanik der Stossvorgänge." *Zeitschrift für Physik* 38: 803. Translated in *Wave Mechanics* by Gunther Ludwig, 206. New York: Pergamon Press, 1968.

———. 1928. "VI. Kongress der Assoziation der russischen Physiker." *Die Naturwissenschaften* 16: 741.

———. 1933. *Optik: Ein Lehrbuch der elektromagnetische Lichttheorie*. Heidelberg: Springer-Verlag.

———. 1934. "On the Quantum Theory of the Electromagnetic Field." *Proceedings of the Royal Society A*. 143: 410.

———. 1935. *The Restless Universe*. Translated by W. M. Deans. London: Blackie & Sons.

———. 1940a. "Prof. J. J. Thomson." *Nature* 146: 356.

———. 1940b. "Prof. Otto Toeplitz." *Nature* 145: 617.

———. 1942. "Arnold Berliner." *Nature* 150: 285.

———. 1949. *Natural Philosophy of Cause and Chance*. Oxford: Oxford University Press.

———. 1955a. "Zum Jahresbeginn." *Physikalische Blätter* 11(1).

———. 1955b. "Remarks at Le Banquet Nobel." *Les Prix Nobel en 1954*. Stockholm: Royale P.A. Norstedt & Söner.

———. 1956. *Experiment and Theory in Physics*. New York: Dover.

———. 1959. "Erinnerungen an Hermann Minkowski zur 50. Wiederkehr seines Todestags." *Die Naturwissenschaften* 46: 501.

———. 1960. *Problems of Atomic Dynamics*. New York: Frederick Ungar.

———. 1964. "Reminiscences of My Work on the Dynamics of Crystal Lattices." *Lattice Dynamics: Proceedings of the International Conference Held at Copenhagen, Denmark, August 5–9, 1963*. New York: Pergamon Press.

———. 1968. *My Life and My Views*. New York: Charles Scribner's Sons.

———. 1969a. *Atomic Physics*. New York: Dover.

———. 1969b. *Physics in My Generation*. New York: Springer-Verlag.

Born, Max, and Hedwig Born. 1969. *Der Luxus des Gewissens: Erlebnisse und Einsichten im Atomzeitalter*. Munich: Nymphenburger Verlagshandler GmbH.

Born, Max, Walter Brandt, and Gustav Born. 1950. "In Memoriam Gustav Born, Experimental Embryologist." *Acta Anatomica* 10: 466.

Born, Max, Werner Heisenberg, and Pascual Jordan. 1926. "Zur Quantenmechanik. II." *Zeitschrift für Physik* 35: 557.

Born, Max, and Kun Huang. 1954. *Dynamical Theory of Crystal Lattices*. Oxford: Clarendon Press.

Born, Max, and Pascual Jordan. 1925a. "Zur Quantentheorie aperiodischer Vorgänge." *Zeitschrift für Physik* 33: 479.

———. 1925b. "Zur Quantenmechanik." *Zeitschrift für Physik* 34: 858.

———. 1930. *Elementare Quantenmechanik (II. Band der Vorlesungen über Atommechanik*. Heidelberg: Springer-Verlag.

Born, Max, and Theodore von Kármán. 1912. "Über Schwingungen in Raumgittern." *Physikalische Zeitschrift* 13: 297.

———. 1913. "Zur Theorie der Specifischen Wärme." *Physikalische Zeitschrift* 14: 15.

Born, Max, and Alfred Landé. 1918a. "Über die absolute Berechnung der Kristalleigenschaften mit Hilfe Bohrscher Atommodelle." *S.B. Preussische Akademie Wissenschaften*: 1048.

———. 1918b. "Über die Berechnung der Kompressibilität regulärer Kristalle aus der Gittertheorie." *Verhandlungen deutschen Physikalische Gesellschaft* 20: 210.

Born, Max, and Erich Oettinger. 1907. "Variationsprinzipe der Wärmetheorie." *Physikalische Zeitschrift* 8: 571.

Born, Max, and Robert Oppenheimer. 1927. "Zur Quantentheorie der Molekeln." *Annalen der Physik* 84: 457.

Born, Max, and Emil Wolf. 1964. *Principles of Optics: Electromagnetic Theory of Propagation, Interference and Diffraction of Light*. Oxford: Pergamon Press.

Born, Stephan. 1978. *Erinnerungen Eines Achtundvierzigers*. Berlin: Verlag J.H.W. Dietz Nachf. GmbH.

Briggs, Asa. 1983. *A Social History of England*. New York: Viking.

Brilling, Bernard. 1936. *Die Vorfahren des Meyer Kauffmann*. Breslau. Private papers.

Cassidy, David C. 1992. *Uncertainty: The Life and Science of Werner Heisenberg*. New York: W. H. Freeman.

Chickering, Roger. 1998. *Imperial Germany and the Great War, 1914–1918*. Cambridge: Cambridge University Press.

Clark, Ronald W. 1971. *Einstein: The Life and Times*. New York: World.

Courant, Richard. 1964. "Reminiscences from Hilbert's Göttingen." Transcript from a tape made at Yale University on January 13. Courant Papers, Bobst Library.

Crawford, Elisabeth, John Heilbron, and Rebecca Ullrich. 1987. *The Nobel Population 1901–1937*. Berkeley: University of California Press.

Cropper, William H. 2001. *Great Physicists: The Life and Times of Leading Physicists from Galileo to Hawking*. New York: Oxford University Press.

Csapo, Arpad. 1958. "Progesterone." *Scientific American*.198: 40.

Dahms, Hans-Joachim. 1987. "Einleitung." In *Die Universität Göttingen unter dem Nationalsozialismus*, edited by Heinrich Becker, Hans-Joachim Dahms, and Cornelia Wegeler. Munich: K. G. Saur.

Dawidowicz, Lucy S. 1975. *The War Against the Jews, 1933–1945*. New York: Holt, Rinehart and Winston.

Debye, Peter. 1964. "The Early Days of Lattice Dynamics." *Lattice Dynamics: Proceedings of the International Conference Held at Copenhagen, Denmark, August 5–9, 1963*. New York: Pergamon Press.

Delbrück, Max. "Erinnerung an Max Born." Max-Born-Gymnasium. CALTECH-D. 37.8.

Dresden, Max. 1987. *H.A. Kramers: Between Tradition and Revolution*. New York: Springer-Verlag.

Eck, Reimer. 1989. "Zur Entstehung des Archivs für berufsständische Rassenstatistik in der Göttinger Universitätsbibliothek." In *Bibliotheken während des Nationalsozialismus*. Part 1, edited by Peter Vodosek and Manfre Korowski. Wiesbaden: Otto Harrassowitz.

Elsasser, Walter M. 1978. *Memoirs of a Physicist in the Atomic Age*. New York: Science History.

"Erklärung der Professoren Grossbritanniens an die Deutschen Akademischen Kreise." DM-W.

Estermann, I. 1976. "Otto Stern." *Dictionary of Scientific Biography*. Vol. 13. Edited by Charles Couston Gillespie. New York: Charles Scribner's Sons.

Feldman, Burton. 2000. *The Nobel Prize: A History of Genius, Controversy, and Prestige*. New York: Arcade.

Flavin, Martin. 1996. *Kurt Hahn's Schools and Legacy*. Wilmington, DE: Middle Atlantic Press.

Fölsing, Albrecht. 1997. *Albert Einstein: A Biography*. New York: Viking.

Forman, Paul. 1970. "Alfred Landé and the Anomalous Zeeman Effect, 1919–1921." *Historical Studies in the Physical Sciences*. Vol. 2. Edited by Russell McCormmach. Philadelphia: University of Pennsylvania Press.

———. 1971. "Weimar Culture, Causality, and Quantum Theory, 1918–1927: Adaptation by German Physicists and Mathematicians to a Hostile Environment." *Historical Studies in the Physical Sciences*. Vol. 3. Edited by Russell McCormmach. Philadelphia: University of Pennsylvania Press.

———. 1986. "Il Naturforscherversammlung B Nauheim del settembre 1920: una introduzione alla vita scientifica nella Republica de Weimar." In *La ristrutturazione delle scienze tra le due guerre mondiali*. Vol. 1. *Proceedings of a Conference in Florence and Rome, 1980*, edited by G. Battimelli, M. De Maria, and A. Rossi, 59–78. Rome: La Goliardica.

Fort, Adrian. 2003. *PROF: The Life of Frederick Lindemann*. London: Jonathan Cape.

Franck, James. 1962. "Ein Geburtstagsgruss am Max Born." *Physikalische Blätter* 18: 541. CHICAGO-F.

Frenkel, Viktor. 1972. "Max Born." *Ideen des exakten Wissens* 5: 289.
———. 1996. *Yakov Ilich Frenkel*. Translated by Alexander S. Silbergleit. Boston: Birkhäuser Verlag.
Friedman, Robert Marc. 2001. *The Politics of Excellence: Behind the Nobel Prize in Science*. New York: W. H. Freeman.
"Der Froschlaich." 1884. Breslau. Unpublished. FAM.
Gay, Ruth. 1992. *The Jews of Germany: A Historical Portrait*. New Haven: Yale University Press.
Gleick, James. 1993. *Genius: The Life and Science of Richard Feynman*. New York: Vintage Press.
Goenner, Hubert, and Guiseppe Castagnetti. 1996. *Albert Einstein as a Pacifist and Democrat During the First World War*. Preprint 35. Berlin: Max-Planck-Institut für Wissenchaftsgeschichte.
Gregor-Dellin, Martin. 1983. *Richard Wagner: His Life, His Work, His Century*. Translated by J. Maxwell Brownjohn. New York: Harcourt Brace Jovanovich.
Hacohen, Malachi Haim. 2000. *Karl Popper: The Formative Years, 1902–1945*. Cambridge: Cambridge University Press.
Hanle, Wilhelm. 1989. *Memoiren*. Giessen: I. Physikalischen Institut der Justus-Liebig-Universität Giessen.
Heilbron, J. L. 1986. *The Dilemmas of an Upright Man: Max Planck as Spokesman for German Science*. Berkeley: University of California Press.
Heisenberg, Werner. 1930. *Die Physikalischen Prinzipien der Quantentheorie*. Leipzig: S. Hirzel-Verlag.
———. 1951. "50 Jahre Quantentheorie." *Die Naturwissenschaften* 38: 49.
———. 1971. *Physics and Beyond: Encounters and Conversations*. Translated by Arnold J. Pomerans. New York: Harper and Row.
———. 1985. "Reminiscences from 1926 and 1927." In *Niels Bohr: A Centenary Volume*, edited by A. P. French and P. J. Kennedy. Cambridge: Harvard University Press.
Hendry, John. 1984. *The Creation of Quantum Mechanics and the Bohr-Pauli Dialogue*. Boston: D. Reidel.
Hentschel, Klaus. 1996. *Physics and National Socialism: An Anthology of Primary Sources*. Boston: Birkhäuser Verlag.
Heppner, Rabbi Dr. A., and Lehrer J. Herzberg. 1929. *Aus Vergangenheit und Gegenwart*. Breslau: Im Selbstverlage.
Hoch, Paul. 1983. "How Britain Lost the Physics War." *London Times Higher Education Supplement*, 21 October, 13.
Hoffmann, Banesh. 1972. *Albert Einstein: Creator and Rebel*. New York: Viking.
Hoffmann, Dieter. 2003. *Pascual Jordan im Dritten Reich—Schlaglichter*. Preprint 248. Berlin: Max-Planck-Institut für Wissenschaftsgeschichte.
Hoffmann, Klaus. 1995. *J. Robert Oppenheimer: Schöpfer der ersten Atombombe*. Berlin: Springer.
Holl, Frank. 1996. *Produktion und Distribution wissenschaftlicher Literatur: der Physiker Max Born und sein Verleger Ferdinand Springer 1913–1970*. Frankfurt am Main: Buchhändler-Vereinigung.
Hopwood, Nick. 1999. "'Giving Body' to Embryos: Modeling, Mechanism, and the Microtome in Late Nineteenth-Century Anatomy." *Isis* 90: 462.

Hund, Friedrich. 1983. "Born, Göttingen und die Quantenmechanik." *James Franck und Max Born in Göttingen: Reden zur akademischen Feier aus Anlass der 100. Wiederkehr ihres Geburtsjahres*. Göttingen: Vandenhoeck and Ruprecht.

———. 1985. "Bohr, Göttingen, and Quantum Mechanics." In *Niels Bohr: A Centenary Volume*, edited by A. P. French and P. J. Kennedy. Cambridge: Harvard University Press.

Infeld, Leopold. 1941. *Quest: The Evolution of a Scientist*. London: Victor Gollancz.

Jacobi, Selma. 1917. *Notes: Recollections of Her Parents' Home and of Her Brother Gustav*. Unpublished. FAM.

Jayaraman, K. S. 1998. "Insult Thwarted 1934 Bid to Raise Profile of Indian Science." *Nature* 392 (12 March): 112.

Jeffrey, Andrew. 1992. *This Present Emergency: Edinburgh, the River Forth, and South-East Scotland and the Second World War*. Edinburgh: Mainstream.

Jungk, Robert. 1958. *Brighter Than a Thousand Suns: A Personal History of the Atomic Scientists*. New York: Harcourt Brace.

Jungnickel, Christa, and Russell McCormmach. 1986. *Intellectual Mastery of Nature: Theoretical Physics from Ohm to Einstein*. Vol 2. *The Now Mighty Theoretical Physics 1875–1924*. Chicago: University of Chicago Press.

Kaganoff, Benzion C. 1996. *A Dictionary of Jewish Names and Their History*. Northvale: Jason Aronson.

Kauffmann, Salomon. 1896. *Erinnerungen*. Unpublished. Private papers.

Kemble, E. C. 1926. "Book Reviews—Problems of Atomic Dynamics." *Physical Review* 28: 423.

Kemmer, Nicholas, and Schlapp, Robert, Robert. 1971. "Max Born." *Biographical Memoirs of Fellows of the Royal Society*. 17 November:17.

Klein, Martin, A.J Kox, Jürgen Renn, and Robert Schulmann. 1993. "Einstein on Length Contraction in the Theory of Relativity." *The Collected Papers of Albert Einstein*. Vol. 3, 478–480. Princeton: Princeton University Press.

Lagiewski, Maciej. 1996. *Breslauer Juden, 1850–1944*. Wroclaw: Muzeum Historyczne.

Laurikainen, K. V. 1985. *Beyond the Atom: The Philosophical Thought of Wolfgang Pauli*. New York: Springer-Verlag.

Lemmerich, Jost. 1983. *Science and Conscience: The World of Two Atomic Scientists*. London: Science Museum.

Lindley, David. 1996. *Where Does the Weirdness Go? Why Quantum Mechanics Is Strange, but Not as Strange as You Think*. New York: Basic Books.

———. 2001. *Boltzmann's Atom: The Great Debate that Launched a Revolution in Physics*. New York: The Free Press.

Livingston, Dorothy. 1973. *The Master of Light: A Biography of Albert A. Michelson*. New York: Charles Scribner's Sons.

Low, Alfred D. 1979. *Jews in the Eyes of the Germans: From the Enlightenment to Imperial Germany*. Philadelphia: Institute for the Study of Human Issues.

Luft, Edward Robert. 1987. *The Naturalized Jews of the Grand Duchy of Posen in 1834 and 1835*. Atlanta: Scholars Press.

MacLane, Saunders. 1995. "Mathematics at Göttingen under the Nazis." *Notices of the American Mathematical Society*. October.

Meissner, Hanna. 1990. "Ernst Hellinger." *Operator Theory: Advances and Applications* 48: 43.

Mets, David. 1987. *Nonnuclear Aircraft Armament: The Search for a Surgical Strike, the United States Air Force and Laser Guided Bombs*. Eglin Air Force Base, FL: Office of History, Armament Division, Air Force Systems Command.

Meyenn, Karl von, and Engelbert Schucking. 2001. "Wolfgang Pauli." *Physics Today* (February): 43.

Meyer Kauffmann Textilwerke A.-G., 1824–1924. Berlin: Ecksteins Biographischer Verlag. STABI.

Meyer, Michael. 1996–1998. *German-Jewish History in Modern Times*. Vols. 2–4. New York: Columbia University Press.

Michelmore, Peter. 1969. *The Swift Years: The Robert Oppenheimer Story*. New York: Dodd, Mead.

Miller, Arthur. 1998. *Albert Einstein's Special Theory of Relativity: Emergence (1905) and Early Interpretation (1905–1911)*. New York: Springer.

Monk, Ray. 2000. *Bertrand Russell: The Ghost of Madness, 1921–1970*. New York: The Free Press.

Moore, Walter. 1989. *Schrödinger: Life and Thought*. New York: Cambridge University Press.

Moritz. 1948. "Physikertagung in Claustal: Tagungsbericht." *Physikalische Blätter* 4: 391.

Nissen, Walter, and Waldemar R. Röhrbein. 1975. *Göttingen: so wie es war*. Düsseldorf: Droste.

Oppenheimer, J. Robert. 1954. *Science and the Common Understanding*. New York: Simon and Schuster.

Pais, Abraham. 1982a. *Subtle Is the Lord: The Science and the Life of Albert Einstein*. New York: Oxford University Press.

———. 1982b. "Max Born's Statistical Interpretation of Quantum Mechanics." *Science* 218: 1193.

Pauli, Wolfgang. 1930. Book review. "Elementare Quantenmechanik." *Die Naturwissenschaften* 18: 602.

Pyenson, Lewis. 1985. *The Young Einstein: The Advent of Relativity*. Boston: Adam Hilger.

Reid, Constance. 1986. *Hilbert-Courant*. New York: Springer-Verlag.

Rigden. John. 2002. *Hydrogen: The Essential Element*. Cambridge: Harvard University Press.

Ringer, Fritz. 1969. *The Decline of the German Mandarins: The German Academic Community, 1890–1933*. Cambridge: Harvard University Press.

Robertson, Peter. 1979. *The Early Years: The Niels Bohr Institute, 1921–1930*. Copenhagen: Akademisk Forlag.

Roos, Julia. 2003. "Weimar's Crisis Through the Lens of Gender: The Case of Prostitution." *Bulletin of the German Historical Institute* 32: 85.

Rosenfeld, Léon. 1950. "Early History of Quantum Mechanics." *Nature* 166(2): 883.

Rowe, David, E. 1986. "'Jewish Mathematics' at Göttingen in the Era of Felix Klein." *Isis* 77: 422.

Rupke, Nicolaas, ed. 2002. *Göttingen and the Development of the Natural Sciences*. Göttingen: Wallstein Verlag.

Schäfer, Hans. 1933. *Erschautes, Erlebtes, Erdachtes, Erstrebtes*. Munich. LBI.

Seth, Suman. 2003. *Principles and Problems: Constructions of Theoretical Physics in Germany, 1890–1918*. Ph.D. diss., Princeton University.

Sigurdsson, Skuli. 1991. *Hermann Weyl, Mathematics and Physics, 1900–1927*. Ph.D. diss., Harvard University.

———. 1996. "Physics, Life, and Contingency." In *Forced Migration and Scientific Change*, edited by Mitchell G. Ash and Alfons Söllner. Washington, DC: German Historical Institute.

Sime, Ruth Lewin. 1996. *Lise Meitner: A Life in Physics*. Berkeley: University of California Press.

Smith, Alice Kimball, and Charles Weiner. 1995. *Robert Oppenheimer: Letters and Recollections*. Stanford: Stanford University Press.

Sommerfeld, Arnold. 1919. *Atombau und Spektrallinien*. Braunschweig: F. Vieweg und Sohn.

Stark, Johannes. 1922. *Die Gegenwärtige Krisis in der deutschen Physik*. Leipzig: J.A. Barth.

Stark, Johannes. 1941. "Jüdische und Deutsche Physik: Vorträge zur Eröffnung des Kolliquiums für theoretische Physik an der Universität München." Herausgegeben von Wilhelm Müller. Leipzig: Helingsche Verlaganstalt.

Staley, Richard. 1992. *Max Born and the German Physics Community: The Education of a Physicist*. Ph.D. diss., University of Cambridge.

Stern, Fritz. 1999. *Einstein's German World*. Princeton: Princeton University Press.

Streiter, Karl Heink. 1985. *Die nationalen Beziehungen im Grossherzogtum Posen (1815–1848)*. Frankfurt am Main: Peter Lang.

Stuewer, Roger. 1975a. *The Compton Effect: Turning Point in Physics*. New York: History.

———. 1975b. *Solid-State and Molecular Theory: A Scientific Biography*. New York: John Wiley & Sons.

Taylor, A. J. P. 1965. *English History 1914–1945*. New York: Oxford University Press.

Trageser, Wolfgang. 1998. "Die Berufung Albert Einsteins nach Frankfurt am Main." *The IPG Preprint Series*. Preprint no. 30. Frankfurt: University of Frankfurt.

Van der Kloot, William Van der. 2004. "April 1915: Five Nobel Prize Winners Inaugurate Weapons of Mass Destruction and the Academic-Industrial-Military Complex." *Notes Rec. Royal Society London* 58: 149.

Van der Waerden, B. L. 1967. *Sources of Quantum Mechanics*. New York: Dover.

Wandycz, Piotr S. 1974. *The Lands of Partitioned Poland, 1795–1918. The History of East Central Europe*. Vol. 7. Seattle: University of Washington Press.

"Wolfenbüttel Speech to the Assembly of Trustees of Samsonschule." 18 March 1883. Julie Ehrenberg Papers. LBI.

Zarchin, Michael. 1939. *Jews in the Province of Posen*. Philadelphia: Dropsie College for Hebrew and Cognate Learning.

INDEX